AF832046

Ganoderma

Ganoderma holds significant traditional importance in various ethnic cultures around the world, particularly in China, Japan, and Korea. Many indigenous traditions have incorporated *Ganoderma* into medicinal practices, being considered a symbol of longevity, vitality, and good health. At present, the taxon is believed to possess various health benefits and is used to treat ailments and promote overall well-being. In this context, the first volume of the book, titled *Ganoderma: Cultivation, Chemistry and Medicinal Applications,* aims to comprehensively cover the taxonomy, morphological features, domestication strategies, structures of secondary metabolites, and therapeutic prospects of *Ganoderma.* It may serve as a definite resource for students, researchers, healthcare professionals, traditional medicine practitioners, and enthusiasts.

FEATURES

- Provides a comprehensive classification system for *Ganoderma* species, highlighting their taxonomy and distinguishing characteristics
- Delves into the techniques and practices involved in cultivating *Ganoderma*, offering detailed guidance for individuals interested in growing this valuable fungus
- Explores the cultural and traditional significance of *Ganoderma* in various ethnic cultures intertwined with customs, beliefs, rituals, myths, and folklore around the world
- Investigates the secondary metabolites of *Ganoderma*, highlighting their implications
- Examines diverse bioactivities associated with *Ganoderma*, including antioxidant, hepatoprotective, antidiabetic, prebiotic, anti-inflammatory, anti-arthritis, anticancer, hypolipidemic, and cholesterol-lowering effects

This book includes relevant illustrations, diagrams, and images to enhance the understanding of concepts associated with *Ganoderma.*

Ganoderma
Cultivation, Chemistry and Medicinal Applications
Volume 1

Edited by
Krishnendu Acharya
Somanjana Khatua

CRC Press
Taylor & Francis Group
Boca Raton London New York

CRC Press is an imprint of the
Taylor & Francis Group, an **informa** business

Designed cover image: Shutterstock

First edition published 2024
by CRC Press
2385 NW Executive Center Drive, Suite 320, Boca Raton FL 33431

and by CRC Press
4 Park Square, Milton Park, Abingdon, Oxon, OX14 4RN

CRC Press is an imprint of Taylor & Francis Group, LLC

© 2024 selection and editorial matter, Krishnendu Acharya and Somanjana Khatua; individual chapters, the contributors

Reasonable efforts have been made to publish reliable data and information, but the author and publisher cannot assume responsibility for the validity of all materials or the consequences of their use. The authors and publishers have attempted to trace the copyright holders of all material reproduced in this publication and apologize to copyright holders if permission to publish in this form has not been obtained. If any copyright material has not been acknowledged please write and let us know so we may rectify in any future reprint.

Except as permitted under U.S. Copyright Law, no part of this book may be reprinted, reproduced, transmitted, or utilized in any form by any electronic, mechanical, or other means, now known or hereafter invented, including photocopying, microfilming, and recording, or in any information storage or retrieval system, without written permission from the publishers.

For permission to photocopy or use material electronically from this work, access www.copyright.com or contact the Copyright Clearance Center, Inc. (CCC), 222 Rosewood Drive, Danvers, MA 01923, 978–750–8400. For works that are not available on CCC please contact mpkbookspermissions@tandf.co.uk

Trademark notice: Product or corporate names may be trademarks or registered trademarks and are used only for identification and explanation without intent to infringe.

ISBN: 978-1-032-39761-0 (hbk)
ISBN: 978-1-032-40807-1 (pbk)
ISBN: 978-1-003-35478-9 (ebk)

DOI: 10.1201/9781003354789

Typeset in Times LT Std
by Apex CoVantage, LLC

Contents

Preface

Noncommunicable diseases (NCDs), or chronic diseases, encompass a vast group of ailments and account for 70% of deaths worldwide. The risk factors include overweight, low physical activity, unhealthy diet, sedentary lifestyle, and so forth, resulting in NCDs to be considered "diseases of lifestyle" or "diseases of civilization." Several synthetic drugs are available in the market to treat and manage NCDs; however, they do have limitations. Such medications not only may cause side effects ranging from mild to severe but can also develop resistance to the treatment over time. In this context, natural drugs have historically been used in various indigenous medicinal systems to cure and manage a wide range of health conditions, including NCDs.

Ganoderma, popularly known as lingzhi and reishi, is one of such matrices that has been revered in healthcare practices, particularly in Asian countries, for centuries. The polyporous Basidiomycetes has traditionally been regarded in China as a herb of spiritual potency that can extend the lifespan, improve intelligence, relieve stress, boost the immune system, and support overall well-being. The growing popularity and demand played a significant role in driving the desire for cultivation of *Ganoderma* to have a more reliable and sustainable supply of the mushroom. Since then, the domestication methods have continued to evolve and improve, making the taxon more readily available for medicinal purposes. At present, *Ganoderma* is often touted for its potent health benefits and has gained attention as a natural remedy for various conditions together with NCDs. Credit goes to the bioactive compounds such as polysaccharides and triterpenoids, which contribute to the antioxidant, hepatoprotective, antidiabetic, gut microbiota regulatory, anti-arthritis, hypolipidemic, anticancer, antimicrobial, cardioprotective, neuroprotective, and immune-regulatory effects. As a result, not only are the number of patents and clinical trials with reishi increasing but also the industrial value of *Ganoderma*-based products is experiencing remarkable progress. An overview on *Ganoderma* research accomplished to date is thus highly needed to direct further developments and applications.

Keeping this in mind, the present book aims to provide a comprehensive guide to *Ganoderma* medicinal prospects, offering readers a deeper understanding. It will delve into a detailed description on morphological and molecular characterization of several valuable *Ganoderma* species, which will help students, researchers, scientists, and industrialists to easily recognize the desired sample. A compilation of information regarding artificial cultivation of the genus has also been provided that will help any person irrespective of his or her academic background to become an entrepreneur. Further chapters divided in two volumes will take the readers on a scientific exploration of the bioactive compounds and medicinal prospects of *Ganoderma*, presenting findings of numerous studies and clinical trials explaining the mechanism of actions. It is our sincere hope that this two-volume set will serve as a definitive resource for researchers, healthcare professionals, traditional medicine practitioners, and enthusiasts.

Briefly, the first volume will initially provide an overview on the hierarchical classification of *Ganoderma*, which has undergone revisions over time. By shedding light on the challenges and complexities encountered in *Ganoderma* taxonomy, we hope to provide a deeper understanding in classifying the species accurately. Further, we will unveil the different cultivation methods employed for *Ganoderma*, such as indoor cultivation on artificial substrates and outdoor cultivation on natural substrates. Additional chapters will provide a comprehensive journey through the traditional history, cultural significance, and modern therapeutic potential of *Ganoderma*. We will explore the rich historical roots of *Ganoderma* in traditional medicine practices across different cultures and civilizations. A fascinating journey will be embarked on the world of *Ganoderma* secondary metabolites, delving into their characterization and potential applications. As we progress through the chapters, we will emphasize the importance of scientific research, clinical studies, and evidence-based approaches in understanding the therapeutic potentials such as antioxidant, hepatoprotective, antidiabetic, prebiotic, anti-inflammatory, anti-arthritis, anticancer, hypolipidemic and cholesterol-lowering effects. All in all, we hope that the first volume will serve as a valuable resource for students, researchers, and enthusiasts, inspiring further research, collaboration, and exploration into the captivating world of *Ganoderma*.

Most humbly and respectfully, we extend our gratitude to all the esteemed researchers, scientists, and contributors for their timely response, meaningful contributions, constant support, and cooperation. Their commitment to unraveling the secrets of *Ganoderma* has brought us closer to understanding and harnessing nature's healing treasure. The dedicated scientific community worldwide, diligently engaged in *Ganoderma* research, is also appreciated. We hope that students, teachers, medical professionals, researchers, and companies involved with macrofungal research will find our two-volume book *Ganoderma: Cultivation, Chemistry and Medicinal Applications* a useful resource.

Krishnendu Acharya
Somanjana Khatua

Editors

Krishnendu Acharya, PhD, earned a master's in botany at the University of Calcutta and then earned an MTech in biotechnology and a PhD at Jadavpur University. Prof. Acharya joined the University of Calcutta as a Lecturer of botany in 2004 and soon acceded to professorship in 2012. He has 25 years of research and teaching experience. Prof. Acharya has supervised 20 PhD students and published 465 research articles and 8 books. He has more than 8,000 and 11,000 Scopus and Google Scholar citations, respectively. His h-index is 47 (Scopus) and 54 (Google Scholar), and his Google Scholar i10 index is 238. He has one patent and two applied. He was ranked first among the top 100 authors of state universities based on the Indian Citation Index 2016 by the Confederation of Indian Industry (CII). He has been listed on Stanford University's list of the world's top 2% of scientists for the last three years. Prof. Acharya has been elected as a Fellow of the West Bengal Academy of Science and Technology, the Linnean Society of London, the Mycological Society of India, the Indian Mycological Society, and the International College of Nutrition, Canada. He has received many awards, including the Sir Edwin John Butler Memorial Award, the Patel Memorial Award, the Global Achievement in the Field of Phyllosphere Biology, and the Prof. K. Natarajan Memorial Award. He is on the editorial boards of many national and international journals. Prof. Acharya is actively and regularly engaged in lecturing at different seminars, universities, and colleges to boost young minds and inspire students.

Somanjana Khatua, PhD, is an Assistant Professor in the Department of Botany, University of Allahabad, Uttar Pradesh, India. Earlier, she was appointed by the West Bengal Education Service and posted at Krishnagar Government College, West Bengal, India. She earned a master's in botany and a PhD at the University of Calcutta. Her areas of interest include drug development, immunology, carbohydrate biology, natural products, medical mycology, and functional food. Dr. Khatua has published 74 papers in different peer-reviewed journals, including *Chemico-Biological Interactions*, *Food and Function*, *Frontiers in Pharmacology*, *Scientific Reports*, *PloS ONE*, *Journal of Pharmacy and Pharmacology*, *PeerJ*, *Carbohydrate Polymers*, *Carbohydrate Research*, and *International Journal of Biological Macromolecules*. She has 63 papers in Scopus and 1,254 Scopus citations, as well as 1,629 Google Scholar citations. Her h-index is 21 (Scopus) and 25 (Google Scholar), and her Google Scholar i10 index is 45. She has received many awards, including the 2020 Woman Botanist Award from the Indian Botanical Society and the Outstanding Paper Award by the government of West Bengal, India. Dr. Khatua delivers invited lectures at various colleges throughout India.

Contributors

Krishnendu Acharya
Molecular and Applied Mycology and Plant Pathology Laboratory
Department of Botany
University of Calcutta
Calcutta, India

Thekkuttuparambil A. Ajith
Department of Biochemistry
Amala Institute of Medical Sciences
Kerala, India

Anik Barman
Department of Microbiology
Bose Institute
Kolkata, India

Jayita De
Molecular and Applied Mycology and Plant Pathology Laboratory
Department of Botany
University of Calcutta
Kolkata, India

M. C. A. Galappaththi
Postgraduate Institute of Science (PGIS)
University of Peradeniya
Peradeniya, Sri Lanka

K. K. Hapuarachchi
Engineering Research Center of Southwest Bio-Pharmaceutical Resource
Ministry of Education
Guizhou University
Guiyang, China

Kainoor K. Janardhanan
Department of Microbiology
Amala Cancer Research Centre
Kerala, India

S. C. Karunarathna
College of Biological Resource and Food Engineering
Qujing Normal University
Qujing, China
and
National Institute of Fundamental Studies (NIFS)
Kandy, Sri Lanka

Somanjana Khatua
Department of Botany
Faculty of Science
University of Allahabad
Prayagraj, India

Anita Klaus
Faculty of Agriculture
University of Belgrade
Belgrade, Serbia

Maja Kozarski
Faculty of Agriculture
University of Belgrade
Belgrade, Serbia

Umah Rani Kuppusamy
Department of Biomedical Science
Faculty of Medicine
and
Mushroom Research Centre
Universiti Malaya
Kuala Lumpur, Malaysia

Narayan Chandra Mandal
Department of Botany
Visva Bharati University
Shantiniketan, India

Supratim Mandal
Department of Microbiology
University of Kalyani
Kalyani, India

Anish Nag
Department of Life Sciences
Christ (Deemed to Be University)
Bangalore, Karnataka

Sudeshna Nandi
Molecular and Applied Mycology and Plant Pathology Laboratory
Department of Botany
University of Calcutta
Kolkata, India

N. M. Patabendige
Institute of Chemistry
Chinese Academy of Sciences
Beijing, China

Annika Marial Paul
Department of Life Sciences
Christ (Deemed to Be University)
Bangalore, Karnataka

Predrag Petrović
Innovation Center of the Faculty of Technology and Metallurgy
University of Belgrade
Belgrade, Serbia

Chia Wei Phan
Department of Pharmaceutical Life Sciences
Faculty of Pharmacy
and
Mushroom Research Centre
Universiti Malaya
Kuala Lumpur, Malaysia

Prakash Pradhan
Molecular and Applied Mycology and Plant Pathology Laboratory
Department of Botany
University of Calcutta
and
West Bengal Biodiversity Board
Kolkata, India

A. K. H. Priyashantha
Nittambuwa, Sri Lanka

Adhiraj Roy
Amity Institute of Molecular Medicine and Stem Cell Research
Amity University Noida
Noida, India

Vikineswary Sabaratnam
Institute of Biological Sciences
Faculty of Science
and
Mushroom Research Centre
Universiti Malaya
Kuala Lumpur, Malaysia

Aloke Saha
Department of Zoology
University of Kalyani
Nadia, India

Kunal Kumar Saha
Department of Botany
Visva Bharati University
Shantiniketan, India

Steven L. Stephenson
Department of Biological Sciences
University of Arkansas
Fayetteville, Arkansas, USA

Jovana Vunduk
Institute of General and Physical Chemistry
and
Ekofungi Ltd.
Belgrade, Serbia

Wan Abd Al Qadr Imad Wan-Mohtar
Institute of Biological Sciences
Faculty of Science
Universiti Malaya
Kuala Lumpur, Malaysia

1 Taxonomy, Phylogeny, and Beneficial Uses of *Ganoderma* (Ganodermataceae, Polyporales)

M. C. A. Galappaththi[1], *A. K. H. Priyashantha*[2], *N. M. Patabendige*[3], *Steven L. Stephenson*[4], *K. K. Hapuarachchi*[5], *and S. C. Karunarathna*[6]

1 Harry Butler Institute, Murdoch University, WA, Australia
2 Chiang Mai University, Thailand
3 Chinese Academy of Sciences, Beijing, China
4 University of Arkansas, Fayetteville, Arkansas, USA
5 College of Biodiversity Conservation, Southwest Forestry University, Kunming 650224, P.R China
6 Qujing Normal University, Qujing, China and National Institute of Fundamental Studies (NIFS), Kandy, Sri Lanka

1.1 INTRODUCTION

The Ganodermataceae is one of the largest families among polypores, with 14 accepted genera: *Amauroderma* (Murrill), *Amaurodermellus* (Costa-Rezende, Drechsler-Santos & Góes-Neto), *Cristataspora* (Robledo & Costa-Rezende), *Foraminispora* (Robledo, Costa-Rez. & Drechsler-Santos), *Furtadoa* (Costa-Rez., Robledo & Drechsler-Santos), *Furtadoella* (B.K. Cui & Y.F. Sun), *Ganoderma* (P. Karst), *Haddowia* (Steyaert), *Humphreya* (Steyaert), *Neoganoderma* (B.K. Cui & Y.F. Sun), *Sanguinoderma* (Y.F. Sun, D.H. Costa & B.K. Cui), *Sinoganoderma* (B.K. Cui, J.H. Xing & Y.F. Sun), *Tomophagus* (Murrill), and *Trachydermella* (B.K. Cui & Y.F. Sun) (Zhou et al., 2015; Hapuarachchi et al., 2019; Costa-Rezende et al., 2020b; Sun et al., 2022). Four hundred ninety four records (494), comprising the majority of species, are included in the genus *Ganoderma* according to Index Fungorum (2024) (accessed on 12 January 2024); however, approximately half of these records have been found to be synonyms. Recently, Sun et al. (2022) confirmed just 181 taxa for this genus based on strong molecular and phylogenetic support.

Ganoderma species have economically significant medicinal and nutritional bioactive components (Dai et al., 2009; Galappaththi et al., 2023). As a matter of fact, Ganoderma has been particularly used Southeast Asian countries as a traditional medicine to maintain health and life expectancy, gaining widespread use as a health supplement (Hapuarachchi et al., 2018). Currently, *Ganoderma* is a hot research topic due to its high potential for use in biotechnology (He et al., 2022). *Ganoderma* species have therapeutic, aesthetic, and pathogenic properties. *Ganoderma* species cause white rot in hardwood by breaking down cellulose and lignin-like polysaccharide components. Furthermore, *Ganoderma* is associated with root decay and the burning of the lower trunk or stems, such as basal stem rot in coconut, oil palm, and areca nut, as well as in many other trees like maple and oak (Kumar et al., 2022). This can lead to tree collapse, posing dangerous conditions for trees and potential property damage cause dangerous tree conditions and property damage (Loyd et al., 2017;

DOI: 10.1201/9781003354789-1

He et al., 2022). Some species are considered important sources of lignin-degrading enzymes, and some have been shown to selectively delignify wood (Otjen et al., 1987; He et al., 2022).

The fame of *Ganoderma* species spans thousands of years and is known in Asia as lingzhi (in China) or reishi (in Japan), which means mushroom of immortality (Loyd and Blanchette, 2019). The Qianlong Emperor of China (1735–1796) compared lingzhi to purple sandalwood or white jade in a poem, highlighting its pricelessness and rarity (Loyd and Blanchette, 2019). In 1781, the mushroom was given to the London mushroom collection by William Curtis and identified as *Boletus lucidus*; he seemed to be excited by his discovery and called it "a handsome mushroom shining as if painted" and "so beautifully polished that I barely know if what I found is natural or artificial" (Loyd and Blanchette, 2019).

Nevertheless, along with the taxonomical works, the Finnish mycologist Karsten (1881) introduced the genus *Ganoderma* to fungal taxonomy, and the only species was *G. lucidum*. Patouillard (1887) transferred other species to *Ganoderma*. Karsten (1889) proposed the monotypic genus *Elfvingia* with the type species *Boletus applanatus* Pers. for nonlaccate *Ganoderma* species, and the type specimen is preserved at the Rijks Herbarium Leiden, Netherlands (Richter et al., 2014). Some mycologists questioned whether *Elfvingia* should be segregated from *Ganoderma* (Imazeki, 1939, 1952; Steyaert, 1980). In the same year, the section *Amauroderma* was created by Patouillard (1889) to accommodate species with spherical/subspherical basidiospores and uniformly thickened walls, and 48 species of the genus were recorded worldwide. Furthermore, *Ganoderma* species are divided into two sections: *Ganoderma* sect. *Ganoderma* and *G.* sect. *Amauroderma*. Murrill (1905) circumscribed the genus *Amauroderma* with *Fomes regulicolor* Berk. ex Cooke as the type species. Bresadola (1979) identified *Ganoderma* with two sections, the same as in Patouillard (1889), but not *Elfvingia*. Moreover, Lloyd (1898–1925) never identified *Amauroderma, Elfvingia*, or *Ganoderma* at the generic level and recombined species epithets in *Polyporus* or *Fomes* (Stevenson, 1933).

Murrill (1902, 1908) published synopses of species emerging in North America, but he omitted *G. lucidum* from his 1908 publication. In 1905, Murrill introduced *Tomophagus* as the type species, designating *Polyporus colossus* Fr. 1851, based on its distinctive morphological features (Furtado, 1965b). Nevertheless, several authors have not accepted the generic segregation of this species, based on its microstructural features and spores (Torrend, 1920; Furtado, 1965a; Steyaert, 1972, 1980; Corner, 1983; Ryvarden, 1991; Wasser et al., 2006; Torres-Torres et al., 2015).

Subsequently, Atkinson (1908), Ames (1913), Torrend (1920), Haddow (1931), Donk (1933), Imazeki (1939), Donk (1948), Nobles (1948, 1958), Overholts (1953), Hansen (1958), Sarkar (1959), Teixeira (1962), Steyaert (1972, 1980), Pegler and Young (1973), Furtado (1981), Bazzalo and Wright (1982), and Corner (1983) reported this genus.

Saccardo and his colleagues (1882–1928) identified *Ganoderma*, not *Elfvingia* and *Amauroderma*. Nonetheless, they had a conflicted and confusing perspective in circumscribing *Ganoderma*. Torrend (1920) monographed both *Ganoderma* and *Amauroderma* in South America. Imazeki (1939, 1952) reviewed the knowledge of *Ganoderma* species in Eastern Asia. *Elfvingia* P. Karst, the new subgenus segregated by Imazeki (1939) in *Ganoderma* based on *G. tsunodae* Yasuda, was later discussed at a generic level (Imazeki, 1952). The segregation of Ganodermataceae as a separate family from other polypores was proposed by Donk (1948) based on morphological characteristics. Steyaert (1972) constructed the genera *Haddowia* and *Humphreya* based on basidiospore morphology. Steyaert (1980) established *Ganoderma* into several subgenera and sections based on the cutis type. The taxonomy of the Ganodermataceae, including species of tropical Asia and America and many new species, was summarized by Corner (1983). However, Corner criticized Steyaert's (1980) basic classification for the subdivisions of *Ganoderma*. Zhao (1989) also provided a summary of the current knowledge of the genus in China, with 84 listed species.

Later on, Donk (1948) introduced the family Ganodermataceae, which includes *Ganoderma* (Curtis) P. Karsten (1881). However, uncertainty about the rank of the family Ganodermataceae should not be neglected. Adaskaveg and Gilbertson (1988) pointed out that Ganodermataceae differs from other families of Polyporaceae in having a peculiar type of double-walled basidiospore. Binder et al. (2013) conducted phylogenomic and phylogenetic analyses of Polyporales, demonstrating that *Ganoderma* is a part of the "core polyporoid clade." Justo et al. (2017) proposed that Ganodermataceae should be considered a synonym of Polyporaceae based on the analyses of ITS, nLSU, and rpb1 sequences. Although He et al. (2019) renamed Ganodermataceae as Polyporaceae based on evolutionary studies, several other authors (Hapuarachchi et al., 2018; Xing et al., 2018; Costa-Rezende et al., 2020a, 2020b; Luangharn et al., 2021; Sun et al., 2022; He et al., 2022; Konara et al., 2022) still accept Ganodermataceae as a separate family from Polyporaceae.

After 1950, several studies on *Ganoderma* were conducted by Adaskaveg and Gilbertson (1986, 1988, 1989), Hseu (1990), Peng (1990), and Wang and Hua (1991). However, the impact of these studies on *Ganoderma* systematics was limited, as they focused on a small number of taxa and made limited comparisons. Consequently, accurate names for many taxa remained unclear in these works (Buchanan and Wilkie, 1995; Gottlieb et al., 1995; Moncalvo et al., 1995a, 1995b, 1995c). Due to its worldwide distribution, primarily found in temperate and tropical regions of Africa, America, Asia, and Europe (Luangharn et al., 2021; Wong et al., 2021), and the fact that many species are phytopathogenic (Pilotti, 2005; Bharudin et al., 2022) and possess medicinal properties (Jung et al., 2005; Hsu et al., 2008; Du et al., 2019), numerous studies have been focused on the genus. As a result, today hundreds of species have been recognized.

Ganoderma comprises three subgenera named *Ganoderma* P. Karst., *Elfvingia* P. Karst., and *Trachyderma* Imazeki (He et al., 2021). The subgenus *Elfvingia* includes all nonlaccate species with a dull upper surface, while laccate species with a cutis surface composed of a palisade of inflated hyphal ends, are found in the subgenera *Phaenema* and *Ganoderma* (Luangharn et al., 2021). Because the genus *Trachyderma* is lichenized, the subgenus *Trachyderma* has been declared invalid and is considered a synonym of *Ganoderma* (Luangharn et al., 2021). According to a recent classification by Sun et al. (2022), *Trachyderma* is an illegitimate name and was renamed *Trachydermella,* which formed an independent clade within the Ganodermataceae.

In this chapter, the authors have attempted to provide an overview of *Ganoderma* taxonomy, based on the latest scientific research. In addition, particular attention is given to highlighting the beneficial usage of the genus. A literature survey was conducted using the databases of Google General, Google Scholar, National Center for Biotechnology Information (NCBI), Scopus, and Web of Science. Furthermore, fungal databases such as the Index Fungorum, GenBank, and MycoBank are also used to obtain additional information.

1.2 TAXONOMIC CHARACTERS USED TO IDENTIFY *GANODERMA* SPECIES

Early studies on the identification of *Ganoderma* species were confusing and contentious due to morphological variation under different environmental conditions, maturity, and a lack of type cultures or fruit bodies (Seo and Kitamoto, 1988; Pilotti et al., 2004; Wang et al., 2009a; Dai et al., 2017). Taxonomic confusion has resulted due to wide variation in the macroscopic traits of the *Ganoderma* basidiomes (He et al., 2022). Recent molecular-based and gene sequencing techniques, however, have resolved many of these constraints, made significant improvements, and provided a plethora of data for the further analysis of the genus (Du et al., 2019).

Several species morphologically similar to *G. lucidum* have been described around the world. These include *G. resinaceum* Boud. 1889 (Patouillard, 1889) in Europe; *G. multipileum* (Hou, 1950), *G. sichuanense* J.D. Zhao and X.Q. Zhang (Zhao et al., 1983), and *G. lingzhi* (; Cao et al., 2012) in China; and *G. oregonense* Murrill, 1908, *G. sessile* (Murrill, 1902), *G. tsugae* (Murrill, 1902), and *G. zonatum* Murrill, 1902 (Murrill, 1902, 1908) in the United States. The taxonomic status of the *Ganoderma* complex is debated, with different ideas expressed regarding the members and their validity in the complex (Galappaththi et al., 2023). Given the confusion over species names, it's understandable that most people simply refer to all *Ganoderma* laccates as *G. lucidum*. According to

Haddow (1931), *G. sessile* was considered a synonym of *G. resinaceum*, and *G. lucidum* is believed to be the correct name for a specimen classified as *G. sessile* (Overholts, 1953). *G. lucidum* was also considered under its previous name before its later synonym *G. tsugae* (Haddow, 1931; Steyaert, 1977). However, Nobles (1965) showed, based on mating evidence that specimens classified as *G. lucidum* in the United States, were in fact, represented by *G. sessile*. Wang et al. (2009b) divided Asian specimens classified as *G. lucidum* into two clades, that represented *G. multipileum*, while the other clade was unknown, and Wang et al. (2012) acknowledged that clade as *G. sichuanense*. However, Cao et al. (2012) found that the holotype of *G. sichuanense* was not conspecific with an unknown clade and suggested it as a new species of *G. lingzhi*. It was reported that the Chinese lingzhi (*G. lucidum*) is different from *G. lucidum* found elsewhere in the world (Cao et al., 2012). Dai et al. (2017) stated that *G. lingzhi* is the true scientific name for the species used in Chinese medicine. Currently, the lingzhi or reishi grown in Asia has been found to represent several species, including *G. lingzhi*, *G. multipileum*, and *G. flexipes* (Loyd and Blanchette, 2019).

As a result of several recent taxonomic and molecular phylogenetic studies on *Ganoderma*, an unexpectedly high level of species diversity has been revealed worldwide, leading to the description of many new species (Cao et al., 2012; Cao and Yuan, 2013; Coetzee et al., 2015; Li et al., 2015; Tchotet Tchoumi et al., 2018, 2019; Xing et al., 2018; Liu et al., 2019; Luangharn et al., 2019; Wu et al., 2020; He et al., 2021, 2022).

A comprehensive study conducted by Sun et al. (2022) highlighted the presence of 180 confirmed *Ganoderma* species. The complete list of accepted *Ganoderma* species presented by Sun et al. (2022) is shown alphabetically in Table 1.1.

1.2.1 Morphological Characteristics

Ganoderma taxonomy is usually based on macro- and micro-morphological characteristics, making it easily distinguished due to its unique appearance, especially its double-walled and truncated basidiospores and pileipellis cells (Xing et al., 2018). Basidiocarp is the most commonly used characteristic to differentiate the *Ganoderma* species and is a sessile sexual structure in *Ganoderma* and other polypores, attached to living or, more commonly, dead or decaying trunks or branches of trees (Mawar et al., 2020). The pileus surface of *Ganoderma* could be laccate (Luangharn et al., 2019) or nonlaccate (Tchotet Tchoumi et al., 2019) and exhibits various colors such as red, black, blue/green, white, yellow, and purple (Chan et al., 2021). More precisely, it can be identified by colors like reddish brown, chocolate brown, dark brown, and yellowish brown (Xing et al., 2018; Tchotet Tchoumi et al., 2019; He et al., 2022).

The coloration of *Ganoderma* species can also change based on their freshness or dryness. For instance, *G. citriporum* appears red when fresh, but turns brown when dry (Gomes-Silva et al., 2011). Similarly, various colors can be seen from young to mature specimens (Loyd and Blanchette, 2019). Figure 1.1 shows a *Ganoderma* species with different growth stages.

Generally, the hyphal system is trimitic: generative, skeletal, and binding. It is also recognized based on the prominent type, diameter, hyaline, thin/thick walled, branched/unbranched, clamped, and septate or not (Coetzee et al., 2015). The shape of basidioles generally varies from pear-shaped to fusiform or clavate, depending on the species (He et al., 2022). Furthermore, the shape of basidiospores can also be almond-shaped, truncate (Xing et al., 2018), ellipsoid, or ovoid (He et al., 2022). Additionally, some research that employed cuticle anatomy as a criterion for determining the taxonomy of *Ganoderma* could not clearly characterize the observed features (Furtado, 1965a). As mentioned, numerous synonyms, species complexes, and potential misidentifications of species have arisen due to the taxonomy of *Ganoderma* species being solely based on the macro- and micromorphology of basidiocarps. This is due to the fact that various species frequently share basidiocarp traits, which makes it challenging to distinguish. For instance, the morphology of *G. lucidum* is similar to that of *G. destructans* (Coetzee et al., 2015). Also, *G. eickeri* and *G. knysnamense* resemble species in the subgenus *Elfvingia* (Tchotet Tchoumi et al., 2019).

TABLE 1.1
Complete Alphabetically Ordered List of *Ganoderma* Species Presented by Sun et al. (2022)

G. acaciicola	*G. fallax*	*G. ochrolaccatum*	*G. testaceum*
G. acontextum	*G. fassii*	*G. oerstedii*	*G. thailandicum*
G. adspersum	*G. fassioides*	*G. orbiforme*	*G. tongshanense*
G. aetii	*G. fici*	*G. oregonense*	*G. tornatum*
G. ahmadii	*G. flexipes*	*G. ostracodes*	*G. torosum*
G. alluaudii	*G. fuscum*		*G. trengganuense*
G. alpinum		*G. parvigibbosum*	*G. tropicum*
G. amazonense	*G. gabonensis*	*G. parvulum*	*G. trulla*
G. angustisporum	*G. ghesquierei*	*G. petchii*	*G. trulliforme*
G. applanatum	*G. gibbosum*	*G. pfeifferi*	*G. tsugae*
G. aridicola	*G. gilletii*	*G. philippii*	*G. tuberculosum*
G. aureolum	*G. guangxiense*	*G. piceum*	*G. turbinatum*
G. australe	*G. guianensis*	*G. platense*	
G. austroafricanum		*G. podocarpense*	*G. umbrinum*
	G. impolitum	*G. polychromum*	
G. bambusicola	*G. insulare*	*G. puerense*	*G. valesiacum*
G. barretoi		*G. puglisii*	*G. vanheurnii*
G. baudonii	*G. knysnamense*	*G. pulchella*	*G. vanmeelii*
G. bilobum	*G. kosteri*	*G. pygmoideum*	
G. boninense			*G. weberianum*
G. brownii	*G. lamaoense*	*G. ramosissimum*	*G. weixiense*
G. bruggemanii	*G. leucocontextum*	*G. ravenelii*	*G. wiiroense*
G. bubalinomarginatum	*G. leucocreas*	*G. resinaceum*	*G. williamsianum*
	G. leytense	*G. reticulatosporum*	
G. dejongii	*G. lingua*	*G. rhacodes*	*G. xylonoides*
G. destructans	*G. lingzhi*	*G. rothwellii*	
G. dianzhongense	*G. lobatoideum*	*G. rufoalbum*	*G. yunlingense*
G. dimidiatum	*G. lobatum*	*G. ryvardenii*	
G. donkii	*G. lobenense*		*G. zonatum*
G. dorsale	*G. longistipitatum*	*G. sanduense*	
G. dubio-cochlear	*G. lucidum*	*G. sarasinii*	
G. dunense	*G. luteicinctum*	*G. sculpturatum*	
G. dussii		*G. septatum*	
	G. magniporum	*G. sessile*	
G. ecuadorense	*G. mangiferae*	*G. sessiliforme*	
G. eickeri	*G. manoutchehrii*	*G. shanxiense*	
G. elegantum	*G. martinicense*	*G. sichuanense*	
G. ellipsoideum	*G. mbrekobenum*	*G. silveirae*	
G. endochrum	*G. megalosporum*	*G. sinense*	
G. enigmaticum	*G. melanophaeum*	*G. soyeri*	
G. esculentum	*G. mexicanum*	*G. steyaertianum*	
	G. miniatocinctum	*G. stipitatum*	
	G. mirabile	*G. subangustisporum*	
	G. mizoramense	*G.*	

1.2.2 Culture Characteristics

The *Ganoderma* species have been extensively studied for their cultural significance throughout history. There are, however, few scientific reports on systematic comparative morphological observation of mycelium, asexual sporulation in the life cycle, and their taxonomic value (Murrill, 1902; Coetzee et al., 2015).

FIGURE 1.1 A *Ganoderma* sp. growing on an underground dead wood at Chiang Mai, Thailand. Photo credit: A.K.H Priyashantha

Ganoderma species can be differentiated based on cultural characteristics, including chlamydospore production, growth rate, and thermophilic properties. These fungi typically exhibit optimum growth between pH 4.5 and 6, and, like many other fungi, they utilize glucose, maltose, and starch as the most suitable carbon sources for growth. As a matter of fact, some species (e.g., *G. lucidum*) are practically reluctant to utilize cellulose and lactose as carbon sources (Subedi et al., 2021). Nevertheless, different selected media have been used for in vitro assay of *Ganoderma*, including potato dextrose agar (PDA), malt extract agar (MEA), tryptone glucose extract agar (TGEA), yeast malt extract agar (YMEA), and mushroom complete medium (MCM) (Suansia and John, 2020). The optimum growth of these fungi requires nitrogen sources such as ammonium acetate, glycine, arginine, and calcium nitrate (Jayasinghe et al., 2008).

When it comes to culture characteristics, one unique trait of *Ganoderma* species is the occurrence of chlamydospores or chlamydospore-like swellings. It has been demonstrated that various *Ganoderma* species differ in their host relationships, production of the terminal and intercalary chlamydospores, average growth rate range (which varies from 2.1 mm/day to 7.8 mm/day), and thermophily range (which ranges from 20°C to 50°C). These differences are the most useful taxonomic criteria to distinguish their mycelial cultures and reveal their phylogenetic relationships among species (Badalyan et al., 2019). The chlamydospores could be hyaline or pigmented, and their shapes vary from ovate to spherical or irregular, or elliptical to obpyriform, to ovate, and even to globose, depending on the species. As mentioned, selected growth media can yield varying results in the culture characteristics, for example, chlamydospores may not be observable on MEA, even after 8 days of growth in the dark under optimum conditions for certain species, such as *G. curtisii*, *G. meredithiae*, *G. ravenelii*, *G. tsugae*, *G. tuberculosum*, and *G. zonatum*. In contrast, they may be visible in *G. martinicense*, *G. sessile*, and *G. weberianum* (Loyd et al., 2019). On the other hand, some studies have observed according to some studies, it has been observed that fast-growing *Ganoderma* cultures, which are also thermophilic, tend to produce a large number of ovoid chlamydospores. In contrast, slow-growing cultures that are not thermophilic are less likely to produce chlamydospores. This association between mycelial features is particularly striking, as demonstrated by Badalyan et al. (2019). Limited research has been conducted on the growth rate and thermophily of *Ganoderma* species. However, early work by Adaskaveg and Gilbertson (1989) found that temperature relationships can be used to distinguish *G. colossum*, *G. meredithiae*, and *G. zonatum*.

More precisely, in a recent study, Coetzee et al. (2015) demonstrated the colony characters, growth rate, and optimal temperature ranges of *G. enigmaticum*. As reported, the fungi show optimum growth at 30°C, reaching 85 mm in the dark in 7 days, followed by at 35°C reaching 75 mm, at 25°C reaching 73 mm, at 20°C reaching 32 mm, at 15°C reaching 9 mm, and zero growth at 10°C on 2% MEA. Meanwhile, *G. destructans* shows the best growth at 25°C, reaching 85 mm in the dark in 7 days, followed by at 30°C reaching 72 mm, at 20°C reaching 42 mm, at 15°C reaching 15 mm, and no growth at 35°C. However, in both cases, the mats are circular with flat edges, white above and creamy reverse at all temperatures, and have a felty, superficial mycelium with medium density. It is also noticeable that chlamydospores are absent (Coetzee et al., 2015). In a similar study, Crous et al. (2014), reported that *G. austroafricanum* shows optimum growth of 82 mm at 25°C in the dark in 8 days, followed by 62 mm at 25°C in the dark in 8 days, 37 mm at 20°C, and no growth at 35°C. Furthermore, *G. austroafricanum* differs from *G. enigmaticum* and *G. destructans*, in that the colonies are circular with a flat felt-like texture, at the entire edge, appearing white at all temperatures with a sporadic tint of yellow at the inner 20 mm circle at 30°C, and chlamydospores are also present.

1.2.3 Sexual Compatibility Tests

Sexual compatibility has also been evaluated throughout the genus *Ganoderma* since the beginning of taxonomic studies, although fewer studies focus on it today (Adaskaveg and Gilbertson, 1986; Pilotti et al., 2002, 2003, 2004, 2021). It has been used to assess whether two isolates belong to the same biological species and/or to identify unknown isolates (Adaskaveg and Gilbertson, 1986; Nieuwenhuis et al., 2013). The complete spectrum of mycelial interactions, ranging from weak to powerful, was observed between closely related species (Adaskaveg and Gilbertson, 1987). Nevertheless, those compatibility tests are not always accurate. For instance, Adaskaveg and Gilbertson (1986) found that the North American *G. lucidum* and the European isolate of *G. resinaceum* belonged to the same biological species, but later studies by Hong and Jung (2004) classified the two as phylogenetically distinct species.

1.2.4 Molecular-Based Identification

Molecular-based identification of *Ganoderma* species provides a reliable approach for taxonomic studies. DNA can be extracted from dried pieces of the pileus with tubes (Ile et al., 2021) and living mycelial cultures (Zhou et al., 2015) for molecular analysis. The genes' internal transcribed spacer (ITS), TEF1-α, and RPB2 are typically amplified using the polymerase chain reaction (PCR) technique, and subsequent sequencing is necessary. The rDNA regions, including ITS1 and ITS2, intergenic spacer (IGS) and RNA polymerase II subunit 2 (RPB2), RPB1, translation elongation factor 1-alpha (TEF1-α) (Kwon et al., 2016; Zhou et al., 2015; Zhang et al., 2017; He et al., 2021), nrLSU, nrSSU, β-tubulin, mtSSU, mtLSU, and *atp6* genes (Li et al., 2016a) have been used for phylogenetic analyses. ITS regions are highly polymorphic and are valuable tools for taxonomic and phylogenetic studies (Gottlieb et al., 2000). rDNA has been extensively used in the taxonomy and phylogeny of *Ganoderma* species (Moncalvo et al., 1995a; Gottlieb et al., 2000; Park et al., 2012; Palanna et al., 2022). This is because these regions, particularly the variability found in their introns, provide adequate resolution to infer phylogenetic relationships among various fungal genera (Gottlieb et al., 2000).

Several well-established molecular marker techniques, such as isozyme variation, DNA sequences in the ITS1 region, mitochondrial small-subunit ribosomal RNAs, random amplified polymorphic DNA (RAPD), restriction-amplified fragment length polymorphism (RFLP), and amplified fragment length polymorphism (AFLP), have been applied to the molecular characterization and investigation of phylogenetic relationships among *Ganoderma* species (Zheng et al., 2009). Sun et al. (2006) successfully used a PCR-based sequence-related amplified polymorphism (SRAP) marker for the first time to study the genetic variation among strains of *Ganoderma* and establish an identification system for these strains.

1.3 BENEFICIAL USES OF *GANODERMA*

Ganoderma products have gained significant interest in various countries, particularly in Asia, North America, and Europe, owing to their valuable properties, including nutritional and medicinal benefits (Hennicke et al., 2016; Loyd et al., 2018; El Sheikha, 2022).

Most importantly, China alone has reported 67 medicinal *Ganoderma* species (Li et al., 2016a). In 2003, the global trade value of *Ganoderma* products was only US$2.5 billion, indicating the popularity of these fungal products (Wang et al., 2012).

In addition, *Ganoderma* species have been used as functional food items and incoporated into several cosmetic ingredients. Furthermore, *Ganoderma* species have a high ornamental value (Du et al., 2019). *G. lucidum*, the most extensively studied species in the genus, has witnessed an increase in consumption for several reasons, including its use as an effective alternative to modern medicine (El Sheikha, 2022). China is the main producer and exporter of *G. lucidum*, and the species plays a significant role in the Chinese economy (Li et al., 2016a). Nowadays, a wide range of *Ganoderma*-based products is commercially available in the market (El Sheikha, 2022). As per the China National Medical Products Administration (NMPA) website, there are currently 1,233 registered *Ganoderma* products including 1,215 domestic and 18 imported products, encompassing capsules, granules, oral liquids, tablets, tea bags, and toothpaste (Li et al., 2019).

1.3.1 Therapeutic Effects of *Ganoderma*

The pharmacological activities of *Ganoderma*, which contains over 400 bioactive ingredients, including polysaccharides and ganoderic acids, have been extensively studied and widely acknowledged for their numerous benefits (Cör et al., 2018). These are the main functional metabolites of *Ganoderma*, and they have been reported to have antioxidant and anti-inflammatory properties (Yin et al., 2019). Furthermore, *Ganoderma* species, contain various glycoproteins, primarily β-glucans, as well as secondary metabolites, including nucleotide analogues, metal chelators, terpenoids, polyphenols, alkaloids, lactones, and sterols. These compounds have been reported to possess significant medicinal properties, as demonstrated by several studies (Xia et al., 2014; Baby et al., 2015; Hapuarachchi et al., 2018).

For instance, recent studies on *G. lucidum* have shown that polysaccharides, proteins, and triterpenoids are the main components that fight against various diseases, such as asthma, cerebral ischemia-reperfusion injury, cancer, hypertension, hypercholesterolemia, liver disorders, and obesity (Yin et al., 2019). Similarly, like the fruiting body of *G. lucidum*, *Ganoderma* spore powder

TABLE 1.2
Some Therapeutic Effects of *Ganoderma* Species

Therapeutic Effects	Species	References
Anticancer	*G. amboinense*	Hsu et al., 2008
	G. applanatum	Elkhateeb et al., 2018
	G. lucidum	Reis et al., 2015; Bryant et al., 2017; Zhao and He, 2018
	G. tsugae	Du et al., 2019
Antidiabetic	*G. applanatum*	Jung et al., 2005; Du et al., 2019
	G. atrum	Zhu et al., 2016; Du et al., 2019
	G. lucidum	Teng et al., 2011; Ma et al., 2015; Du et al., 2019
Antihyperglycemic	*G. lucidum*	Li et al., 2011
	G. applanatum	Vazirian et al., 2014; Du et al., 2019
Anti-inflammatory	*G. atrum*	Du et al., 2019
	G. capense	Zhou et al., 2014; Du et al., 2019
	G. colossus	El Dine et al., 2009; Du et al., 2019
	G. lucidum	Du et al., 2019; Yang et al., 2019
	G. sichuanense	Du et al., 2019
	G. sinense	Du et al., 2019
	G. tsugae	Ko et al., 2008
Antioxidant	*G. applanatum*	Liu et al., 2015
	G. capense	Jiang et al., 2016
	G. cochlear	Dou et al., 2014
	G. hainanense	Li et al., 2016b
	G. lucidum	Abdullah et al., 2012; Kana et al., 2015; Zeng et al., 2017
	G. pfelfferi	Du et al., 2019
	G. tsugae	Tseng et al., 2008
Antimicrobial	*G. boninense*	Abdullah et al., 2020
	G. lucidum	Karwa and Rai 2012; Sa-Ard et al., 2015
Cardiovascular problems	*G. lucidum*	Du et al., 2019
Hepatoprotective activity	*G. lucidum*	Pham et al., 2016
Immunomodulatory activity	*G. lucidum*	Wang et al., 2018
Neuroprotective activity	*G. lucidum*	Qin et al., 2019
Wound healing activity	*G. lucidum*	Cheng et al., 2013

has various therapeutic effects such as antitumor, antioxidation, immune regulation, liver function improvement, neuroprotection, antiradiation and chemotherapy damage, and free radical scavenging (Xu and Li, 2019). Table 1.2 summarizes the therapeutic effects of *Ganoderma* species based on recent findings.

1.3.2 *Ganoderma* as Functional Foods

As mentioned earlier, *Ganoderma* is being incorporated into various health foods to enhance human well-being and promote longevity. Several *Ganoderma* species are considered functional foods in a number of countries. For example, the Chinese government has approved over 1,000 types of *Ganoderma* health foods, and more recently, *G. tsugae* has also been recognized as a healthy food (Chen et al., 2016; Li et al., 2019). Valuable components can be extracted from various parts of *Ganoderma*, including the fruiting body, mycelium, and spores, to create a range of products such as soup, yogurt (Dong and Han, 2015), coffee, dietary supplements, drinks, powder, spore products, and syrups (Du et al., 2019). Additionally, *Ganoderma* (e.g., *G. lucidum*) is used alone or mixed with other herbs to produce medicinal wines and teas, which can help regulate the immune system and combat aging (Dong and Han, 2015). However, capsules are the most common formulation among these dietary supplements. In general, those *Ganoderma* extract–based foods act as an immunity booster, insomnia improvement agent, fatigue relief agent, and blood lipid and glucose lowering auxiliary agent (Li et al., 2019).

1.3.3 *Ganoderma* as Cosmetic Products

Ganoderma extracts are incorporated into several cosmetic products, particularly those manufactured in the United States, Southeast Asian countries, and some Asian and European countries. Many of these products are used for skin lightening (Jiang, 2015) and whitening (Hyde et al., 2010). In this regard, *G. lucidum* has demonstrated the highest inhibition of tyrosinase activity compared to other basidiomycetes, which is the most common way to lighten or whiten the skin and reduce melanin formation (Chien et al., 2008). Also, when combined with three other herbs and zinc, *G. lucidum* can promote hair growth in human males by reducing levels of dihydrotestosterone (DHT) or prostatic hyperplasia (Meehan, 2015). According to the Chinese NMPA website, there are approximately 3,000 cosmetic products containing *Ganoderma* extract. Currently, there are only around 350 cosmetic products on the market containing *Ganoderma* extract, including moisturizers, serums, masks, liquid or powder foundations, shampoos, skin repair creams, and sunscreens (Li et al., 2019). Table 1.3 shows some of the cosmetic products that include the derivatives of *G. lucidum* and their corresponding ingredients.

1.3.4 Other Uses of *Ganoderma*

In response to the environmental degradation due to the accumulation of highly toxic pollutants, scientists have started to explore various eco-friendly solutions, one of which is the utilization of microbes. In this aspect, Ipeaiyeda et al. (2020) found that *G. lucidum* has tremendous bioremediation potential for removing heavy metals lead, zinc, nickel, copper, and cadmium. In a similar study, Chang et al. (2020) reported significant findings regarding the potential application of *G. lucidum* to remove Pb^{2+} and Cd^{2+} from water.

Several studies have reported on *Ganoderma* as a biocontrol agent. For instance, Shahid et al. (2016) reported the effect of methanolic extracts of *G. lucidum* against phytopathogenic fungi, including *Fusarium oxysporum*, and *Alternaria alternata* which were isolated from *Calendula officinalis* (marigold). Asif et al. (2022) concluded that the mycelium and crude protein extracts from *G. lucidum* have the potential to control the growth of the tomato early blight pathogen *Alternaria solani*. Thus, *G. lucidum* is a potential candidate as a biological control agent to manage tomato early blight, offering a viable and eco-friendly substitute to chemical pesticides. The mycelium growth filtrate and extract of *G. lucidum* significantly reducedthe severity of Septoria leaf spot disease in tomato plants by over 90% (da Cruz et al., 2022). These results suggest that *G. lucidum* can be used in the treatment of plant diseases through the potential formulation of bioproducts.

Ganoderma is famous in Chinese history for its importance. Besides its therapeutic and health benefits, *Ganoderma* holds significant cultural value in China (Li et al., 2019). *G. lucidum* and *G. sinense* were used as bonsai to embellish gardens, decorations, and many other art products due to their texture and vibrent colors (Wang et al., 2014; Hapuarachchi et al., 2018). *Ganoderma* jade ornaments and *Ganoderma* porcelain enjoy great popularity among both domestic and foreign customers. In addition, the residue of *Ganoderma* contains chitin, and the biomaterial sacchachitin prepared from the fruiting body of *G. tsugae* can be used as artificial skin, blood vessels, dialysis membranes, synthetic contact lenses, and more (Li et al., 2019).

TABLE 1.3
Certain Cosmetics Containing *G. lucidum* and Their Ingredients

Product Name and Country	Uses	References
CV Skinlabs (Body Repair Lotion), USA	Anti-inflammatory and wound-healing ability	Wu et al., 2016
Dr. Andrew Weil for ORIGINS (Mega-Mushroom Skin Relief Face Mask), USA	Anti-inflammatory activity	Wu et al., 2016
FOUR SIGMA FOODS (Instant Reishi Herbal Mushroom Tea), UK	Immunity promoter	Wu et al., 2016
Hankook-Sansim Firming Cream (Tan Ryuk SANG), Korea	Make vitalized and tight skin	Wu et al., 2016
Kat Burki (Form Control Marine Collagen Gel), UK	Enhance elasticity, provide hydration, and boost collagen	Wu et al., 2016
LA BELLA FIGURA (Gentle Enzyme Cleanser), Italy	Antioxidant	Wu et al., 2016
EMBELLIR (Refresh Massage cream), France	Antiaging	Wu et al., 2016
Moon Juice (Spirit Dust), USA	Immune system promoter	Wu et al., 2016
Tela Beauty Organics (Encore, Texture and Style Paste), UK	Provides sun protection to hair and prevents discoloration	Wu et al., 2016
Yves Saint Laurent (Temps Majeur Elixir De Nuit), France	Antiaging	Wu et al., 2016
DXN Products (*Ganoderma* skin cleanser: DXN Ganozhi E Deep Cleansing Cream), UK	Removes impurities and dead skin cells	Hapuarachchi et al., 2018
Guangzhou Maycare cosmetics (Collagen crystal facial mask), China	Skin renewing and whitening	Hapuarachchi et al., 2018
Paris Skin Institute (Derma Sublime-Eye Crème Suprême), France	Delivers rejuvenating results while reducing the appearance of dark lines, dark circles under the eyes, and puffiness	Hapuarachchi et al., 2018
Guangzhou Ocean Cosmetic Beauty (*Ganoderma* Moisturizing Cream), China	Antiaging, antiwrinkle, anti-acne, removing dark circles, moisturizing, firming, nourishing, skin rejuvenation, sunscreen, and whitening agent	Hapuarachchi et al., 2018
DXN Products (*Ganoderma* cleansing milk: DXN Ganozhi E Hydrasoft Toner), UK	Purifies and minimizes pores; penetrates and revitalizes the skin	Hapuarachchi et al., 2018
DXN Products (*Ganoderma* day cream: DXN Ganozhi E UV Defense Day Cream), UK	Firms and moisturizes, hydrates, and protects the skin from UV rays	Hapuarachchi et al., 2018
DXN (Ganozhi Moisturizing Micro Emulsion), Malaysia	Hydrates and nourishes the skin	Hapuarachchi et al., 2018
Guangzhou Bocaly Bio-Tec. (*Ganoderma* Cells Repairing Anti-aging Face Mask), China	Firming, antiwrinkle, moisturizer, lightening, nourishing, pore cleaner, pigmentation correctors, and whitening	Hapuarachchi et al., 2018
MAVEX (AHA/BHA Peeling), Hong Kong	Skin antiaging and revitalizing, effective exfoliating, keratolytic, and biostimulating properties	Hapuarachchi et al., 2018

1.4 CONCLUSION

The taxonomy of *Ganoderma* has evolved over the past several centuries, and confusion among the species has largely been resolved through advanced phylogenetic studies based on molecular analysis. However, recent findings have highlighted the necessity of further extensive studies, as misinterpretation among the genera could be due to environmental plasticity, geographic isolation, and the transition wild-to-artificial cultures. *Ganoderma* is widely recognized for its exceptional medicinal value in treating a wide range of diseases. The global consumption of *Ganoderma* is rising, and an increasing number of commercially available products incorporate fungi as an active ingredient in food supplements, cosmetic products, and other items. Furthermore, the recent understanding of fungi in addressing the environmental pollution scenarios demonstrates an unfolding aspect, emphasizing the need for additional research to explore novel applications of fungi.

REFERENCES

Abdullah, N., S.M. Ismail, N. Aminudin, A.S. Shuib, B.F. Lau. 2012. Evaluation of selected culinary-medicinal mushrooms for antioxidant and ACE inhibitory activities. *Evid. Based Complementary Altern. Med.* 2012: 464238.

Abdullah, S., S.-E. Jang, M.-K. Kwak, K.P. Chong. 2020. *Ganoderma boninense* mycelia for phytochemicals and secondary metabolites with antibacterial activity. *J Microbiol.* 58, no. 12: 1054–1064.

Adaskaveg, J.E., R.L. Gilbertson. 1986. Cultural studies and genetics of sexuality of *Ganoderma lucidum* and *G. tsugae* in relation to the taxonomy of the *G. lucidum* complex. *Mycologia* 78, no. 5: 694–705.

Adaskaveg, J.E., R.L. Gilbertson. 1987. Vegetative incompatibility between intraspecific dikaryotic pairings of *Ganoderma lucidum* and *G. tsugae*. *Mycologia* 79, no. 4: 603–613.

Adaskaveg, J.E., R.L. Gilbertson. 1988. Basidiospores, pilocystidia, and other basidiocarp characters in several species of the *Ganoderma lucidum* complex. *Mycologia* 80, no. 4: 493–507.

Adaskaveg, J.E., R.L. Gilbertson. 1989. Cultural studies of four North American species in the *Ganoderma lucidum* complex with comparisons to *G. lucidum* and *G. tsugae*. *Mycol. Res.* 92, no. 2: 182–191.

Ames, A. 1913. A consideration of structure in relation to genera of the Polyporaceae. *Ann. Mycol.* 11, no. 3: 211–253.

Asif, M., A.A. Shahid, N. Ahmad. 2022. *Ganoderma lucidum* as a biocontrol agent for management of *Alternaria solani*, a pathogen of early blight of tomato. *Sarhad J. Agric.* 38, no. 2: 734–741.

Atkinson, G.F. 1908. Observations of *Polyporus lucidus* leys and some of its allies from Europe and North America. *Bot. Gaz.* 46, no. 5: 321–338.

Baby, S., A.J. Johnson, B. Govindan. 2015. Secondary metabolites from *Ganoderma*. *Phytochemistry* 114: 66–101.

Badalyan, S.M., N.G. Gharibyan, M. Iotti, A. Zambonelli. 2019. Morphological and ecological screening of different collections of medicinal white-rot bracket fungus *Ganoderma adspersum* (Schulzer) Donk (Agaricomycetes, Polyporales). *Ital. J. Mycol.* 48, no. 1: 1–15.

Bazzalo, M.E., J.E. Wright. 1982. Survey of the Argentine species of *Ganoderma lucidum* complex. *Mycotaxon* 16, no. 1: 293–325.

Bharudin, I., A.F.F. Ab Wahab, M.A. Abd Samad, N. Xin-Yie, M.A. Zairun, F.D. Abu Bakar, A.M. Abdul Murad. 2022. Review update on the life cycle, plant-microbe interaction, genomics, detection and control strategies of the oil palm pathogen *Ganoderma boninense*. *Biology* 11, no. 2: 251.

Binder, M., A. Justo, R. Riley, A. Salamov, F. Lopez-Giraldez, E. Sjökvist, A. Copeland, B. Foster, H. Sun, E. Larsson, K.H. Larsson, J. Townsend, I.V. Grigoriev, D.S. Hibbett. 2013. Phylogenetic and phylogenomic overview of the Polyporales. *Mycologia* 105, no. 6: 1350–1373.

Bresadola, G. 1979. *Omnia Bresadoliana extracta in unum collecta*. Trento: Gruppo Micologico G. Bresadola.

Bryant, J.M., M. Bouchard, A. Haque. 2017. Anticancer activity of Ganoderic acid DM: Current status and future perspective. *J. Clin. Cell Immunol.* 8, no. 6: 535.

Buchanan, P.K., J.P. Wilkie. 1995. Taxonomy of New Zealand *Ganoderma*: Two non-laccate species. In *Ganoderma: Systematics, phytopathology and pharmacology; proceedings of contributed symposium 59A,B, 5th international mycological congress, Vancouver*, ed. P.K. Buchanan, R.S. Hseu, J.M. Moncalvo, 7–17. Taipei: Hseu Ruey-Shyang, Applied Microbiology Laboratory, Agricultural Chemistry Department, National Taiwan University.

Cao, Y., S.H. Wu, Y.C. Dai. 2012. Species clarification of the prize medicinal *Ganoderma* mushroom "Lingzhi". *Fungal Divers.* 56: 49–62.

Cao, Y., H.S. Yuan. 2013. *Ganoderma mutabile* sp. nov. from Southwestern China based on morphological and molecular data. *Mycol. Prog.* 12: 121–126.

Chan, S.W., B. Tomlinson, P. Chan, C.W.K. Lam. 2021. The beneficial effects of *Ganoderma lucidum* on cardiovascular and metabolic disease risk. *Pharm. Biol.* 59, no. 1: 1161–1171.

Chang, J., H. Zhang, H. Cheng, Y. Yan, M. Chang, Y. Cao, F. Huang, G. Zhang, M. Yan. 2020. Spent *Ganoderma lucidum* substrate derived biochar as a new bio-adsorbent for Pb^{2+}/Cd^{2+} removal in water. *Chemosphere* 241: 125121.

Chen, R.Y., J. Kang, G.H. Du. 2016. Construction of the quality control system of *Ganoderma* products. *Edi. and Med. Mushroom* 24, no. 6: 339–344.
Cheng, P.G., C.W. Phan, V. Sabaratnam, N. Abdullah, M.A. Abdulla, U.R. Kuppusamy. 2013. Polysaccharides-rich extract of *Ganoderma lucidum* (M.A. Curtis:Fr.) P. Karst. accelerates wound healing in streptozotocin-induced diabetic rats. *Evid. Based Complementary Altern. Med.* 2013: 671252.
Chien, C.C., M.L. Tsai, C.C. Chen, S.J. Chang, C.H. Tseng. 2008. Effects on tyrosinase activity by the extracts of *Ganoderma lucidum* and related mushrooms. *Mycopathologia* 166, no. 2: 117–120.
Coetzee, M.P.A., S. Marincowitz, V.G. Muthelo, M.J. Wingfield. 2015. *Ganoderma* species, including new taxa associated with root rot of the iconic *Jacaranda mimosifolia* in Pretoria, South Africa. *IMA Fungus* 6, no. 1: 249–256.
Cör, D., Ž. Knez, M. Knez Hrnčič. 2018. Antitumour, antimicrobial, antioxidant and antiacetylcholinesterase effect of *Ganoderma lucidum* terpenoids and Polysaccharides: A review. *Molecules* 23, no. 3: 649.
Corner, E.J.H. 1983. Ad Polyporaceae I, *Amauroderma* and *Ganoderma*. *Nova Hedwigia* 75: 1–182.
Costa-Rezende, D.H., A. Góes-Neto, E.R. Drechsler-Santos. 2020a. Studies on Brazilian *Amauroderma* s. str. reveal a new species from the Atlantic forest, *Amauroderma robledoi* sp. nov. (Polyporales, Ganodermataceae). *J. Torrey Bot. Soc.* 147, no. 2: 199–205.
Costa-Rezende, D.H., G.L. Robledo, E.R. Drechsler-Santos, M. Glen, G. Gates, B.R. de Madrignac Bonzi, O.F. Popoff, E. Crespo, A. Góes-Neto. 2020b. Taxonomy and phylogeny of Polypores with Ganodermatoid basidiospores (Ganodermataceae). *Mycol. Progress* 19, no. 8: 725–741.
Crous, P.W., M.J. Wingfield, R.K. Schumacher, B.A. Summerell, A. Giraldo, J. Gené, J. Guarro, et al. 2014. Fungal planet description sheets: 281–319. *Persoonia* 33, no. 1: 212–289.
da Cruz, M.P., R.B. Felipini, M.M. Cardozo, S.M. Mazaro, R.M. Di Piero. 2022. *Ganoderma lucidum* mycelial growth filtrate and the mycelial extract increase defense responses against Septoria leaf spot in tomato. *Biol. Control* 173: 105002.
Dai, Y.C., Z.L. Yang, B.K. Cui, C.J. Yu, L.W. Zhou. 2009. Species diversity and utilization of medicinal mushrooms and fungi in China (review). *Int. J. Med. Mushrooms* 11, no. 3: 287–302.
Dai, Y.C., L.W. Zhou, T. Hattori, Y. Cao, J.A. Stalphers. 2017. *Ganoderma lingzhi* (Polyporales, Basidiomycota): The scientific binomial for the widely cultivated medicinal fungus Lingzhi. *Mycol. Prog.* 16: 1051–105.
Dong, C., Q. Han. 2015. *Ganoderma lucidum* (Lingzhi, *Ganoderma*): Fungi, algae, and other materials. In *Dietary Chinese herbs chemistry: Pharmacology and clinical evidence*, ed. Y. Liu, Z. Wang, J. Zhang, 759–765. London: Springer.
Donk, M.A. 1933. *Revision der Niederländischen Homobasidiomycetae-Aphyllophoraceae II*. Amsterdam-Haarlem: De Technische Boekhandel H. Stamquot.
Donk, M.A. 1948. Notes on Malesian fungi I. *Bull. du Jard. Bot. Buitenzorg* 17: 473–482.
Dou, M., L. Di, L.L. Zhou, Y.M. Yan, X.L. Wang, F.J. Zhou, Z.L. Yang, R.T. F.F. Hou, Y.X. Cheng. 2014. Cochlearols A and B, polycyclic meroterpenoids from the fungus *Ganoderma cochl*ear that have renoprotective activities. *Org. Lett.* 16, no. 23: 6064–6067.
Du, Z., C.H. Dong, K. Wang, Y.J. Yao. 2019. Classification, biological characteristics and cultivations of *Ganoderma*. In *Ganoderma and health: Advances in experimental medicine and biology*, ed. Z. Lin, B. Yang, 15–58. Singapore: Springer.
El Dine, R.S., A.M. El Halawany, C.M. Ma, M. Hattori. 2009. Inhibition of the dimerization and active site of HIV-1 protease by secondary metabolites from the Vietnamese mushroom *Ganoderma colossum*. *J. Nat. Prod.* 72, no. 11: 2019–2023.
Elkhateeb, W.A., G.M. Zaghlol, I.M. El-Garawani, E.F. Ahmed, M.E. Rateb, A.E. Abdel Moneim. 2018. *Ganoderma applanatum* secondary metabolites induced apoptosis through different pathways: In vivo and in vitro anticancer studies. *Biomed. Pharmacother.* 101: 264–277.
El Sheikha, A.F. 2022. Nutritional profile and health benefits of *Ganoderma lucidum* "Lingzhi, Reishi, or Mannentake" as functional foods: Current scenario and future perspectives. *Foods* 11, no. 7: 1030.
Furtado, J.S. 1965a. Relation of microstructure of the taxonomy of the Ganodermataceae (Polyporaceae) with special reference to the structure of the cover of the pilear. *Mycologia* 57, no. 4: 588–611.
Furtado, J.S. 1965b. *Ganoderma colossum* and the status of *Tomophagus*. *Mycologia* 57, no. 6: 979–984.
Furtado, J.S. 1981. *Taxonomy of Amauroderma (Basidiomycetes, Polyporaceac)*. Volume 34. New York: Botanical Garden.
Galappaththi, M.C.A., N.M. Patabendige, B.M. Premarathne, K.K. Hapuarachchi, S. Tibpromma, D.-Q. Dai, N. Suwannarach, S. Rapior, S.C. Karunarathna. 2023. A review of *Ganoderma* triterpenoids and their bioactivities. *Biomolecules* 13, no. 1: 24.
Gomes-Silva, A.C., L. Ryvarden, T.B. Gibertoni. 2011. New records of Ganodermataceae (Basidiomycota) from Brazil. *Nova Hedwigia* 92, no. 1–2: 83–94.

Gottlieb, A.M., E. Ferrer, J.E. Wright. 2000. rDNA analyses as an aid to the taxonomy of species of *Ganoderma*. *Mycol. Res.* 104, no. 9: 1033–1045.

Gottlieb, A.M., B.O. Saidman, J.E. Wright. 1995. Characterization of six isoenzymatic systems in Argentine representatives of two groups of *Ganoderma*. In *Ganoderma: Systematics, phytopathology and pharmacology; proceedings of contributed symposium 59A,B, 5th international mycological congress, Vancouver*, ed. P.K. Buchanan, R.S. Hseu, J.M. Moncalvo, 25–29. Taipei: Hseu Ruey-Shyang, Applied Microbiology Laboratory, Agricultural Chemistry Department, National Taiwan University.

Haddow, W.R. 1931. Studies in *Ganoderma*. *J. Arnold Arbor Harv. Univ.* 12, no. 1: 25–46.

Hansen, L. 1958. On the anatomy of the Danish species of *Ganoderma*. *Bot. Tidsskr.* 54: 333–352.

Hapuarachchi, K.K., W.A. Elkhateeb, S.C. Karunarathna, C.R. Cheng, A.R. Bandara, P. Kakumyan, K.D. Hyde, G.M. Daba, T.C. Wen. 2018. Current status of global *Ganoderma* cultivation, products, industry and market. *Mycosphere* 9, no. 5: 1025–1052.

Hapuarachchi, K.K., S.C. Karunarathna, P. Phengsintham, H.D. Yang, P. Kakumyan, K.D. Hyde, T.C. Wen. 2019. Ganodermataceae (Polyporales): Diversity in greater Mekong Subregion countries (China, Laos, Myanmar, Thailand and Vietnam). *Mycosphere* 10, no. 1: 221–316.

He, J., X. Han, Z.-L. Luo, E.-X. Li, S.-M. Tang, H.-M. Luo, K.-Y. Niu, X.-J. Su, S.-H. Li. 2022. Species diversity of *Ganoderma* (Ganodermataceae, Polyporales) with three new species and a key to *Ganoderma* in Yunnan Province, China. *Front. Microbiol.* 13: 1035434.

He, J., Z.L. Luo, S.M. Tang, Y.J. Li, S.H. Li, H.Y. Su. 2021. Phylogenetic analyses and morphological characters reveal two new species of *Ganoderma* from Yunnan Province, China. *MycoKeys* 84: 141–162.

He, M.-Q., R.-L. Zhao, K.D. Hyde, D. Begerow, M. Kemler, A. Yurkov, E.H. McKenzie, et al. 2019. Notes, outline and divergence times of Basidiomycota. *Fungal Divers.* 99, no. 1: 105–367.

Hennicke, F., Z. Cheikh-Ali, T. Liebisch, J.G. Maciá-Vicente, H.B. Bode, M. Piepenbring. 2016. Distinguishing commercially grown *Ganoderma lucidum* from *Ganoderma lingzhi* from Europe and East Asia on the basis of morphology, molecular phylogeny, and triterpenic acid profiles. *Phytochemistry* 127: 29–37.

Hong, S.G., H.S. Jung. 2004. Phylogenetic analysis of *Ganoderma* based on nearly complete mitochondrial small-subunit ribosomal DNA sequences. *Mycologia* 96, no. 4: 742–755.

Hou, D. 1950. A new species of *Ganoderma* from Taiwan. *Quat. J. Taiwan Mus.* 3: 101–105.

Hseu, R.S. 1990. *An identification system for cultures of Ganoderma species*. PhD diss., National Taiwan University, Taipei, Taiwan. (in Chinese).

Hsu, C.C., K.Y. Lin, Z.H. Wang, W.L. Lin, M.C. Yin. 2008. Preventive effect of *Ganoderma amboinense* on acetaminophen-induced acute liver injury. *Phytomedicine* 15, no. 11: 946–950.

Hyde, K.D., A.H. Bahkali, M.A. Moslem. 2010. Fungi—an unusual source for cosmetics. *Fungal Divers.* 43: 1–9.

Imazeki, R. 1939. Studies on *Ganoderma* of Nippon. *Bull. Nat. Sci. Mus. Tokyo* 1: 29–52. (in Japanese).

Imazeki, R. 1952. A contribution to the fungus flora of Dutch New Guinea. *Bull. Govt. Forest. Exp. St. Tokyo* 57: 87–128.

Index Fungorum 2024. https://www.indexfungorum.org/

Ipeaiyeda, A.R., C.O. Adenipekun, O. Oluwole. 2020. Bioremediation potential of *Ganoderma lucidum* (Curt:Fr) P. Karsten to remove toxic metals from abandoned battery slag dumpsite soil and immobilisation of metal absorbed fungi in bricks. *Cogent Environ. Sci.* 6, no. 1: 1847400.

Jayasinghe, C., A. Imtiaj, H. Hur, G.W. Lee, T.S. Lee, U.Y. Lee. 2008. Favorable culture conditions for mycelial growth of Korean wild strains in *Ganoderma lucidum*. *Mycobiology* 36, no. 1: 28–33.

Jiang, J., F. Kong, N. Li, D. Zhang, C. Yan, H. Lv. 2016. Purification, structural characterization and in vitro antioxidant activity of a novel Polysaccharide from Boshuzhi. *Carbohydr. Polym.* 147: 365–371.

Jiang, L. 2015. *Ganoderma lucidum* (Reishi mushroom): Potential application as health supplement and cosmeceutical ingredient. *Glob. J. Res. Anal.* 4, no. 9: 124–125.

Jung, S.H., Y.S. Lee, S.H. Shim, S. Lee, K.H. Shin, J.S. Kim, Y.S. Kim, S.S. Kang. 2005. Inhibitory effects of *Ganoderma applanatum* on rat lens aldose reductase and sorbitol accumulation in streptozotocin-induced diabetic rat tissues. *Phytother. Res.* 19, no. 6: 477–480.

Justo, A., O. Miettinen, D. Floudas, B. Ortiz-Santana, E. Sjökvist, D. Lindner, K. Nakasone, T. Niemelä, K.H. Larsson, L. Ryvarden, D.S. Hibbett. 2017. A revised family-level classification of the Polyporales (Basidiomycota). *Fungal Biol.* 121, no. 9: 798–824.

Kana, Y., T. Chen, Y. Wu, J. Wu. 2015. Antioxidant activity of Polysaccharide extracted from *Ganoderma lucidum* using response surface methodology. *Int. J. Biol. Macromol.* 72: 151–157.

Karsten, P.A. 1881. Enumeratio Boletinearum et Polyporearum Fennicarum, systemate novo dispositarum. *Rev. Mycol. Toulouse* 3, no. 9: 16–19.

Karsten, P.A. 1889. *Kritisk öfversigt af Finlands Basidsvampar (Basidiomycetes; Gastero- & Hymenomycetes)*. Volume 48. Finland: Finska litteratursällskapets tryckeri.

Karwa, A., M. Rai. 2012. Naturally occurring medicinal mushroom-derived antimicrobials: A case-study using Lingzhi or Reishi *Ganoderma lucidum* (W. Curt.: Fr.) P. Karst. (higher basidiomycetes). *Int. J. Med. Mushrooms* 14, no. 5: 481–490.

Ko, H.H., C.F. Hung, J.P. Wang, C.N. Lin. 2008. Anti-inflammatory triterpenoids and steroids from *Ganoderma lucidum* and *G. tsugae*. *Phytochemistry* 69, no. 1: 234–249.

Konara, U.A., K.M. Thambugala, K.K. Hapuarachchi. 2022. *Ganoderma* (Ganodermataceae, Polyporales): Historical perspectives, recent advances, and future research in Sri Lanka. *Stud. Fungi* 7: 17.

Kumar, H.M.A., M. Sarkar, K. Darshan, T. Ghoshal, B.S. Kavya, B.M. Bashayl, A.J.K. Asaiya, N. Berry. 2022. The *Ganoderma*: Biodiversity and significance. In *Fungal diversity, ecology and control management: Fungal biology*, ed. V.R. Rajpal, I. Singh, S.S. Navi, 255–291. Singapore: Springer.

Kwon, O.-C., Y.-J. Park, H.-I. Kim, W.-S. Kong, J.-H. Cho, C.-S. Lee. 2016. Taxonomic position and species identity of the cultivated Yeongji "*Ganoderma lucidum*" in Korea. *Mycobiology* 44, no. 1: 1–6.

Li, F., Y. Zhang, Z. Zhong. 2011. Antihyperglycemic effect of *Ganoderma lucidum* Polysaccharides on streptozotocin-induced diabetic mice. *Int. J. Mol. Sci.* 12, no. 9: 6135–6145.

Li, S., C. Dong, H. Wen, X. Liu. 2016a. Development of Lingzhi industry in China — emanated from the artificial cultivation in the institute of microbiology, Chinese academy of sciences (IMCAS). *Mycology* 7, no. 2: 74–80.

Li, T.H., H.P. Hu, W.Q. Deng, S.H. Wu, D.M. Wang, T. Tsering. 2015. *Ganoderma leucocontextum*, a new member of the *G. lucidum* complex from Southwestern China. *Mycoscience* 56, no. 1: 81–85.

Li, W., L.L. Lou, J.Y. Zhu, J.S. Zhang, A.A. Liang, J.M. Bao, G.H. Tang, S. Yin. 2016b. New lanostane-type triterpenoids from the fruiting body of *Ganoderma hainanense*. *Fitoterapia* 115: 24–30.

Li, Z., J. Zhou, Z. Lin. 2019. Development and innovation of *Ganoderma* industry and products in China. In *Ganoderma and health: Advances in experimental medicine and biology*, ed. Z. Lin, B. Yang, Volume 1181, 187–204. Singapore: Springer.

Liu, H., L.J. Guo, S.L. Li, L. Fan. 2019. *Ganoderma shanxiense*, a new species from Northern China based on morphological and molecular evidence. *Phytotaxa* 406, no. 2: 129–136.

Liu, H., X.G. Hou, J.H. Zhao, L. He. 2015. Liquid fermentation of *Ganoderma applanatum* and antioxidant activity of exopolysaccharides. *Open Biomed. Eng. J.* 9: 224–227.

Loyd, A.L., R.A. Blanchette. 2019. Glossy with grandeur: The laccate *Ganoderma* of North America. *Fungi* 12, no. 1: 31–36.

Loyd, A.L., E.R. Linder, M.E. Smith, R.A. Blanchette, J.A. Smith. 2019. Cultural characterization and chlamydospore function of the Ganodermataceae present in the Eastern United States. *Mycologia* 111, no. 1: 1–12.

Loyd, A.L., B.S. Richter, M.A. Jusino, C. Truong, M.E. Smith, R.A. Blanchette, J.A. Smith. 2018. Identifying the "mushroom of immortality": Assessing the *Ganoderma* species composition in commercial Reishi products. *Front. Microbiol.* 9: 1557.

Loyd, A.L., J.A. Smith, B.S. Richter, R.A. Blanchette, M.E. Smith. 2017. The laccate *Ganoderma* of the Southeastern United States: A cosmopolitan and important genus of wood decay fungi. *UF/IFAS Extension* 2017, no. 1: 333.

Luangharn, T., S.C. Karunarathna, A.K. Dutta, S. Paloi, I. Promputtha, K.D. Hyde, J. Xu, P.E. Mortimer. 2021. *Ganoderma* (Ganodermataceae, Basidiomycota) species from the Greater Mekong subregion. *J. Fungi.* 7, no. 10: 819.

Luangharn, T., S.C. Karunarathna, P.E. Mortimer, K.D. Hyde, N. Thongklang, J. Xu. 2019. A new record of *Ganoderma tropicum* (Basidiomycota, Polyporales) for Thailand and first assessment of optimum conditions for mycelia production. *MycoKeys* 51: 65–83.

Ma, H.T., J.F. Hsieh, S.T. Chen. 2015. Anti-diabetic effects of *Ganoderma lucidum*. *Phytochemistry* 114: 109–113.

Mawar, R., L. Ram, N.A. Deepesh, T. Mathur. 2020. *Ganoderma*. In *Beneficial microbes in agro-ecology*, ed. N. Amaresan, M. Senthil-Kumar, K. Annapurna, K. Kumar, A. Sankaranarayanan, 625–649. Amsterdam: Academic Press.

Meehan, K. 2015. *Composition to promote hair growth in humans*. U.S. Patent US9144542.

Moncalvo, J.M., H.F. Wang, R.S. Hseu. 1995a. Gene phylogeny of the *Ganoderma lucidum* complex based on ribosomal DNA sequences: Comparison with traditional taxonomic characters. *Mycol. Res.* 99, no. 12: 1489–1499.

Moncalvo, J.M., H.F. Wang, H.H. Wang, R.S. Hseu. 1995b. The use of ribosomal DNA nucleotide sequence data for species identification and phylogeny in the Ganodermataceae. In *Ganoderma: Systematics, phytopathology and pharmacology; proceedings of contributed symposium 59A,B, 5th international mycological congress, Vancouver*, ed. P.K. Buchanan, R.S. Hseu, J.M. Moncalvo, 31–44. Taipei: Hseu

Ruey-Shyang, Applied Microbiology Laboratory, Agricultural Chemistry Department, National Taiwan University.

Moncalvo, M., H.H. Wang, R.S. Hseu. 1995c. Phylogenetic relationships in *Ganoderma* inferred from the internal transcribed spacers and 25s ribosomal DNA sequences. *Mycologia* 87, no. 2: 223–238.

Murrill, W.A. 1902. The Polyporaceae of North America, genus I *Ganoderma*. *J. Torrey Bot. Soc.* 29, no. 10: 599–608.

Murrill, W.A. 1905. *Tomophagus* for *Dendrophagus*. *Torreya* 5: 197.

Murrill, W.A. 1908. Agaricales (Polyporaceae). *North Amer. Flora* 9.

Nieuwenhuis, B.P.S., S. Billiard, S. Vuilleumier, E. Petit, M.E. Hood, T. Giraud. 2013. Evolution of uni and bifactorial sexual compatibility systems in fungi. *Heredity* 111, no. 6: 445–455.

Nobles, M.K. 1948. Studies in forest pathology: VI. Identification of cultures of wood-rotting fungi. *Can. J. Res.* 26, no. 3: 281–431.

Nobles, M.K. 1958. Cultural characters as a guide to the taxonomy and phylogeny of the Polyporaceae. *Can. J. Bot.* 36, no. 6: 883–926.

Nobles, M.K. 1965. Identification of cultures of wood-inhabiting Hymenomycetes. *Can. J. Bot.* 43, no. 9: 1097–1139.

Otjen, L., R. Blanchette, M. Effland, G. Leatham. 1987. Assessment of 30 white rot basidiomycetes for selective lignin degradation. *Holzforschung* 41, no. 6: 343–349.

Overholts, L.O. 1953. *Polyporaceae of the United States, Alaska, and Canada*. Ann Arbor: University of Michigan Press.

Palanna, K.B., P.S. Koti, S. Basavaraj, B. Boraiah, T. Narendrappa. 2022. Morphological and molecular diversity of *Ganoderma* spp. causal agent of basal stem rot of coconut in Southern dry tracts of Karnataka. *J. Hortl. Sci.* 17, no. 2.

Park, Y.J., O.C. Kwon, E.S. Son, D.E. Yoon, W. Han, Y.B. Yoo, C.S. Lee. 2012. Taxonomy of *Ganoderma lucidum* from Korea based on rDNA and partial β-tubulin gene sequence analysis. *Mycobiology* 40, no. 1: 71–75.

Patouillard, N.T. 1887. Notes sur quelques champignons de l'Herbier du Mus6um d'Histoire naturelle de Paris. *J. Bot.* 1: 169–171.

Patouillard, N.T. 1889. Le genre *Ganoderma*. *Bull. Soc. Mycol. Fr.* 5: 64–80.

Pegler, D.N., T.W.K. Young. 1973. Basidiospore form in the British species of *Ganoderma* Karst. *Kew Bull.* 28, no. 3: 351–364.

Peng, J.T. 1990. *Identification and culture conservation of the wild Ganoderma species in Taiwan*. Wufeng, Taichung, Taiwan: Taiwan Agricultural Research Institute (TARI). (in Chinese).

Pham, H.N., L.S. Hoang, V.T. Phung. 2016. Hepatoprotective activity of *Ganoderma lucidium* (Curtis) P. Karst. against cyclophosphamide-induced liver injury in mice. *Cogent Biol.* 2, no. 1: 1267421.

Pilotti, C.A. 2005. Stem rots of oil palm caused by *Ganoderma boninense*: Pathogen biology and epidemiology. *Mycopathologia* 159, no. 1: 129–137.

Pilotti, C.A., G. Killah, D. Rama, E.A. Gorea, A.M. Mudge. 2021. A preliminary study to identify and distinguish Southern tropical populations of *Ganoderma boninense* from oil palm via mating assays, sequence data, and microsatellite markers. *Mycologia* 113, no. 3: 574–585.

Pilotti, C.A., F.R. Sanderson, E.A.B. Aitken. 2002. Sexuality and interactions of monokaryotic and dikaryotic mycelia of *Ganoderma boninense*. *Mycol. Res.* 106, no. 11: 1315–1322.

Pilotti, C.A., F.R. Sanderson, E.A.B. Aitken. 2003. Genetic structure of a population of *Ganoderma boninense* Pat. on oil palm. *Plant Pathol.* 52, no. 4: 455–463.

Pilotti, C.A., F.R. Sanderson, E.A.B. Aitken, W. Armstrong. 2004. Morphological variation and host range of two *Ganoderma* species from Papua New Guinea. *Mycopathologia* 158, no. 2: 251–265.

Qin, L.H., C. Wang, L.W. Qin, Y.F. Liang, G.H. Wang. 2019. Spore powder of *Ganoderma lucidum* for Alzheimer's disease: A protocol for systematic review. *Medicine* 98, no. 5: e14382.

Reis, F.S., R.T. Lima, P. Morales, I.C.F.R. Ferreira, M.H. Vasconcelos. 2015. Methanolic extract of *Ganoderma lucidum* induces autophagy of AGS human gastric tumor cells. *Molecules* 20, no. 10: 17872–17882.

Richter, C., K. Wittstein, P.M. Kirk, M. Stadler. 2014. An assessment of the taxonomy and chemotaxonomy of *Ganoderma*. *Fungal Divers.* 71, no. 1: 1–15.

Ryvarden, L. 1991. *Genera of Polypores: Nomenclature and taxonomy*. Synopsis Fungorum 5, Fungiflora, Oslo, Norway.

Sa-Ard, P., R. Sarnthima, S. Khammuang, W. Kanchanarach. 2015. Antioxidant, antibacterial and DNA protective activities of protein extracts from *Ganoderma lucidum*. *J. Food Sci. Technol.* 52, no. 5: 2966–2973.

Saccardo, P.A. 1882–1928. *Sylloge Fungorum omnium hurscusque cognitorum*. Volume I–XXIV. New York: Johnson Reprint Corporation.

Sarkar, A. 1959. Developmental anatomy and differentiation of tissue systems of *Ganoderma lucidum* (Leyss. ex Fries) Karst. *Phyton*. 13: 89–104.

Seo, G.S., Y. Kitamoto. 1988. Morphological features and morphogenesis in the *Ganoderma lucidum* complex. *Int. J. Med. Mushrooms* 6, no. 2: 43–54.

Shahid, A.A., M. Asif, M. Shahbaz, M. Ali. 2016. *Antifungal potential of Ganoderma lucidum extract against plant pathogenic fungi of Calendula officinalis L.* 5th International Conference on Biological, Chemical and Environmental Sciences (BCES-2016), London, 24–25 March.

Stevenson, J.A. 1933. *General index to the mycological writings of C.G. Lloyd (1898–1925)*. Cincinnati, OH: Lloyd Library and Museum.

Steyaert, R.L. 1972. Species of *Ganoderma* and related genera mainly of the Bogor and Leiden herbaria. *Persoonia* 7, no. 1: 55–118.

Steyaert, R.L. 1977. Basidiospores of two *Ganoderma* species and others of two related genera under the scanning electron microscope. *Kew Bull.* 31, no. 3: 437–442.

Steyaert, R.L. 1980. Study of some *Ganoderma* species. *Bull. Jard. Bot. Nat. Belg.* 50, no. 1–2: 135–186.

Suansia, A., P. John. 2020. Morphological and cultural characteristics of different collections of medicinal white-rot bracket fungi *Ganoderma* P. Karst. *Int. J. Curr. Microbiol. App. Sci.* 9, no. 1: 2636–2644.

Subedi, K., B.B. Basnet, R. Panday, M. Neupane, G.R. Tripathi. 2021. Optimization of growth conditions and biological activities of Nepalese *Ganoderma lucidum* strain Philippine. *Adv. Pharmacol. Sci.* 2021: 4888979.

Sun, S.-J., W. Gao, S.-Q. Lin, J. Zhu, B.-G. Xie, Z.-B. Lin. 2006. Analysis of genetic diversity in *Ganoderma* population with a novel molecular marker SRAP. *Appl. Microbiol. Biotechnol.* 72, no. 3: 537–543.

Sun, Y.F., J.H. Xing, X.L. He, D.M. Wu, C.G. Song, S. Liu, J. Vlasák, G. Gates, T.B. Gibertoni, B.K. Cui. 2022. Species diversity, systematic revision and molecular phylogeny of Ganodermataceae (Polyporales, basidiomycota) with an emphasis on Chinese collections. *Stud. Mycol.* 101: 287–415.

Tchotet Tchoumi, J.M., M.P.A. Coetzee, M. Rajchenberg, J. Roux. 2019. Taxonomy and species diversity of *Ganoderma* species in the garden route national park of South Africa inferred from morphology and multilocus phylogenies. *Mycologia* 111, no. 5: 1–18.

Tchotet Tchoumi, J.M., M.P.A. Coetzee, M. Rajchenberg, M.J. Wingfield, J. Roux. 2018. Three *Ganoderma* species, including *Ganoderma dunense* sp. nov., associated with dying *Acacia cyclops* trees in South Africa. *Australas. Plant Pathol.* 47: 431–447.

Teixeira, A.R. 1962. The taxonomy of the Polyporaceac. *Biol. Rev.* 37, no. 1: 51–81.

Teng, B.-S., C.-D. Wang, H.-J. Yang, J.-S. Wu, D. Zhang, M. Zheng, Z-.H. Fan, D. Pan, P. Zhou. 2011. A protein tyrosine phosphatase 1B activity inhibitor from the fruiting bodies of *Ganoderma lucidum* (Fr.) Karst and its hypoglycemic potency on streptozotocin induced type 2 diabetic mice. *J. Agric. Food Chem.* 59, no. 12: 6492–6500.

Torres-Torres, M.G., L. Ryvarden, L. Guzmán-Dávalos. 2015. *Ganoderma* subgenus *Ganoderma* in Mexico. *Rev. Mex. Micol.* 41: 27–45.

Torrend, C. 1920. Les Polyporacées du Brésil. *Brotéria Bot.* 18, no. 1: 23–142.

Tseng, Y.H., J.H. Yang, J.L. Mau. 2008. Antioxidant properties of Polysaccharides from *Ganoderma tsugae. Food Chem.* 107, no. 2: 732–738.

Vazirian, M., S. Dianat, A. Manayi, R. Ziari, A. Mousazadeh, E. Habibi, S. Saeidnia, Y. Amanzadeh. 2014. Anti-inflammatory effect, total Polysaccharide, total phenolics content and antioxidant activity of the aqueous extract of three basidiomycetes. *RJP* 1, no. 1: 13–19.

Wang, B.C., J. Hua. 1991. *A cultural atlas of some Ganoderma cultures*. Hsinchu, Taiwan: Food Industry Research and Development Institute (FIRDI).

Wang, C., S. Shi, Q. Chen, S. Lin, R. Wang, S. Wang, C. Chen. 2018. Antitumor and immunomodulatory activities of *Ganoderma lucidum* Polysaccharides in glioma-bearing rats. *Integr. Cancer Ther.* 17, no. 3: 674–683.

Wang, D.M., S.H. Wu, T.H. Li. 2009a. Two records of *Ganoderma* new to mainland China. *Mycotaxon* 108: 35–40.

Wang, D.M., S.H. Wu, C.H. Su, J.T. Peng, Y.H. Shih, L.C. Chen. 2009b. *Ganoderma multipileum*, the correct name for "*G. lucidum*" in tropical Asia. *Bot. Stud.* 50: 451–458.

Wang, X.-C., R.-J. Xi, Y. Li, D.-M. Wang, Y.-J. Yao. 2012. The species identity of the widely cultivated *Ganoderma*, "*G. lucidum*" (Lingzhi), in China. *PLOS One* 7, no. 7: e40857.

Wang, X.D., Y. Zhang, H.Y. Jiang, Y. Luo, X. Zhang, Y. Li. 2014. Cultivation of Zi-Zhi and bonsai production. *Edible Fungi* 36, no. 4: 51–52. (in Chinese).

Wasser, S.P., I.V. Zmitrovich, M.Y. Didukh, W.A. Spirin, V.F. Malysheva. 2006. *Morphological traits of Ganoderma lucidum complex highlighting G. tsugae var. jannieae: The current generalization*. Ruggel, Germany: A.R.A. Gantner Verlag K.-G.

Wong, W.C., H.J. Tung, M.N. Fadhilah, F. Midot, S.Y.L. Lau, L. Melling, S. Astari, D. Hadziabdic, R.N. Trigiano, K.J. Goh, Y.K. Goh. 2021. Genetic diversity and gene flow amongst admixed populations of *Ganoderma boninense*, causal agent of basal stem rot in African oil palm (*Elaeis guineensis* Jacq.) in Sarawak (Malaysia), Peninsular Malaysia, and Sumatra (Indonesia). *Mycologia* 113, no. 5: 902–917.

Wu, S.H., C.L. Chern, C.L. Wei, Y.P. Chen, M. Akiba, T. Hattori. 2020. *Ganoderma bambusicola* sp. nov. (Polyporales, Basidiomycota) from Southern Asia. *Phytotaxa* 456, no. 1: 75–85.

Wu, Y., M.-H. Choi, J. Li, H. Yang, H.-J. Shin. 2016. Mushroom cosmetics: The present and future. *Cosmetics* 3, no. 3: 22.

Xia, Q., H. Zhang, X. Sun, H. Zhao, L. Wu, D. Zhu, G. Yang, Y. Shao, X. Zhang, X. Mao, L. Zhang, G. She. 2014. A comprehensive review of the structure elucidation and biological activity of triterpenoids from *Ganoderma* spp. *Molecules* 19, no. 11: 17478–17535.

Xing, J.-H., Y.-F. Sun, Y.-L. Han, B.-K. Cui, Y.-C. Dai. 2018. Morphological and molecular identification of two new *Ganoderma* species on *Casuarina equisetifolia* from China. *MycoKeys* 34: 93–108.

Xu, J., P. Li. 2019. Researches and application of *Ganoderma* spores powder. In *Ganoderma and health: Advances in experimental medicine and biology*, ed. Z. Lin, B. Yang, Volume 1181, 157–186. Singapore: Springer.

Yang, Y., H. Zhang, J. Zuo, X. Gong, F. Yi, W. Zhu, L. Li. 2019. Advances in research on the active constituents and physiological effects of *Ganoderma lucidum*. *Biomed. Dermatol.* 3: 6.

Yin, Z., B. Yang, H. Ren. 2019. Preventive and therapeutic effect of *Ganoderma* (Lingzhi) on skin diseases and care. In *Ganoderma and health: Advances in experimental medicine and biology*, ed. Z. Lin, B. Yang, Volume 1182, 311–321. Singapore: Springer.

Zeng, Q., F. Zhou, L. Lei, J. Chen, J. Lu, J. Zhou, K. Cao, L. Gao, F. Xia, S. Ding, L. Huang, H. Xiang, J. Wang, Y. Xiao, R. Xiao, J. Huang. 2017. *Ganoderma lucidum* Polysaccharides protect fibroblasts against UVB-induced photoaging. *Mol. Med. Rep.* 15, no. 1: 111–116.

Zhang, X., Z. Xu, H. Pei, Z. Chen, X. Tan, J. Hu, B. Yang, J. Sun. 2017. Intraspecific variation and phylogenetic relationships are revealed by ITS1 secondary structure analysis and single-nucleotide polymorphism in *Ganoderma lucidum*. *PLOS One* 12, no. 1: e0169042.

Zhao, J.D. 1989. *The Ganodermataceae in China*. Bibliotheca Mycologica, Band 132. Berlin, Stuttgart: J. Cramer.

Zhao, J.D., L.W. Xu, X.Q. Zhang. 1983. Taxonomic studies on the family Ganodermataceae of China II. *Acta Mycol. Sin.* 2: 159–167.

Zhao, R.-L., Y.-M. He. 2018. Network pharmacology analysis of the anti-cancer pharmacological mechanisms of *Ganoderma lucidum* extract with experimental support using Hepa1-6-bearing C57 BL/6 mice. *J. Ethnopharmacol.* 210: 287–295.

Zheng, L., D. Jia, X. Fei, X. Luo, Z. Yang. 2009. An assessment of the genetic diversity within *Ganoderma* strains with AFLP and its PCR-RFLP. *Microbiol. Res.* 164, no. 3: 312–321.

Zhou, L.W., Y. Cao, S.H. Wu, J. Vlasák, D.W. Li, M.J. Li, Y.C. Dai. 2015. Global diversity of the *Ganoderma lucidum* complex (Ganodermataceae, Polyporales) inferred from morphology and multilocus phylogeny. *Phytochemistry* 114: 7–15.

Zhou, Y., S. Chen, R. Ding, W. Yao, X. Gao. 2014. Inflammatory modulation effect of glycopeptide from *Ganoderma capense* (Lloyd) Teng. *Mediators Inflamm.* 2014: 691285.

Zhu, K.X., S.P. Nie, L.H. Tan, C. Li, D.M. Gong, M.Y. Xie. 2016. A Polysaccharide from *Ganoderma atrum* improves liver function in type 2 diabetic rats via antioxidant action and short-chain fatty acids excretion. *J. Agric. Food Chem.* 64, no. 9: 1938–1944.

2 Cultivation Strategies of *Ganoderma* or the Reishi Mushroom

Prakash Pradhan[1], *Jayita De, and Krishnendu Acharya*[2]

1 University of Calcutta, Kolkata, India and West Bengal Biodiversity Board, Kolkata, India

2 University of Calcutta, Kolkata, India

2.1 INTRODUCTION

The twenty-first century has seen many scientific developments as well as human health catastrophes. While medicines and vaccines have been developed to support human health, there has been continuous pressure from rise of novel human diseases and viruses, including COVID-19. With the aim to maintain wellness through natural means and negate suppression of immunity and resistivity towards certain synthetic drugs due to prolonged usage, scientists are in search of foods with functional and nutraceutical properties (Lin, 2009). Besides higher plants, mushrooms and algae are two important alternative natural food sources (Ghosh, 2004). Being endowed with artistic beauty, exotic taste, delectable texture, and a wide range of biologically active compounds, wild edible mushrooms have received wide attention from both laymen and the pharmacological and scientific fraternity alike (Wasser, 2002; Cha and Yoo, 2009; Kakon et al., 2012).

Since time immemorial, mushrooms have been known for their culinary, nutritional, medicinal, and gastronomic features. Many bioactive compounds found in the mushrooms have demonstrated effective results in combating several chronic and acute human health disorders (Zhou et al., 2020). They are popularly known as "food of the gods" and the "elixir of life" (Thakur and Singh, 2013). Major bioactive metabolites of mushrooms can be broadly grouped into proteins, polysaccharides, fats, alkaloids, glycosides, tocopherols, flavonoids, phenolics, folates, carotenoids, terpenoids, organic acids, enzymes, etc. Overpopulation with accompanying malnutrition is a serious problem faced by the people of countries like India, where there is an urge to develop a successful functional food. Cultivation of mushrooms can be regarded as a source of income for the coming generations to fight poverty and venture into the entrepreneurship sectors (Sana et al., 2017; Deshmukh et al., 2018).

Many mushroom genera that can be used as food or supplementary aids have been explored (Cha and Yoo, 2009). Among such genera, *Ganoderma*, under the division Eumycota, sub-division Basidiomycotina, class Hymenomycetes, order Aphyllophorales, and family Ganodermataceae (Polyporaceae), is a notable fungal resource with human utility. Polyporaceae is one of the most important families of Basidiomycota, consisting of some of the major fungal components in forest ecosystems, especially as saprophytes and parasites (Hibbett and Donoghue, 1995). Besides their ecological role, some of them are valued as medicine and indigenous food (Cui et al., 2019; Silva-Neto and Pinto, 2021). *Ganoderma lucidum* was first mentioned in 221–227 BC, during the reign of the Chinese emperor, Shin-Huang of the Ch'in Dynasty (Wagner et al., 2003). The genus *Ganoderma* P. Karst. of Polyporaceae contains facultative parasites causing white rot wood decay in a range of tree species mostly distributed in tropical areas, though also distributed in temperate Europe (Adaskaveg et al., 1988, 1991; Xing et al., 2018; GBIF, n.d.). Mycobank.org (accessed on 03.06.2021) has 502 listed records of sub-generic taxa under *Ganoderma*, out of which 5 are illegitimate names, 9 are invalid names, and 36 orthographic variants (Alam et al., 2010). While

DOI: 10.1201/9781003354789-2

Ganoderma exhibits laccate (upper, shiny surface) or non-laccate (upper dull surface) type macro-morphological diversity (Smith and Sivasithamparam, 2003; Pilotti et al., 2004), its unique microscopic features include truncated double-walled basidiospores with pillar-like ornamentation on their inner walls (Cao et al., 2013).

Members of *Ganoderma* such as *G. lucidum* (reishi in Japanese), *G. lingzhi* (Líng zhī in Chinese), and *G. sinense* (Zǐ zhī in Chinese) have been an invaluable part of Chinese traditional medicine for over 2000 years (Yang and Liau, 1998; Wagner et al., 2003; Hyde et al., 2019). Aș per Chinese and Japanese indigenous beliefs, *Ganoderma* spp. possesses strong healing power and is regarded as a "Celestial herb", "Mushroom of immortality", and "Auspicious herb" indicating spiritual potency, sanctity, happiness, longevity, well-being, and success (Babitskaya et al., 2003; Podgornik et al., 2007; Sanodiya et al., 2009; Baby et al., 2015; Hapuarachchi et al., 2016; Bijalwan et al., 2021). In addition, *G. mbrekodenum* and *G. enigmaticum* have been the focus of domestication and commercial cultivation (Ofodile et al., 2022).

Among the mushrooms used by humans, *Ganoderma* spp. are unique in the sense that they are valued for their pharmaceutical rather than nutritional properties (Wachtel-Galor, 2011). Wide scientific studies on notable species of *Ganoderma*, such as *G. lucidum*, have provided scientific support to some of the ancient claims of the therapeutic health benefits ascribed to the mushroom. Some of the health benefits credited to this *Ganoderma* spp. are presented in Figure 2.1 (Harsh et al., 1993; Ling et al., 1997; Chang and Buswell, 1999; Chang and Miles, 2004; Aydemir, 2002; Wagner et al., 2003; Sliva, 2006; Wachtel-Galor, 2011; Cao et al., 2006, 2012; Russell and Paterson, 2006; Jeong et al., 2008; Deepalakshmi and Mirunalini, 2011; Hung and Nhi, 2012; De Silva et al., 2012; Rawat et al., 2012; Agarwal et al., 2013; Boh, 2013; Celık et al., 2014; Kaur et al., 2015; Roy et al., 2015; Wińska et al., 2019). As a widely distributed fungus, *Ganoderma* has attracted extensive attention for its economic value. *Ganoderma* has been developed into a variety of products such as powders, teas, and dietary supplements, with global sales of these products estimated to exceed $2.5 billion, 34 with sales growing at an annual rate of 18% (Wang et al., 2020).

Traditional cultivation methods are known to be tedious and time-consuming, while the artificial cultivation process, which utilizes commonly available substrates like wood logs, sawdust, cereal grains, rice bran, wheat bran, and agricultural waste, is ideal for large-scale cultivation, as it produces *Ganoderma* fruitbodies in relatively less time. To surpass the traditional cultivation methods, researchers have drawn their attention to the artificial cultivation methods for growth (Dadwal and Jamaluddin, 2004; Tripathy, 2010). Research is going on to develop suitable techniques for higher productivity with minimal expenditure (Yu et al., 2003; Wasser, 2004). For the very first time, *G. lucidum* was traditionally cultivated in China in the year 1969 and artificially cultivated in 1971 with successful results (Yu et al., 2003; Singh et al., 2014). *G. lucidum*, which is in demand as a

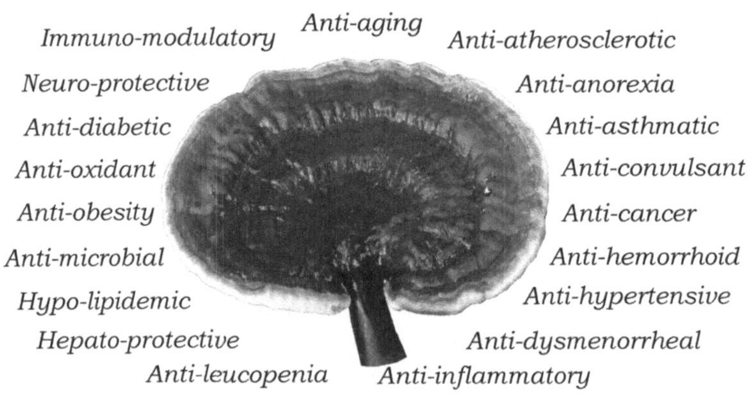

FIGURE 2.1 Medicinal properties of *Ganoderma* spp.

medicine in most Asian countries like China, South Korea, and Japan, presently has a market of around $2.5 billion (Hapuarachchi et al., 2016). As natural sources of *Ganoderma* are limited and rampant natural collection would be ecologically unsound, the sustainable option of cultivating *Ganoderma* spp., especially *G. lucidum*, has been adopted globally (Wasser, 2004; Barros et al., 2008; Banuelos and Lin, 2009; Bishop et al., 2015).

In order to achieve high yields on a broad scale, we have explored the growth parameters of *G. lucidum* and attempted to synthesize the work that has been done in relation to its commercial production. In the current work, only solid-state fermentation techniques have been reported together with their current prospects. The discussion also includes a number of supplements that are employed in the artificial culture of *G. lucidum*. This section seeks to engage readers by providing clear and straightforward methods for studying and putting into practice the most practical *Ganoderma* cultivation strategy.

2.2 FUNGAL STRAIN ISOLATION AND PURE CULTURE PREPARATION

Fruitbodies of *Ganoderma* may be collected from the wild (Figure 2.2), from various host trees such as *Picea abies, Betula pubescens, Eucalyptus* spp., *Acacia* spp., *Quercus ilex* subsp. *ballota* etc., or its mycelial cultures may be procured from type culture collection centers. While the direct mycelial source would not need much preprocessing, fruitbodies collected from the field have to be first sterilized with 70% ethanol, preferably under laminar airflow. After sterilization, a sterilized scalpel is used to scrape tissue just beneath the upper surface of the cap. This cut portion is taken out and aseptically transferred to a Petri plate with desirable culture media e.g., 2% potato dextrose agar (PDA; 20 g/L) or malt extract (ME; 20 g/L; bacteriological agar 15 g/L) (Suberu et al., 2013), and incubated at 24–30°C and regularly checked every 7 days. After 10–18 days, a colony

FIGURE 2.2 Fruitbodies of *Ganoderma lucidum* growing in their natural habitat in Eastern India.

diameter of around 8.5 cm is observed in Petri plates (Cortina-Escribano et al., 2020), and it may be sub-cultured at an interval of 3 weeks and further stored at a temperature of 4°C (Wagner et al., 2003; Roy et al., 2015; Cortina-Escribano et al., 2020; Bijalwan et al., 2021). In addition, molecular techniques may be followed for species and strain verification by studying the rDNA-ITS sequences (White et al., 1990; Zheng et al., 2009; Nithya et al., 2014). White et al. (1990) have amplified ITS sequences in a thermal cycler using forward ITS1–F (5'-CTTGGTCATTTAGAGGAAGTAA-3') as well as reverse ITS4 (5'-TCCTC- CGCTTATTGATATGC-3') primers. This protocol involves an initial cycle of denaturation at 95°C for 3 minutes and 35 successive cycles of subsequent denaturation of 30 seconds at 95°C, annealing of 30 seconds at 52°C, and extension of 1 minute at 72°C (Cortina-Escribano et al., 2020). The generated nucleotide sequence is at first compared to those deposited in GenBank with the help of the NCBI-BLAST program. Alignment of DNA sequences is performed by MAFFT 7 and optimization by BioEdit v7.2.3. Phylogenetic analyses are also carried out using PAUP v4.0b 10. This is followed by identification of the obtained sequence to species of *Ganoderma* (Manavalan et al., 2012). Ofodile et al. (2022) has also provided details of DNA isolation, ITS amplification, and sequencing of *Ganoderma* isolates. Radial growth (mm) in fungal pure cultures can be studied for comparison of available commercial strains (Kaliyaperumal, 2013).

However, the selection of a strain must be prioritized before undertaking full-scale production. As evident from the work of Cortina-Escribano et al. (2020), who studied six strains of *G. lucidum* (MUS9, MUS75, MUS19, MUS12, MUS192, and MUS6), there are strain-specific mycelial growth differences in terms of growth rate and mycelial density (Al-Obaidi et al., 2016; Alheeti et al., 2020).

2.3 PREPARATION OF SPAWN

Spawn may be either solid type or liquid type. While solid spawn may be prepared using grains or a sawdust-bran mix, the liquid spawn may be prepared in potato dextrose broth. Solid spawn can be classified according to the variety of substrates used, such as poplar billets (10–15 cm lengths and 4–5 cm thick segments), wheat grains (59.1% w/w), or other substrates like bran sawdust. In the case of wheat grain spawn, the grains weighing 1 kg are boiled until they become translucent. It has been investigated by the Bangladesh Council of Scientific and Industrial Research (BCSIR) that a ratio of 9:1, with 9 parts sawdust and 1 part wheat bran/rice bran, has elevated the production of *G. lucidum* fruiting bodies (Roy et al., 2015). About 400 g of grains are filled in polypropylene bags (18 cm × 25 cm) along with 1 g of lime with 0.5 g gypsum powder and 0.8% w/w $CaSO_4$ to maintain optimum pH (González-Matute et al., 2002; Roy et al., 2015). The necks of these bags are prepared using polyvinyl chloride (PVC) tubes, which are heat-resistant. With the help of a sharp-ended stick, a hole measuring 2–3 cm deep is made to put the inoculum in the center. These bags are sealed using cotton plugs and packed using sterilized brown paper and tied with rubber bands. Autoclaving of these bags is performed at 121°C at 15 pounds inch^2 pressure and ultimately cooled for 24 hours in an aseptic chamber. The sterilized bags are inoculated by removal of the cotton plugs, with pure *G. lucidum* culture in front of laminar airflow. These inoculated bags are again plugged with cotton and kept in an incubator at 25 ± 1°C for 15 days (Ajith and Janardhanan, 2006; Uddin et al., 2011; Azizi et al., 2012; Singh et al., 2014). Besides plastic bags, spawn substrate of sorghum grains and calcium carbonate (3:1), having a moisture content of 70%, may be autoclaved in glass bottles. From 12-day-old PDA culture, 5 agar blocks of 1 cm diameter may be inoculated to such bottled, sterilized, and cooled spawn substrate. Colonization of the substrate is reported to take around 14 days, after which it may be used in the inoculation of cultivation substrates (Ofodile et al., 2022).

Preparation of liquid spawn involves the following steps, (i) culturing of mycelia in PDA Petri plates at 25°C for a duration of 7–10 days in the dark; (ii) a 1-cm diameter mycelial plug is generated from a colony aged about 1 week, is taken out, is incubated in a 500-mL flask containing 150 mL PDB media, and kept at 25–30°C at 150–200 rpm in rotary shaker incubator; (iii) after 8–10 days, mycelial inoculum for the fermenter would be ready, and the ideal quantity of this inoculum is added to the sterilized fermenter; and (iv) 20–30 minutes of fermentation at 25°C would yield around 10–15% (v/v) of inoculum, which may be harvested and used as planting spawn (Zhou, 2017).

2.4 GROWTH CONDITIONS FOR FRUITBODIES OF *GANODERMA*

Different stages of development of *Ganoderma* fruitbodies are the mycelial stage, followed by primordia formation, the appearance of young and mature fruiting bodies towards the time of harvest, and each stage has particular requirements for nutritional and environmental factors, which are discussed as follows.

2.4.1 Nutritional Factors

2.4.1.1 Sources of Carbon and Nitrogen and the C:N Ratio

Uptake of carbon and nitrogen are necessary for the supply of energy and protein synthesis in *Ganoderma* (Kashangura, 2008). In terms of carbon, mycelial growth requires fructose and glucose, while sawdust, rice, wheat straw, bran, cottonseed hull, corncob powder, and a variety of agricultural byproducts are important carbon sources for the cultivation of fruitbodies (Zhou, 2017). Nitrogen sources of peptone and yeast powder are considered useful for mycelial growth, while for fruitbody production, wheat bran, rice bran, corn powder, ammonium sulfate, urea, etc., are reported to be beneficial (Kashangura, 2008). Optimum C:N ratios of 20:1 and 30–40:1 should be maintained for mycelial growth and fruitbody production, respectively (Zhou, 2017).

2.4.1.2 Inorganic Salts and Vitamins

Many elements, including calcium, magnesium, sodium, potassium, phosphorus, zinc, and sulfur, are important for *Ganoderma* cultivation. One percent gypsum or calcium sulfate or 3% calcium carbonate is vital for the pH adjustment of the substrate, increasing ventilation, fixing nitrogen levels, and elevating the levels of sulfur and calcium of the substrate (Kashangura, 2008; Ofodile et al., 2022). Vitamins such as biotin, B_1, B_2, and B_6 to the minimum of 10 mg/L are known to be essential for the proper growth and development of the mushroom (Zhou, 2017).

2.4.1.3 Moisture Content

About 60–65% moisture is required for *Ganoderma* cultivation, as water availability meets the oxygen demands of mycelia and plays an important role in its growth. Substrates with large voids demand a higher moisture content of 70% to avoid any oxygen deficiencies, which can interfere with the growth mechanism. In the case of sawdust substrates, 80–90% humidity is required to develop primordia into stalk and then develop into the cap (Kashangura, 2008; Zhou, 2017).

2.4.2 Environmental Parameters

2.4.2.1 Temperature

Temperature is crucial for any enzymatic activity in mushrooms (Kashangura, 2008). For *G. lucidum*, the optimum temperature for mycelial growth and fruitbody development is reported as 25–30°C. Temperature exceeding 35°C is reported to negatively affect the health of fruitbodies, and temperature below 20°C is reported to yield yellow, rigid fruitbodies (Setiawan, 2001; Kibar and Peksen, 2008; Zhou, 2017; Sakamoto, 2018).

2.4.2.2 Carbon Dioxide Concentration

Ganoderma is an aerobic mushroom; hence, uptake of oxygen and release of CO_2 is a routine process (Kashangura, 2008). However, the enclosed cultivation environment may lead to the accumulation of CO_2. It has been found that if the CO_2 concentration exceeds 0.1%, then the growth of branched antler-shaped mushrooms is observed (Zhou, 2017). However, CO_2 below 0.1% results in the development of large fruitbodies, with thick round caps and a short stipe (Sakamoto, 2018).

2.4.2.3 Light

The sensitivity of *Ganoderma* towards light and dark conditions has been observed during its growing period (Kashangura, 2008). Light inhibits the growth of mycelia, and spawn-running is preferred in the dark; that's why there is the practice of burying inoculated substratum or wood logs. Brief exposure to weak light can induce primordial differentiation and formation of pileus in the early stage of fruitbody differentiation (Sakamoto, 2018). It has been observed that light intensity ranging from 20 to 100 lux can produce antler-shaped mushrooms, whereas intensities of 300–1000 lux and 3000–10,000 lux promote small caps with slender stipes and large, round caps with thick stripes, respectively (Kibar and Peksen, 2008; Zhou, 2017).

2.5 CULTIVATION PATTERN

Selection of proper raw materials is a basic necessity to be maintained throughout the cultivation procedure, and types of substrates may differ for region-specific cultivation (González-Matute et al., 2002; Gurung et al., 2013; Tesfaw et al., 2015). The biomass conversion rate for the substrate is reported to be around 100–120 g of a dry fruitbody from 1 kg of the substrate (Zhou, 2017). There are two major patterns of *Ganoderma* cultivation using a solid-state and a liquid-state media. While the solid-state cultivation pattern focuses on the production and isolation of fruit bodies and spores, the main aim of the liquid-state cultivation pattern is the development of mycelial biomass and bioactive constituents like ganoderic acid, polysaccharides, triterpenoids, peptides, etc. (Dawit, 1998). Different cultivation strategies of *G. lucidum* cultivation are elaborated in the following context (Stamets, 2000; de Carvalho et al., 2015).

2.5.1 Solid-State Cultivation

Solid lignocellulosic substrates rich in phenols and tannins are known to supplement the growth of *Ganoderma* (Ozcariz-Fermoselle et al., 2018). Solid-state cultivation may further be divided into wood log cultivation and substitute cultivation, which are described next.

2.5.1.1 Wood Log–Based Cultivation

Fruitbody production involves various steps, such as strain selection followed by spawn preparation, selection of wood log species, and substrate preparation using raw materials (Erkel, 2009; Bernabé-González et al., 2015). Artificial cultivation of *G. lucidum* requires wood logs of broadleaf trees cut into short lengths, bagging, tying, sterilization, and inoculation, followed by stump cultivation. Tree species, especially hardwoods of Fagaceae like *Castanopsis fargessi*, *C. eyrei*, *C. sclerophylla*, *C. carlessi*, mountain olive and peach trees of Elaeocarpaceae and Rosaceae, respectively, and *Liquidambar* spp. belonging to Hamamelidaceae are suitable for *G. lucidum* cultivation. Trees are felled, and blocks of 12–15 cm lengths with a diameter of 10–13 cm are cut and taken to the inoculation site. During operation, the wood segments are cut clean of side branches and burrs to prevent the puncturing of plastic bags. Wood segments are dried for around 2 weeks to lower the moisture content to an optimum level of 45–55% and then packed, compacted with 0.15 MPa pressure for 1.5–2 hours at 100°C, sterilized for 10–12 hours, and finally inoculated (Azizi et al., 2012). Aseptic inoculation must be done in a clean, dry place with optimum spawn inoculation volume. An estimated 80–100 bottles of solid spawn are required for inoculation of the log with an area of 1 square hectometer. Primordial initiation of *G. lucidum* has been recorded in 60–70 days of wood log cultivation (Zhou, 2017). The wood log inoculation may be carried out either with the cross-sectional method, hole-bore method, or bag method.

2.5.1.1.1 Cross-Sectional Method of Inoculation

The cross-sectional method of inoculation involves uniform spawn deposition in a 5-mm layer on the sides and top of sterilized wooden blocks, followed by stacking with another wooden stump on top of it. The inoculated wooden blocks/logs are covered with paper and are kept in shade. After proper spawn rooting, the paper covering is discarded and the wood blocks are separated and buried in the soil. When soil burial is not performed, these logs are spaced at an interval of 15 cm and left for further mycelial growth. The logs buried in alignment must have a distance of 6–8 cm between them. After this, a covering layer of 1–2 cm is placed on top of the logs (Zhou, 2017).

2.5.1.1.2 Hole-Bore Method of Inoculation

However, in the hole-bore method of inoculation, holes of 2–4 cm deep and 12–15 mm diameter are drilled on the wooden stumps, partially filled with solid spawn to the level of the bark, and followed by wax sealing. Then burial of prepared logs in the soil is done within a simple greenhouse to ensure proper drainage, aeration, and water retention. After the burial, digging of a trench or building of a soil mound for log burial is done. Temperature of about 23–34°C is maintained along with humidity and shade of 80–90% (Mayzumi et al., 1997).

2.5.1.1.3 Plastic Bag Method of Inoculation

In the bag cultivation method, the packaged ingredients in the heat-resistant plastic bags, containing a substrate of wood log/wood blocks, are wetted at first to enhance their moisture content to about ~65%. Then the substrate is mixed with 2% K_2HPO_4, gypsum and lime in a 1:1 ratio (w/w). The bags are properly plugged using cotton wool, covered with brown paper, the neck prepared using PVC rings, autoclaved at 121°C under 15 pounds inch^2 pressure, cooled, and then inoculated (Zhou, 2017; Bijalwan et al., 2021). The bag cultivation method is advantageous in having a short growth cycle and high yield (Zhou, 2017).

2.5.1.2 Substitute Cultivation

The log-based method is the primary mode of fruitbody cultivation, and substitutes for a woody substrate like sawdust, agricultural by-products, bagasse, corncob, and cottonseed husk with certain additives like wheat bran, rice bran, soybean powder, finger millet, corn flour, and *Eleusine corcana* or ragi powder, etc., are used for cultivation. Wastes of *Carya illinoinensis* (Ozcariz–Fermoselle et al., 2018), sunflower (*Helianthus annuus*) seed hulls (Curvetto et al., 2002), and tea (*Camellia sinensis*) are known to be used as a substrate for *G. lucidum* cultivation, and especially tea is reported to have better biological efficiency in terms of yield (Peksen and Yakupoglu, 2009). However, it has been often observed that *G. lucidum* grown using sawdust yielded high-quality fruiting bodies (Zhou, 2017).

Sawdust of plants like *Albizzia richardiana, A. procera, Alnus nepalensis, Borassus flabellifer, Bombax ceiba, Dalbergia sissoo, Mangifera indica, Eucalyptus camaldulensis, Shorea robusta* (Singh et al., 2014), *Artocarpus heterophyllus, Hevea brasiliensis, Melia dubia* (Jeewanthi et al., 2017), *Betula pubescens, B. pendula, Populus tremula, Picea abies, Pinus sylvestris, Alnus incana* (Cortina-Escribano et al., 2020), *Swietenia mahagoni, Tectona grandis, Michelia champaca, Dipterocarpus turbinates*, and *Gmelina arborea* (Roy et al., 2015) can be used as basic raw materials. Supplementation of sawdust substrate with water hyacinth is also reported to give one of the highest biological efficiencies in the conversion of substrate dry weight to the *Ganoderma* fruitbody (Ofodile et al., 2022). The raw materials are mixed properly with other substitutes, bottled or bagged, sterilized, and inoculated. The bag method is mostly preferred over bottling due to ease of transport, more substrate utilization, and obtaining large fruitbodies (Zhou, 2017).

In the bag method of cultivation, the moisture from spawn needs to be reduced to avoid unnecessary movement of spawn and inoculated to the substrate by removing the cotton plugs from the

TABLE 2.1
Composition of Substrate Formulation and Types

Formulation Types	Composition	Percentage
Type 1	Sawdust	78
	Wheat bran	20
	Sucrose	1
	Gypsum	1
Type 2	Cottonseed hull	42
	Sawdust	45
	Wheat bran	15
	Gypsum	1
Type 3	Cottonseed hull	89
	Wheat bran	10
	Gypsum	1
Type 4	Corncob	50
	Wheat bran	35
	Sawdust	15
Type 5	Corncob	74.5
	Cornmeal	24.5
	Gypsum	0.5
	Plant ash	0.5

mouth of the plastic bags and injecting with 5–10% of inoculation volume. This inoculation must be done aseptically, preferably in front of a laminar to reduce the chances of contamination. After inoculation, a spawn run is done followed by soil embedment with harvesting (Zhou, 2017).

Several formulations for substitute cultivation are listed in Table 2.1, which are subject to changes according to the need of a grower (Hossain, 2009; Chen, 2014). These formulations with several additives are developed initially on a laboratory scale and implemented after trials to a large-scale operational mode.

2.5.2 Effect of Process Variables on Growth and Yield

Different components used in solid-state fermentation, including the composition of agar media, wood substrate, and fungal strains, govern cell growth and product yield of *G. lucidum*.

2.5.2.1 Effect of Agar Media Composition

Different agar media, including PDA, water agar (WA), and malt extract agar (MEA), are used for the artificial cultivation of *G. lucidum*. Statistical analyses are amended to study the effect of agar media composition on cell growth. The mean of colony diameters is measured initially when PDA, WA, and MEA are inoculated for 10–18 days. A colony diameter of 8.5 cm is observed in PDA and MEA plates. This confirms that PDA and MEA supplement the mycelial growth of *G. lucidum*. MEA accounts for the best media for growth, followed by PDA and WA. Low-density weak and thin mycelia are obtained from WA. Mycelial distribution in the case of WA resembles the root system of a tree, and sometimes hyphae are difficult to see with the naked eye. Evidence of aerial hyphae is recorded in PDA but not in MEA. Colonies take up a dark red-orange color when grown in MEA. Hyphal exudates are often observed in colonies on PDA and MEA but are absent in WA (Cortina-Escribano et al., 2020).

2.5.2.2 Effect of Strain

Cortina-Escribano et al. (2020) has reported the use of six different *G. lucidum* strains, namely MUS9, MUS75, MUS19, MUS12, MUS192, and MUS6, for cultivation. Strains MUS6 and MUS192 exhibited higher growth than others after inoculation of 10 days and 18 days, respectively. MUS9 exhibited the lowest rate of growth, followed by MUS75 after the same period of inoculation. However, after 24 days, all strains attained a maximum growth of 8.5 cm colony diameter. Strains MUS9 and MUS192, with higher replicates, can colonize the entire plate towards the end of the experiment. Contrary to this, strains MUS12 and MUS75 have lower replicates to colonize the entire Petri plate. MUS6 and MUS19 had high mycelial density, followed by MUS12 and MUS192, which had moderate density, while MUS9 and MUS75 had low mycelial density. All strains, except MUS19 and MUS6, had a glabrous texture with scarce granulation. Most strains exhibited aerial hyphae, except MUS19 and MUS12. Hyphal color ranged from white to orange or dark red, especially in the case of MUS75. Irregular mycelial growth can be defined if the hyphae grows in a submerged manner at the inoculated point. Therefore, colonization of mycelia starts from the plate extremities towards the center. This type of colonization results in low-density aerial hyphae with a granular texture. A study of Ofodile et al. (2022) suggests the primordia production and the pace of development of fruitbodies are not only dependent upon *Ganoderma* isolates but also the combination of substrate used. To study the strain-specific mycelial growth differences, measurement of the mean growth on specific substrate combinations and their statistical analyses are suggested (Cortina-Escribano et al., 2020; Ofodile et al., 2022).

2.5.2.3 Effect of Wood Substrate

It is pertinent before going for full-scale cultivation to screen the efficiency of sawdust source(s) for the particular strain. Observations recorded for strain response to agar media composition govern the effect of wood substrates on *G. lucidum*. Media containing *Betula pubescens* and *B. pendula* with *Populus tremula* sawdust exhibited the highest growth rate of mycelia. After inoculation of 10 days and 18 days, sawdust of *Pinus sylvestris, Larix* spp., *Picea abies*, and *Alnus incana* stimulated the mycelial growth compared to those without sawdust amended in media. Media composed of *A. incana* and *P. tremula* sawdust have the most replicates, whereas those with *P. sylvestris* sawdust have lower replicates to colonize the respective medium. Media with *A. incana* and *Betula* spp. yield high-density mycelial growth, whereas *P. sylvestris*–implemented media and those devoid of any sawdust exhibited low-density mycelial growth. Mycelia grown on all sawdust-amended culture media, except *Betula* spp., achieved a granular and glabrous texture. Media devoid of any sawdust yields mycelia with a flat topography. Aerial hyphae are commonly found in media with *Larix* spp. and *Betula* spp. sawdust. Hyphae appear white in the case of *P. abies* and *P. tremula*, whereas they attain an orange-red color in *Betula* spp.–supplemented media. Secretion of exudates is also evident in sawdust-implemented media. Irregular colony growth is observed in the case of *Larix* spp., *A. incana*, and *P. abies* amended media (Cortina-Escribano et al., 2020).

A report using *Swietenia mahagoni, Tectona grandis, Michelia champaca, Dipterocarpus turbinatus*, and *Gmelina arborea* sawdust as the substrate has been confirmed. *D. turbinatus* and *S. mahagoni* give biological efficiencies of 6.8% and 7.6%, respectively. Yields of 210.9 g/kg and 235.2 g/kg are recorded for *D. turbinatus* and *S. mahagoni* supplemented media, respectively. However, yields of 110.4 g/kg and 132.9 g/kg are observed when rice bran is used as a substrate along with *D. turbinatus* and *S. mahagoni* sawdust, respectively. Here, biological efficiencies of 3.6% and 4.3%, respectively, are measured for *D. turbinatus* and *S. mahagoni* with rice bran. Other sawdust of *Tectona grandis, Michelia champaca*, and *Gmelina arborea* showed stunted growth of *G. lucidum* (Roy et al., 2015).

A yield record of 570 g/100 kg using commonly available sawdust with maize flour, wheat bran, rice bran, and bagasse as the substrate is reported (Kamra and Bhatt, 2013). Another record of 102.58 g/kg yield with 12.89% efficiency is confirmed which uses hornbeam sawdust, 10% wheat bran, and 5% malt extract as substrate (Azizi et al., 2012). Oak sawdust supplemented with wheat bran exhibited a high yield and biological efficiency of 63.66 g/kg and 18.63%, respectively (Erkel, 2009).

2.5.2.4 Effect of Wood Substrates on Strains

Reports suggest that sawdust of *P. tremula* and *Betula* spp. stimulate the mycelial growth of *G. lucidum*. Sawdust of *Betula* spp. is better suited for MUS192, MUS9, MUS19, and MUS12 compared to *P. sylvestris*. Mycelial growths of strains MUS75 and MUS6 on *P. sylvestris* and *Betula* spp. media do not show many differences. Sawdust of *A. incana* favors the growth of MUS12. *P. abies* sawdust has an affinity for strain MUS19 (Cortina-Escribano et al., 2020).

A spawn run of 14 cm is observed in *S. mahagoni* sawdust with wheat bran as well as rice bran after 6 days and 9 days of inoculation. After 18 days of inoculation, mycelial growth in *D. turbinatus* sawdust attained 13.5 cm and 13 cm spawn runs with rice bran and wheat bran, respectively (Roy et al., 2015). Primordial initiation of *G. lucidum* has been recorded after inoculation of 35 days using a composition of sawdust, maize flour, wheat bran, rice bran, and bagasse (Kamra and Bhatt, 2013).

2.6 HARVEST AND POST-HARVEST PROCESSES

Time taken for primordia formation to the harvest of the first flush fruitbody takes around 25–30 days in the case of substitute culture. The fruitbodies can be harvested when the cap becomes completely red and the white margins disappear. They are harvested by cutting the stalk with a knife. Under optimal growth parameters, second and third flushes of production may also be recovered from the inoculated substrate. However, during harvesting of the first flush, the addition of nutrient solution is recommended for continued spawn running, and the second flush can be picked after 2 weeks (Zhou, 2017).

After harvesting, the moisture from the fruitbodies is removed through sun drying or with heat (60°C), preferably with the upper cap surface facing downwards. If the aim is to prepare a powder from the fruitbody, then the cap may be shredded radially and then processed for drying. However, drying improperly and in less heat for a prolonged period may degrade the quality of the mushroom and increase the chance of contamination by molds (Chen, 2002).

2.7 DISEASES AND PESTS

During cultivation, *G. lucidum* may be infected by many pathogens as well as infested by insects leading to reduction of yield, abnormal morphology, and metabolic disorders. The primary fungal pathogens include *Trichoderma* spp. (causing mildew), *Mycogone perniciosa* (causing brown rot disease), *Neurospora* spp., and *Aspergillus* spp. Insect pests of *G. lucidum* mainly belong to Lepidoptera, Coleoptera, Isoptera, and Collembola. For prevention of disease and pest incidences, selection of healthy substrates, proper sterilization, environmental sanitation, soil disinfection, pest control, and management are recommended (Zhou, 2017).

2.8 GENERAL FACILITIES

The equipment required for large-scale cultivation of *Ganoderma* is a bioreactor, fermentation tank, incubation tank, autoclave, working table, weighing machine or digital balance, inoculation box, culture shelf, sack packer, mixer, etc. It also requires a culture room, inoculation room, preservation room, mushroom house, sterilizing room, waste disposal room, house-packing room, and laboratory. Along with these, proper aeration, humidity, temperature, and light should be available to create the suitable microenvironment needed by *Ganoderma* for a large-scale commercial cultivation strategy (Zhou, 2017).

2.9 RECENT TRENDS IN STRAIN IMPROVEMENT

The prerequisite for *G. lucidum* cultivation is the generation of a suitable strain superior to other available strains in the wild. To improve the fungal strain, several breeding methods are undertaken, classified as primitive and advanced techniques (Oei, 2003).

2.9.1 Selective Breeding

Artificial selection, popularly known as selective breeding, is one of the most primitive techniques which choose the superior strain over the inferior ones (Rolim et al., 2014). It can be done both artificially and naturally. Selective breeding involves single-spore and tissue isolation techniques. This is followed by a screening of clones, purification of strain, and rejuvenation. During this procedure, tissue separation is often implemented, as *G. lucidum* spores are difficult to germinate (Lin and Zhou, 1999). Then cultivation is done and selection of a superior strain is made (Chen and Su, 2008; Zhou, 2017).

2.9.2 Cross-Breeding

In the 1980s, hybrid strains were evolved using the cross-breeding technique in most Asian countries (Zhao et al., 1993). It is aimed to obtain genetically recombinant strains by a selection of superior traits of the parental generation. This method gives rise to genetic variation to inculcate desirable traits from strains of different origins through the method of haploid mating. However, spores of *G. lucidum* face difficulties during germination, and mono-karyotic strains are generally not obtained, which is desirable for this breeding technique. Hence, the cross-breeding technique is of limited use in this regard. As a result, the protoplast mono-karyo-genesis technique is undertaken for practice (Wu et al., 2009; Zhou et al., 2012). Some reports of cross-breeding have been documented which state the use of protoplasts for the selection of traits (Chiu et al., 2005).

2.9.3 Mutation Breeding

Next comes mutation breeding, which targets the genes and brings about genic recombination. A population with desirable strains generated from induced mutation or natural mutation is selected and screened. This procedure involves strain selection, preparation of protoplast or spore suspension, viable counting, and mutagenizing, followed by spreading of plates, picking a desirable strain, and inoculation. Superior strains are selected through post-screening, slope culturing, and rescreening (Zhou et al., 2012). Several chemical or physical agents act as mutagenizing agents. Parental strains undergo mutation; then the mutated mycelia are selected and used for mother flasks, fruiting bags, and spawn preparation. Generally, protoplasts are preferred in mutational breeding experiments (Li et al., 2001). By-products of cultivation such as polysaccharides, triterpenoids, organic germanium, etc., are regarded as the objectives of this breeding (Gao et al., 2008). This breeding method provides genetic markers for further protoplast as well as cross-breeding experiments. However, it comes with certain disadvantages, including complications in working with mutants, and this mutational process is quite randomized.

2.9.4 Cell-Fusion Breeding

An advanced breeding technique known as cell-fusion breeding is used to genetically manipulate intergeneric and intrageneric traits through protoplast fusion (Zhou, 2017). This method involves parental strain selection, isolation of a genetically marked protoplast, regeneration, and culture of a protoplast, followed by protoplast fusion, regeneration, culture, detection, and selection of a fusion (Zhou et al., 2012). This technique has been used since the 1970s and is a dependable breeding method for inducing the exchange of genetic contents of auxotrophic mutants and generating new individuals by protoplast fusion using polyethylene glycol (Ferenczy et al., 1974; Lin and Zhou, 1999) and applied in the development of new strains following the fusion of inter-genus (Yoo et al., 2002) and intra-genus (Park et al., 1988) protoplasts and mono-karyogenesis of the protoplast (Wu et al., 2009).

2.9.5 Genetic Engineering Breeding

Transgenic *Ganoderma* may also be obtained through various genetic engineering techniques (Zhou, 2017). Desirable genes such as those responsible for increased production of ganoderic acid ($C_{30}H_{44}O_7$) can be incorporated within the host strain from the donor strain. This technique involves donor strain selection, gene separation, in vitro gene construction, transfer into the host cell, reproduction, recombinant DNA expression, and selection of the host. Hence, the DNA sequence of an unknown wild strain can be isolated and transformed into commercially available strains. This method revolves around six major gene transformation techniques like biolistic, *Agrobacterium*-mediated, electron transfer, restriction enzyme–mediated, lithium acetate, and protoplast-mediated processes (Zhou et al., 2012). This method brings new vigor to the organism using several molecular biology techniques (Park et al., 1991; Sun et al., 2001; Kim et al., 2004; Zhou, 2017).

2.10 CONCLUSION

Backed by over 2000 years of sustained indigenous usage in the traditional systems of medicine, and given the rising global demand for its medicinal values, the artificial cultivation of *Ganoderma* spp. is a vocation with great prospects. The distribution ranging from tropical to temperate regions as well as its simple and cost-effective substrate requirements make it ideal for cultivation in wide areas with varied geo-climatic setups. The *Ganoderma* cultivation aim may vary from suitable solid-state cultivation-based fruitbody production or the liquid-state culture-based production of mycelia and bioactive metabolites such as ganoderic acid, triterpenoids, polysaccharides, and peptides. Relevant protocols are available for these. However, a large-scale venture would require integrated pest and disease management, screening of strains, and their specificity for substrate sources to maximize returns. Recent developments in modern strain improvement methods and improved substrate formulations would enhance the prospects of the *Ganoderma* cultivation industry.

REFERENCES

Adaskaveg, J.E., R.A. Blanchette, R.L. Gilbertson. 1991. Decay of date palm wood by white-rot and brown-rot fungi. *Can. J. Bot.* 69: 615–629. https://doi.org/10.1139/b91-083

Adaskaveg, J.E., R.L. Gilbertson. 1988. Basidiospores, pilocystidia, and other basidiocarp characters in several species of the *Ganoderma lucidum* complex. *Mycologia* 80: 493–507. https://doi.org/10.1080/00275514.1988.12025571

Agarwal, K., G.S. Chakarborthy, S. Verma. 2013. In vitro antioxidant activity of different extract of *Ganoderma lucidum*. *Int. J. Pharm. Sci.* 3: 48–54.

Ajith, T.A., K.K. Janardhanan. 2006. Indian medicinal mushrooms as a source of antioxidant and antitumor agents. *J. Clin. Biochem. Nutr.* 40: 157–162.

Alam, N., J. Lee, T. Lee. 2010. Mycelial growth conditions and phylogenetic relationships of *Pleurotus ostreatus*. *World App. Sci. J.* 9(8): 928–937.

Alheeti, A.A., M.M. Muslat, L.M. Ayyash, R.M. Theer. 2020. Isolation, identification and organic production of mushrooms *Ganoderma lucidum* (Curt.:Fr) Karst (Reishi). *Indian J. Ecol.* 47: 231–235.

Al-Obaidi, J.R., N.B. Saidi, S.R.A. Usuldin, et al. 2016. Comparison of different protein extraction methods for gel-based proteomic analysis of *Ganoderma* spp. *Protein J.* 35: 100–106. http://dx.doi.org/10.1007/s10930-016-9656-z

Aydemir, G. 2002. Research on nutrition and cancer: The importance of the standardized dietary assessments. *Asian Pac. J. Cancer Prev.* 3: 177–180.

Azizi, M., M. Tavana, M. Farsi, F. Oroojalian. 2012. Yield performance of Lingzhi or Reishi medicinal mushroom, *Ganoderma lucidum* (W.Curt.:Fr.) P. Karst. (higher Basidiomycetes), using different waste materials as substrates. *Int. J. Med. Mushrooms* 14: 521–527.

Babitskaya, V., N.A. Bisko, V. Scherba, et al. 2003. Some biologically active substances from medicinal mushroom *Ganoderma lucidum* (W. Curt.:Fr.) P. Karst. (Aphyllophoromycetideae). *Int. J. Med. Mushrooms* 5: 301–306.

Baby, S., A.J. Johnson, B. Govindan. 2015. Secondary metabolites from *Ganoderma*. *Phytochem.* 114: 66–101.

Banuelos, G.S., Z.Q. Lin (eds). 2009. *Development and uses of biofortified agricultural products*. Boca Raton, FL: CRC Press. https://doi.org/10.1201/9781420060065

Barros, L., T. Cruz, P. Baptista, L.M. Estevinho, et al. 2008. Wild and commercial mushrooms as a source of nutrients and nutraceuticals. *Food Chem. Toxicol.* 46: 2742–2747.

Bernabé-González, T., M. Cayetano-Catarino, G. Bernabé-Villanueva, et al. 2015. Cultivation of *Ganoderma lucidum* on agricultural by-products in Mexico. *Micologia Aplicada Int.* 27: 25–30.

Bijalwan, A., K. Bahuguna, A. Vasishth, et al. 2021. Growth performance of *Ganoderma lucidum* using billet method in Garhwal Himalaya, India. *Saudi J. Biol. Sci.* 28(5): 2709–2717. https://doi.org/10.1016/j.sjbs.2021.03.030

Bishop, K.S., C.H.J. Kao, Y. Xu, et al. 2015. From 2000 years of *Ganoderma lucidum* to recent developments in nutraceuticals. *Phytochem.* 114: 56–65. https://doi.org/10.1016/j.phytochem.2015.02.015

Boh, B. 2013. *Ganoderma lucidum*: A potential for biotechnological production of anti-cancer and immunomodulatory drugs. *Recent Pat. Anticancer. Drug Discov.* 8(3): 255–287. https://doi.org/10.2174/1574891x113089990036

Cao, Q., Z. Lin. 2006. *Ganoderma lucidum* polysaccharides peptide inhibits the growth of vascular endothelial cells and the induction of VEGF in the human lung cancer cell. *Life Sci.* 78: 1457–1463.

Cao, Y., S.H. Wu, Y.C. Dai. 2012. Species clarification of the prize medicinal *Ganoderma* mushroom "Lingzhi". *Fungal Divers.* 56: 49–62. https://doi.org/10.1007/s13225-012-0178-5

Cao, Y., H.S. Yuan. 2013. *Ganoderma mobile* sp. nov. from Southwestern China based on morphological and molecular data. *Mycol. Prog.* 12: 121–126.

Celık, G.Y., D. Onbasil, B. Altinsoy, et al. 2014. In vitro antimicrobial and antioxidant properties of *Ganoderma lucidum* extracts grown in Turkey. *European J. Medicinal Plants* 4: 709–722. https://doi.org/10.9734/EJMP/2014/8546

Cha, D.Y., Y.B. Yoo. 2009. IV. Cultivation of Reishi (*Ganoderma lucidum*). *Food Rev. Int.* 13: 378–382. https://doi.org/10.1080/87559129709541121

Chang, S.T., J. Buswell. 1999. *Ganoderma lucidum* (Curt.: Fr.) P. Karst. (Aphyllophoromycetideae): A mushrooming medicinal mushroom. *Int. J. Med. Mushrooms* 1: 139–146.

Chang, S.T., P.G. Miles. 2004. *Mushrooms: Cultivation, nutritional value, medicinal effect, and environmental impact*, 2nd edition. Boca Raton: CRC Press.

Chen, A.W. 2002. Natural log cultivation of medicinal mushroom, *Ganoderma lucidum* (Reishi). *Mushroom Grower's Newslett.* 3: 2–6.

Chen, A.W. 2014. Mushrooms worldwide. Part III. Mushrooms for the tropics. Growing *Ganoderma* mushrooms. In: *Mushroom grower's handbook*, vol. 1, 224–334. https://www.scribd.com/document/265831668/mushroom-growers-handbook-1-mushworld-com-chapter-11-pdf#

Chen, J., K.M. Su. 2008. Proceeding of heredity, breeding, and identification of edible fungi. *Edible Fungi Chin.* 27: 3–8.

Chiu, S.W., V.Y. Luk, S. Yu, et al. 2005. Artificial hybridization of *Ganoderma lucidum* and *G. tsugae* Murrill by protoplast fusion for sustainability. *Int. J. Med. Mushroom* 7: 263–280.

Cortina-Escribano, M., P. Veteli, R. Linnakoski, et al. 2020. Effect of wood residues on the growth of *Ganoderma lucidum*. *Karstenia*. 58: 16–28. https://doi.org/10.29203/ka.2020.486

Cui, B.K., H.J. Li, X. Ji, et al. 2019. Species diversity, taxonomy and phylogeny of Polyporaceae (Basidiomycota) in China. *Fungal Divers.* 97: 137–392. https://doi.org/10.1007/s13225-019-00427-4

Curvetto, N.R., D. Figlas, R. Devalis, et al. 2002. Growth and productivity of different *Pleurotus ostreatus* strains on sunflower seed hulls supplemented with N-NH4+ and/or Mn (II). *Bioresour. Technol.* 84: 171–176.

Dadwal, V.S., J. Jamaluddin. 2004. Cultivation of *Ganoderma lucidum* (Fr.) Karst. *Indian For.* 130: 435–440.

Dawit, A. 1998. *Mushroom cultivation: A practical approach.* Ethiopia: Addis Ababa University Press.

de Carvalho, C.S.M., C. Sales-Campos, L.P. de Carvalho, et al. 2015. Cultivation and bromatological analysis of the medicinal mushroom *Ganoderma lucidum* (Curt.: Fr.) P. Karst. cultivated in agricultural waste. *Afr. J. Biotechnol.* 14: 412–418.

Deepalakshmi, K., S. Mirunalini. 2011. Therapeutic properties and current medicinal usages of medicinal mushroom: *Ganoderma lucidum*. *Int. J. Pharm. Sci. Res.* 2: 1922–1929. http://dx.doi.org/10.13040/IJPSR.0975-8232.2(8).1922-29

Deshmukh, A.S., S.S. Deshmukh, V.S. Pathak, et al. 2018. A review of mushroom bioactive metabolites responsible for antioxidant and anti-cancerous effects. *Asian J. Microbiol. Biotechnol. Environ. Sci.* 20: 225–230.

De Silva, D.D., S. Rapior, F. Fons, et al. 2012. Medicinal mushrooms in supportive cancer therapies: An approach to anti-cancer effects and putative mechanisms of action. *Fungal Divers.* 55: 1–35. https://doi.org/10.1007/s13225-012-0151-3

Erkel, E. 2009. Yield performance of *Ganoderma lucidum* (Fr.) Karst cultivation on substrates containing different protein and carbohydrate sources. *Afr. J. Agric. Res.* 4: 1331–1333. https://doi.org/10.5897/AJAR.9000768

Ferenczy, L., F. Kevei, J. Zsolt. 1974. Fusion of fungal protoplasts. *Nature* 248: 793–794. https://doi.org/10.1038/248793a0

Gao, M.X., J.Z. Miao, Z.H. Cao, et al. 2008. Study on breeding *Ganoderma* strains producing high polysaccharides by protoplast electrofusion technique. *Jiangsu Agr. Sci.* 6: 89–91.

GBIF. n.d. www.gbif.org/occurrence/search?q=ganoderma (accessed June 4, 2021).

Ghosh, D. 2004. Search for future viands: Algae and fungi as food. *Resonance J. Sci. Educ.* 9(5): 33–40.

González-Matute, R., D. Figlas, R. Devalis, et al. 2002. Sunflower seed hulls as a main nutrient source for cultivating *Ganoderma lucidum*. *Micologia Applicada Int.* 14: 1–6.

Gurung, O.K., U. Budathoki, G. Parajuli. 2013. Effect of different substrates on the production of *Ganoderma lucidum* (Curt.:Fr.) Karst. *Our Nat.* 10: 191–198. https://doi.org/10.3126/on.v10i1.7781

Hapuarachchi, K.K., T.C. Wen, R. Jeewon, et al. 2016. Mycosphere essays 15. *Ganoderma lucidum*—are the beneficial medical properties substantiated? *Mycosphere* 7: 687–715. https://doi.org/10.5943/mycosphere/7/6/1

Harsh, N.S.K. B.K. Rai, D.P. Tiwari. 1993. Use of *Ganoderma lucidum* in folk medicine. *J. Trop. Biodivers.* 1: 324–326.

Hibbett, D.S., M.J. Donoghue. 1995. Progress toward a phylogenetic classification of the Polyporaceae through parsimony analysis of mitochondrial ribosomal DNA sequences. *Can. J. Bot.* 73(Suppl. 1): S853–S861.

Hossain, K., N. Sarker, A. Kakon, et al. 2009. Cultivation of Reishi mushroom (*Ganoderma lucidum*) on sawdust of different tree species. *Bangladesh J. Mushrooms* 3: 1–5.

Hung, P.V., N.N.Y. Nhi. 2012. Nutritional composition and antioxidant capacity of several edible mushrooms grown in Southern Vietnam. *Int. Food Res. J.* 19: 611–615.

Hyde, K.D., J. Xu, S. Rapior, et al. 2019. The amazing potential of fungi: 50 ways we can exploit fungi industrially. *Fungal Divers.* 97: 1–136. https://doi.org/10.1007/s13225-019-00430-9

Jeewanthi, L.A.M.N., K. Ratnayake, P. Rajapakse. 2017. Growth and yield of Reishi mushroom [*Ganoderma lucidum* (Curtis) P. Karst.] in different sawdust substrates. *J. Food Agric.* 10: 8–16.

Jeong, Y.T., B.K. Yang, S.C. Jeong, et al. 2008. *Ganoderma applanatum*: A promising mushroom for antitumor and immunomodulating activity. *Phytother. Res.* 22: 614–619.

Kakon, A.J., M.B.K. Choudhury, S. Saha. 2012. Mushroom is an ideal food supplement. *J. Dhaka Natl. Med. Coll. Hosp.* 18: 58–62. https://doi.org/10.3329/jdnmch.v18i1.12243

Kaliyaperumal, M. 2013. Molecular taxonomy of *Ganoderma cupreum* from Southern India inferred from ITS rDNA sequences analysis. *Mycobiol.* 41(4): 248–251.

Kamra, A., A.B. Bhatt. 2013. First attempt of organic cultivation of red *Ganoderma lucidum* under subtropical habitat and its economics. *Int. J. Pharm. Pharm. Sci.* 5: 94–98.

Kashangura, C. 2008. *Optimisation of the growth conditions and genetic characterization of Pleurotus species.* PhD Thesis, University of Zimbabwe, Harare.

Kaur, H., S. Sharma, P.K. Khanna, et al. 2015. Evaluation of *Ganoderma lucidum* strains for the production of bioactive components and their potential use as antimicrobial agents. *J. Appl. Nat. Sci.* 7: 298–303. https://doi.org/10.31018/jans.v7i1.605

Kibar, B., A. Peksen. 2008. Modelling the effects of temperature and light intensity on the development and yield of different *Pleurotus* species. *Agric. Trop. Subtrop.* 41: 68–73.

Kim, J.H., D.H. Lee, S.H. Lee, et al. 2004. Effect of *Ganoderma lucidum* on the quality and functionality of Korean traditional rice wine, Yakju. *J. Biosci. Bioeng.* 97: 24–28.

Li, G., F. Yang, R. Li, et al. 2001. A study on the breeding of new *Ganoderma* varieties by UV-induced mutagenesis. *Acta. Microbiol. Sin.* 41: 229–233.

Lin, J., X.W. Zhou. 1999. Artificial cultivation of organizational separation of *Ganoderma lucidum*. *Edible Fungi.* 2: 10–11.

Lin, Z.B. 2009. *Lingzhi: From mystery to science*. Beijing: Peking University Medical Press.
Ling, Z.X., A.F. Chen, Z.B. Lin. 1997. *Ganoderma lucidum* polysaccharides enhance the function of immunological effector cells in immunosuppressed mice. *J. Ethnopharmacol.* 111: 219–226.
Manavalan, T., A. Manavalan, K.P. Thangavelu, et al. 2012. Secretome analysis of *Ganoderma lucidum* cultivated in sugarcane bagasse. *J. Proteomics.* 77: 298–309. https://doi.org/10.1016/j.jprot.2012.09.004
Mayzumi, F., H. Okamoto, T. Mizuno, T. 1997. IV. Cultivation of Reishi (*Ganoderma lucidum*). *Food Rev. Int.* 13: 365–370. https://doi.org/10.1080/87559129709541118
Nithya, V., M. Ambikapathy, A. Panneerselvam. 2014. Collection, identification, phytochemical analysis, and phytotoxicity test of wood-inhabiting fungi *Ganoderma lucidum* (Curt.Fr.) P. Karst. *Hygeia* 6: 31–39. https://doi.org/10.15254/H.J.D.Med.6.2014.120
Oei, P. 2003. *Mushroom cultivation: Appropriate technology for mushroom growers*. Leiden, The Netherlands: Backhuys Publishers.
Ofodile, L.N., O.S. Isikhuemhen, F.N. Anike, et al. 2022. The domestication and cultivation of *Ganoderma* (Agaricomycetes) medicinal mushroom species from Nigeria. *Int. J. Med. Mushrooms* 24(6): 69–78. https://doi.org/10.1615/IntJMedMushrooms.2022043906
Ozcariz-Fermoselle, M.V., R. Fraile-Fabero, T. Girbés-Juan, et al. 2018. Use of lignocellulosic wastes of pecan (*Carya illinoinensis*) in the cultivation of *Ganoderma lucidum*. *Rev. Iberoam. Micol.* 35: 103–109. https://doi.org/10.1016/j.riam.2017.09.005
Park, S.H., E.C. Choi, B.K. Kim. 1991. Studies on intergeneric protoplast fusion and nuclear transfer between *Ganoderma lucidum* and *Corilus versicolor*. *Arch. Pharm. Res.* 14: 282–283.
Park, Y.D., J.S. Lee, Y.B Yoo. 1988. Interspecific protoplast fusion of *Ganoderma applanatum* and *Ganoderma lucidum* and fruit body formation of the fusants. *Kor, J. Mycol.* 16: 79–86.
Peksen, A., G. Yakupoglu. 2009. Tea waste as a supplement for the cultivation of *Ganoderma lucidum*. *World J. Microbiol. Biotechnol.* 25: 611–618. https://doi.org/10.1007/s11274-008-9931-z
Pilotti, C.A., F.R. Sanderson, E.A.B. Aitken, et al. 2004. Morphological variation and host range of two *Ganoderma* species from Papua New Guinea. *Mycopathologia* 158: 251–265. https://doi.org/10.1023/B:MYCO.0000041833.41085.6f
Podgornik, B.B., M. Berovic, J. Zhang, et al. 2007. *Ganoderma lucidum* and its pharmaceutically active compounds. *Biotechnol. Annu. Rev.* 13: 265–301.
Rawat, A., M. Mohsin, A.M. Sah, et al. 2012. Biochemical estimation of wildly collected *Ganoderma lucidum* from Central Himalayan Hills of India. *Adv. Appl. Sci. Res.* 3: 3708–3713.
Rolim, L.N., C. Sales-Campos, M.A.Q. Cavalcanti, et al. 2014. Application of Chinese Jun-Cao technique for the production of Brazilian *Ganoderma lucidum* strains. *Braz. Arch. Biol. Technol.* 57: 367–375.
Roy, S., M.A.A. Jahan, K.K. Das, et al. 2015. Artificial cultivation of *Ganoderma lucidum* (Reishi medicinal mushroom) using different sawdusts as substrates. *Am. J. Bios.* 3: 178–182. https://doi.org/10.11648/j.ajbio.20150305.13
Russell, R., M. Paterson. 2006. *Ganoderma*: A therapeutic fungal factory. *Phytochem.* 67: 1985–2001.
Sakamoto, Y. 2018. Influences of environmental factors on fruiting body induction, development and maturation in mushroom-forming fungi. *Fungal Biol. Rev.* 32: 236–248. https://doi.org/10.1016/j.fbr.2018.02.003
Sana, S., N. Mahmud, S. Rana, et al. 2017. Mushroom: Enroll plants in natural bioactive compounds in biological research including their pharmacological properties. *Int. J. Biol. Med. Res.* 8(4): 6170–6176.
Sanodiya, B.S., G.S. Thakur, R.K. Baghel, et al. 2009. *Ganoderma lucidum*: A potent pharmacological macrofungus. *Curr. Pharm. Biotechnol.* 10(8): 717–742.
Setiawan, B. 2001. Designing temperature control system for mushroom cultivation. *IFAC Proceedings.* 34: 158–161.
Silva-Neto, C.D.M., D.D.S. Pinto. 2021. Food production potential of *Favolus Brasiliensis* (Basidiomycota: Polyporaceae), an indigenous food. *Food. Sci. Technol.* 41: 183–188.
Singh, S., N.S.K. Harsh, P.K. Gupta. 2014. A novel method of economical cultivation of medicinally important mushroom, *Ganoderma lucidum*. *Int. J. Pharm. Sci. Res.* 5: 2033–2037.
Sliva, D. 2006. *Ganoderma lucidum* in cancer research. *Leuk. Res.* 30: 767–768. https://doi.org/10.1016/j.leukres.2005.12.015
Smith, B.J., K. Sivasithamparam. 2003. Morphological studies of *Ganoderma* (Ganodermataceae) from the Australasian and Pacific regions. *Aust. Syst. Bot.* 16: 487–503. https://doi.org/10.1071/SB02001
Stamets, P. 2000. *Growing gourmet, and medicinal mushrooms*, 3rd edition. Berkeley: Ten Speed Press.
Suberu, H.A., A.A. Lateef, I.M. Bello, et al. 2013. Mycelia biomass yield of *Ganoderma lucidum* mushroom by submerged culture. *Nigerian J. Technol. Res.* 8: 64–67.
Sun, L., H.Q. Cai, W.H. Xu, et al. 2001. Efficient transformation of the medicinal mushroom *Ganoderma lucidum*. *Plant. Mol. Biol. Rep.* 19: 383a–383j.

Tesfaw, A., A. Tadesse, G. Kiros. 2015. Optimization of oyster (*Pleurotus ostreatus*) mushroom cultivation using locally available substrates and materials in Debre Berhan, Ethiopia. *J. Appl. Biol. Biotechnol.* 3: 15–20. https://doi.org/10.7324/JABB.2015.3103

Thakur, M.P., H.K. Singh. 2013. Mushrooms, their bioactive compounds, and medicinal uses: A review. Medicinal plants. *Int. J. Phytomed. Relat. Ind.* 5(1): 1–20.

Tripathy, A. 2010. Yield evaluation of paddy straw mushrooms (*Volvariella* spp.) on various lignocellulosic wastes. *Int. J. Appl. Agric. Res.* 5: 317–326.

Uddin, M.N., S. Yesmin, M.A. Khan, et al. 2011. Production of oyster mushrooms in different seasonal conditions of Bangladesh. *J. Sci. Res.* 3: 161–167.

Wachtel-Galor, S., J. Yuen, J.A. Buswell, et al. 2011. *Ganoderma lucidum* (Lingzhi or Reishi): A medicinal mushroom. In: I.F.F. Benzie, S. Wachtel-Galor (eds) *Herbal medicine: Biomolecular and clinical aspects*, 2nd edition. Boca Raton, FL: CRC Press/Taylor & Francis.

Wagner, R., D.A. Mitchell, G.L. Sassaki, et al. 2003. Current techniques for the cultivation of *Ganoderma lucidum* for the production of biomass, ganoderic acid, and polysaccharides. *Food Technol. Biotechnol.* 41: 371–382.

Wang, L., J.Q. Li, J. Zhang, et al. 2020. Traditional uses, chemical components and pharmacological activities of the genus *Ganoderma* P. Karst.: A review. *RSC Adv.* 10(69): 42084–42097, November 18. https://doi.org/10.1039/d0ra07219b

Wasser, S.P. 2002. Medicinal mushrooms as a source of antitumor and immunomodulating polysaccharides. *Appl. Microbiol. Biotechnol.* 60: 258–274.

Wasser, S.P. 2004. Reishi or Lingzhi (*Ganoderma lucidum*). In: *Encyclopedia of dietary supplements*. New York: Marcel Dekker. https://doi.org/10.1081/E-EDS-120022119

White, T.J., T. Bruns, S. Lee, et al. 1990. Amplification and direct sequencing of fungal ribosomal RNA genes for phylogenetics. In: M.A. Innis, D.H. Gelfand, J.J. Sninsky, et al. (eds) *PCR protocols*. San Diego: Academic Press, 315–322. https://doi.org/10.1016/B978-0-12-372180-8.50042-1

Wińska, K., W. Mączka, K. Gabryelska, et al. 2019. Mushrooms of the genus *Ganoderma* used to treat diabetes and insulin resistance. *Molecules* 24: 4075. https://doi.org/10.3390/molecules24224075

Wu, S.Q., X.B. Guo, X. Zhou, et al. 2009. AFLP analysis of genetic diversity in main cultivated strains of *Ganoderma* spp. *Afr. J. Biotechnol.* 8: 3448–3454.

Xing, J.H., Y.F. Sun, Y.L. Han, et al. 2018. Morphological and molecular identification of two new *Ganoderma* species on *Casuarina equisetifolia* from China. *MycoKeys* 34: 93–108. https://doi.org/10.3897/mycokeys.34.22593

Yang, F.C., C.B. Liau. 1998. Effect of cultivating conditions on the mycelial growth of *Ganoderma lucidum* in submerged flask cultures. *Bioprocess Eng.* 19: 233–236. https://doi.org/10.1007/PL00009014

Yoo, Y.B., K.H. Lee, B.G. Kim. 2002. Characterization of somatic hybrids with compatible and incompatible species by protoplast fusion in genera *Pleurotus* (Fr.) P. Karst. and *Ganoderma* P. Karst. by RAPD-PCR analysis. *Int. J. Med. Mushroom* 4: 147–157.

Yu, Y.N., M.Z. Shen. 2003. The history of Lingzhi (*Ganoderma* spp.) cultivation. *Mycosyst.* 22: 3–9.

Zhao, J., S.T. Chang. 1993. Monokaryotization by protoplasting heterothallic species of edible mushrooms. *World J. Microbiol. Biotechnol.* 9: 538–554.

Zheng, L.Y., D.H. Jia, X.F. Fei, et al. 2009. An assessment of the genetic diversity within *Ganoderma* strains with AFLP and ITS PCR-RFLP. *Microbiol Res.* 164: 312–321. https://doi.org/10.1016/j.micres.2007.02.002

Zhou, J., M. Chen, S. Wu, et al. 2020. A review on mushroom—derived bioactive peptides: Preparation and biological activities. *Food Rev. Int.* 134.

Zhou, X.W. 2017. Cultivation of *Ganoderma lucidum*. In: C.Z. Diego, A. Pardo-Giménez (eds) *Edible and medicinal mushrooms: Technology and applications*. New York: John Wiley & Sons Ltd. https://doi.org/10.1002/9781119149446.ch18

Zhou, X.W., K.Q. Su, Y.M. Zhang. 2012. Applied modern biotechnology for the cultivation of *Ganoderma* and development of their products. *Appl. Microbiol. Biotechnol.* 93(3): 941–963. https://doi.org/10.1007/s00253-011-3780-7

3 *Ganoderma* in Traditional Culture

Anita Klaus[1] *and Wan Abd Al Qadr Imad Wan-Mohtar*[2]
1 University of Belgrade, Belgrade, Serbia
2 Universiti Malaya, Kuala Lumpur, Malaysia

3.1 INTRODUCTION

Very likely, even before there was archaeological evidence, mushrooms were an integral part of the lives of ancient people around the world. The attitude towards these unusual creatures varied a lot—some feared and avoided them; others respected, worshiped and used specially selected specimens for ritual purposes. They were exceptional ingredients in the menu of many human communities and highly valued medicines that were initially available only to very important and privileged members of the communities.

Considering their delicate and soft structure, it is not surprising that there is only scant archaeological evidence to indicate the use of mushrooms in various areas of human activities. Therefore, for the oldest mushroom fossil remains that come from the Mesozoic and Cenozoic (Table 3.1), it cannot be said with certainty whether and for what purposes they were used by humans. Nevertheless, these archaeological findings are of inestimable value in themselves, and most often these are specimens of wooden structures from which fossil remnants have survived to the present days due to the favorable environmental conditions in which they have been trapped for millions of years. Probably the oldest fossil remains are those belonging to the species *Phellinites degiustoi* (according to the valid classification *Fomitopsis pinicola*), estimated to be approximately 250 million years old (Singer & Archangelsky, 1958). Among the oldest are also the remnants of mushrooms belonging to the genus *Fomes*, as well as *Ganoderma lucidum*, which are believed to be over millions of years old (Brown, 1940; Chaney & Mason, 1936; Mason, 1934).

One of the oldest pieces of evidence of the presence of mushrooms in people's lives is the fossil remains of a woman named the Red Lady in which using the most modern techniques, spores of different types of mushrooms were detected in dental plaque. The age of this fossil is estimated at 18,700 years. It was found in the El Mirón cave in the mountainous region of Cantabria in Spain, and it is significant because it testifies to the use of mushrooms for food or even some special power long ago (Power et al., 2015). Throughout thousands of years of human history, the use of mushrooms for nutrition and health by the Egyptian, Greek, Roman, Chinese, and Mexican civilizations has been documented (Gargano et al., 2017). Thus, Egyptian pharaohs considered mushrooms to be a gift from the gods, and consequently, only members of the royal family could use them for food, while ordinary people were not allowed to touch them (Abdel-Azeem et al., 2016). One of the most impressive examples of the importance of mushrooms in human life is the perfectly preserved Simalaun mummy, better known as Ötzi (3350–3100 BC), which originates from Chalcolithic Europe (3500–1700 BC) and was found in alpine glaciers (Table 3.3). Among the various objects found with this mummy, three mushrooms were also found—two differently shaped parts of a polypore *Fomitopsis betulina* carpophore and a larger quantity of *Fomes fomentarius* used as tinder (Peinter et al., 1998). Considering various medical effects, such as antiparasitic, antiviral, antibacterial, anticancer, immunomodulatory and neuroprotective (Pleszczyńska et al., 2017), and also against whipworm infections (*Trichuris trichiura*) and for bowel cleansing (Dickson et al., 2000), and based on comprehensive ethnomycological and pharmacological data, as well as forensic

DOI: 10.1201/9781003354789-3

TABLE 3.1
Mesozoic and Cenozoic Remains of Mushrooms

Archeological Site	Estimated Age	Identified Mushroom Species	References
Patagonia, petrified forests of Santa Cruz, Argentina	250 million years	*Phellinites degiustoi* (*Fomitopsis pinicola*)	Singer & Archangelsky, 1958
Idaho, USA	23–2.6 million years	*Fomes idahoensis*	Brown, 1940
Alaska, USA	1,81 million–11.550 years	*Bovista plumbea* *Fomes applanatus* *Fomes fomentarius* *Fomes pinicola* *Ganoderma lucidum*	Chaney and Mason (1936)
California, USA	1,81 million–11.550 years	*Fomes* spp.	Mason, 1934

TABLE 3.2
More Famous Records of Mushrooms in Rock Art

Archeological Site	Estimated Age	Identified Mushroom Species	Assumed Purpose	References
Tassili, Sahara desert, Algeria	7000–5000 BC	*Psilocybe mairei*	Ritual use	Samorini, 1992
Selva Pascuala, Spain	4000 BC	*Psilocybe hispanica*	Ritual use	Akers et al., 2011
Cave 2961, village Baylovo, Bulgaria	3000 BC	*Suillellus luridus*	Religious rites	Angelov Uzunov & Petrova Stoyneva-Gärtner, 2015
Tuva, Asia	1500–1000 BC	*Amanita muscaria*	Shamanistic significance	Molodin & Cheremisin, 1999
Pegtymel River region, Siberia, Russia	1000 BC–1000 AD	*Amanita muscaria*	Shamanistic significance	Dikov, 1971

reports, it is believed that at that time *F. betulina* was used as medicine (Peinter et al., 1998). As Wilford (1998) reports, *F. betulina* is thought to have been a prized laxative by prehistoric peoples living in northern Europe. The importance of mushrooms used for religious and ritual purposes by numerous human communities is evidenced by petroglyphs found in different locations worldwide (Table 3.2).

According to the data so far, the oldest painted mural depicting mushrooms, which is believed to represent *Psilocybe mairei*, a species known in Algeria and Morocco, is the one found in a cave from Tassili, in the Sahara desert, an area belonging to southeastern Algeria. This rock art record indicates the importance of mushrooms in rituals as early as 7000–5000 BC (Samorini, 1992). Of great importance is the Selva Pascuala mural, estimated to 4000 BC, which most likely represents the neurotropic mushroom *Psilocybe hispanica* used by shamans in religious ceremonies (Akers et al., 2011). Other petroglyphs date from the time period from 3000 BC to 1000 AD, including those that testify to the importance of the mushrooms *Suillellus luridus* and *Amanita muscaria*, which were discovered in Cave 2961, village Baylovo, Bulgaria, then Tuva, Asia, and the Pegtymel River region, Siberia, Russia (Angelov Uzunov & Petrova Stoyneva-Gärtner, 2015; Molodin & Cheremisin, 1999; Dikov, 1971).

Fossil remains of mushrooms that date back to 13,000 BC are also very significant because they were still preserved enough that it was possible to estimate with great certainty their purpose in

TABLE 3.3
Some of the Identified Remains of Mushrooms Found around the World

Archeological Site	Estimated Age	Identified Mushroom Species	Assumed Purpose	Ref.
Star Carr, England	3000–13000 BC	*Fomes fomentarius* *Phellinus igniarius* *Piptoporus betulinus*	Tinder for lighting fires Chew-ash Sharpening razors, for transporting fire	Robson, 2018
La Draga, Neolithic site, Spain	5324–4977 BC	*Coriolopsis gallica* *Daedalea quercina* *Daldinia concentrica* *Ganoderma adspersum* *Lenzites warnieri* *Skeletocutis nivea*	Catches fire quickly when ground Tinder, a comb, for flavoring, a hemostatic Tinder, for transporting fire tinder to light a fire	Roussel, 2005 Roussel, 2005; Berihuete-Azorín et al., 2018 Berihuete-Azorín et al., 2018 Roussel, 2005 Roussel, 2005; Berihuete-Azorín et al., 2018
La Marmotta, Neolithic village	5000 BC	*Daedaleopsis tricolor*	During ritual ceremonies, pharmacological properties	Bernicchia et al., 2006
Zhejiang Province, China	4800 BC	*Ganoderma lucidum*	Rituals of witch doctors	Yuan et al., 2018
Neolithic man, Iceman (Ötzi) in an alpine glacier at the Hauslabjoch, 92 m south of the Austrian/Italian border	3350–3100 BC	*Fomes fomentarius* *Piptoporus betulinus* two species	Medical and spiritual purposes, for lighting a fire A knife–sharpener, as food while it is immature, i.e., until it becomes woody, some medical (vermifuge) and even spiritual properties	Peinter et al., 1998
In swamp areas of the Neolithic sites of Clairvaux and Charavines in France	2900–2399 BC	*Fomes fomentarius*	For lighting a fire	Montroux & Lundstrom-Baudais, 1979
Vindolanda, England	85–125 AD	*Bovista nigrescens* *Calvatia utriformis*	Delicacies, hemostatic agents, tinder	Watling & Seaward, 1976
Hontoon Island, Volusia County, Florida, USA	1190 AD 1220 AD	*Polyporus hydnoides* *Fomes sclerodermeus*	Certain ceremonies or everyday needs of natives Tinder	Purdy & Purdy, 1982

human settlements around the world (Table 3.3). Such assessments are possible mainly due to modern scientific methods and techniques (O'Regan et al., 2016; Power et al., 2015).

Archaeological finds of mushrooms are most often found in flooded places where the moisture content is high enough to preserve organic material, either in mud that prevented the erosive action of water currents or in ice that enabled the preservation of mushroom tissue to this day. In order to assess as precisely as possible the way mushrooms were used, ethnomycological data, physical characteristics of fossil remains and distribution analysis are taken into account (Berihuete-Azorín et al., 2018). As can be seen in Table 3.3., during those almost 15,000 years, human communities around the world were using various types of mushrooms for nutrition, in medicinal purposes, to light fires and transport flames for tinder, for sharpening knives, or in ritual ceremonies.

3.2 ARCHAEOLOGICAL RECORDS OF *GANODERMA*

According to the archaeological data available so far, species from the genus *Ganoderma* have been found all over the globe, and the age of the fossilized but still sufficiently preserved specimens to be studied is estimated at 6800 to even 1.81 million years (Figure 3.1).

The oldest fossil remains of the mushroom *G. lucidum*, found in Alaska, come from the Pleistocene era (before 1.81 million to 11,550 years ago). This is a very valuable archaeological find considering that mushrooms, due to their sensitive structure, are rarely preserved as fossil records. However, this fruiting body was placed in frozen mud for thousands of years, which enabled the complete preservation of its structure (Chaney & Mason, 1936). Although this fossil remains important in its own right, due to the lack of other relevant data, it cannot be stated with certainty whether and in what way this species was used by human communities.

Among the oldest archaeological findings of *Ganoderma* remains for which it was possible to assume a certain purpose are those that come from the early Neolithic site of La Draga, located on the eastern shore of Lake Banjoles, Spain (Table 3.3). Research conducted by Berihuete-Azorín et al. (2018) documents that out of a total of 86 identified remains, as many as 51 specimens were classified as *G. adspersum*, which are believed to originate from the period 5324–4977 BC. When classifying these taxa in a certain taxonomic category, several parameters were taken into account, such as the consistency of the surface characteristics, the number of pores per unit area of the hymenium, and the structure of the fruitbody hyphae. Identification to the species of these artifacts was possible due to their position completely below the water level, which preserved organic material for thousands of years. As proposed by Roussel (2005), *G. adspersum* was probably used as tinder in this Neolithic settlement, and since this species is highly flammable when ground, it is assumed that this was its primary purpose.

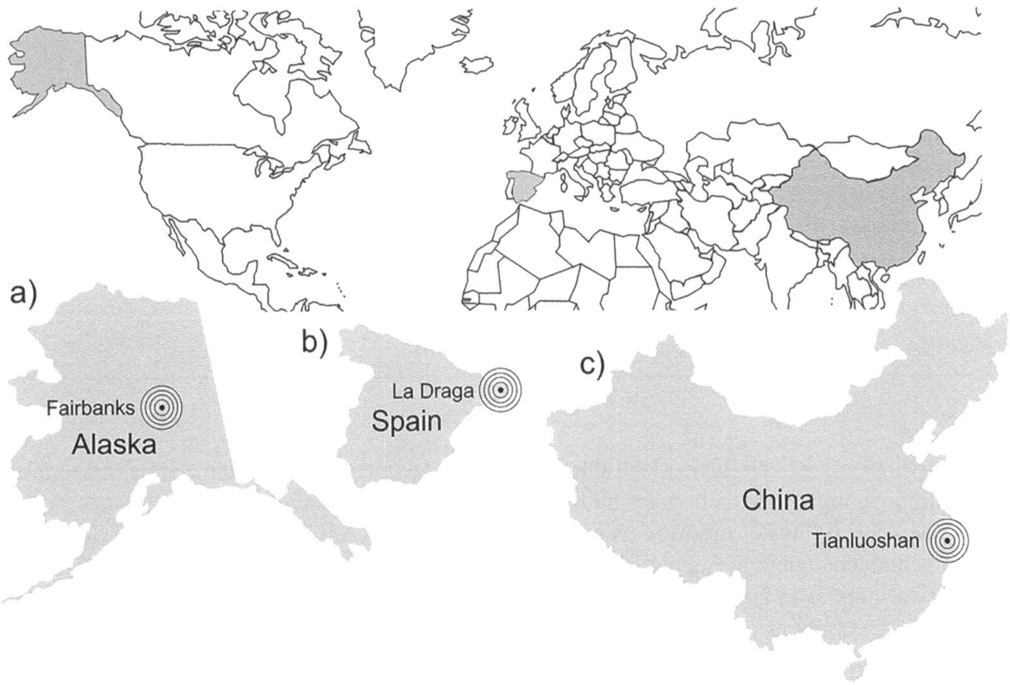

FIGURE 3.1 The most famous archaeological sites where the fossil remains of *Ganoderma* spp. were identified: a) Alaska, USA, *G. lucidum*, 1.81 million–11.550 years; b) La Draga, Spain, Europe, *G. adspersum*, 5324–4977 BC; c) Zhejiang Province, China, Asia, *G. lucidum*, 4800 BC.

The importance of mushrooms from the genus *Ganoderma* in the life of people in the Neolithic era is also evidenced by five artifacts found in three localities in Zhejiang province, China. Fossil remains from 4800 BC found in the Tianloushan site indicate the role of *G. lucidum* in the ritual rites performed by witch doctors. Such important prehistoric artifacts were unearthed alongside numerous cultural relics, including wood carvings, ornaments and head ornaments. These archaeobotanical records contribute to the understanding of the importance that *G. lucidum* acquired in the earliest Neolithic period in the settlements of the lower areas of the Yangtze River (Yuan et al., 2018).

Over time, people began to engage in agriculture, and with such a way of life, to explore the power of plants and mushrooms. The ancient Chinese people accepted the exceptional values of *Ganoderma*, so in their perception it became an extremely valued gift from God, which is also the subject of the legendary events "Shennong gathers *Ganoderma*" and "Xuanyuan gifted with *Ganoderma*", originating from the mythological era of China (Lin, 2019).

3.3 *GANODERMA* THROUGH HISTORY

Although there are certain doubts regarding the taxonomic categories of the genus *Ganoderma*, the development of molecular biology contributed to a more accurate classification, so that today it is considered that 131 species of mushrooms belonging to this genus live around the world (Wang et al., 2020). Various *Ganoderma* species can be found in tropical, subtropical and temperate regions of Africa, America, Asia, Oceania and Europe. In particular, numerous species of this genus are present in the southern tropical and subtropical regions of China (Sun et al., 2020). Considering the existence of suitable natural conditions for the growth of a large number of *Ganoderma* species, these mushrooms have always been available to people in these regions, and there was great interest in them. As a consequence, it is not surprising that the specific attitude towards these exceptional organisms has been developed by humans over the centuries.

Information regarding the use of *G. lucidum* as a health-promoting agent dates back as far as 2,400 years, but it is believed that people throughout the Eastern Hemisphere, including Indian and Korean cultures, have been using it as an important part of folk medicine for more than 4,000 years. Considering bimillennial beliefs, *G. lucidum* has a positive effect on health and contributes to longevity; also it is highly appreciated for its ability to enhance spiritual power and as a source of immortality (Money, 2016; Wasser, 2005). There are different names for this mushroom. It is known as lingzhi in China, which means "mushroom of immortality" or "celestial herb", suggesting that it brings happiness, good health and even immortality. In Korea, the common name is mannentake, or "10,000-year-old mushroom", while in Japan, the name reishi is used, which means "divine mushroom" (Pegler, 2002). Its importance is also confirmed by the fact that in the Orient *Ganoderma* was used as a talisman that protects people and homes from evil. As reported by Tan (2015), ancient tribes in Malaysia traditionally used the fruiting body of *G. neo-japonicum*. They cut these mushrooms into pieces like beads and strung them into necklaces, which they then put around the necks of children suffering from epilepsy. Also, they use this mushroom to treat various diseases, including diabetes.

The respect that people in the Orient had for *G. lucidum* is also evidenced by the depictions that are often a motif in Chinese and Japanese artworks (Wasser, 2005).

Archaeological evidence confirms the importance of *Ganoderma* as a part of Chinese culture for probably more than 2,000 years, but of particular interest is the fact that this extraordinary mushroom was also an indispensable ingredient in traditional Chinese medicine, and since ancient times it has been considered a valuable medicine. However, there are opinions that the basic concept of *Ganoderma* originated in India as an expression of "soma", soma-haoma, the Vedic plant (Pegler, 2002).

The respect and adoration of *Ganoderma* was so strong in China that the emperors of various dynasties sent servants to distant mountains to look for this sacred mushroom, believing that its consumption would provide them with good health and eternal youth. However, due to the remote location, it was very rare to find it, especially high-quality specimens, so the use of this treasure was reserved primarily for rulers and rich families. The unusual and characteristic shape of its fruiting body is also very fondly used as a decorative design by emperors and nobility (Lin, 2019).

At the time of the unification of China and the construction of the well-known part of the Great Wall during the reign of the Qin Dynasty (221–207 BC), first mention of *Ganoderma* as a mystical mushroom with supernatural powers was recorded. Over time, there were more and more depictions of *Ganoderma* in Chinese literature and art, as a reflection of the belief in its supernatural powers (Wasser, 2005).

According to legend, in 109 BC, Emperor Wu discovered a mushroom of an unusual shape inside the Imperial Palace. Everything indicates that it was a branched form of *Ganoderma* growing in the form of "deer antlers", which arises when there is not enough light, which was probably the case behind the walls of the Imperial Palace. This form of *Ganoderma*, which does not develop a spore-producing pileus, retains its original glossy, varnished, burgundy-brown surface for a very long time, and as such it has been valued for centuries as a precious material for making statues and various decorations (Pegler, 2002).

Since the time of the Han Dynasty (202 BC–9 AD, 25–220 AD), Confucian scientists have called *Ganoderma* "fortune herb" due to the circular lines on its cap, which were considered an auspicious symbol. In the prominent Taoist work Ge Hong's Bao Pu Zi, written in 317 AD during the reign of the Dong Jin Dynasty (265–420 AD), *Ganoderma* is ranked as a medicine. According to his theory a person can achieve immortality by taking *Ganoderma*, which is supported by stories of such cases. The followers of this theory believed that *Ganoderma* is among the best remedies, whose power is so great that regular intake prevents aging and even death. This is how the names xiancao (magic grass) and shenzhi (heavenly herb) were born, which contributed even more to the mystification of this mushroom. Famous Taoists, such as Ge Hong, Lu Xiu–Jing (406–477 AD), Tao Hong–Jing and Sun Si–Miao (541–682 AD), understood its importance and power, so they greatly influenced the promotion of *Ganoderma* culture in China. Thus, in their quest for immortality, they contributed to the development of Taoist medical practices that promote good health and welfare (Lin, 2019). Belief in the unlimited powers of *Ganoderma* grew, and the mushroom itself was shrouded in mystery, largely due to the lack of methods to determine its true effects (Lin, 2019).

The importance of this mushroom in the thousands of years of tradition and culture of the Chinese people is also witnessed by the stylized depiction of *Ganoderma* on the walls of the Forbidden City in Beijing. This gift from God has always been considered ritual food for the Emperor (Pegler, 2002).

3.3.1 *Ganoderma* in Art

The consequence of accepting *Ganoderma* as a gift from God, and thus worshiping this exceptional mushroom, is the expression of artistic impulses through many different directions. Various artifacts such as paintings, balustrades, carpet and furniture designs, deer's antlers and jade carvings, jewelry, perfume bottles and women's hair combs are centuries old, and some of them date back to the Yuan Dynasty (1280–1368 AD) (Wasser, 2005).

One of the evidences that testify to the importance of these mushrooms is a picture from the Liao Dynasty (916 and 1125 AD) painted on the wall of a wooden pagoda in Shanxi. It represents a barefoot man walking on rocky mountains. He is covered with fur and leaves, has a plump face and bare stomach and carries bamboo baskets on his shoulders. Taking into account all scriptures, legends and tangible evidence, it is believed that this is a representation of Shennong, the god of agriculture. According to legend, he tried 100 plants in one day and came across 70 poisonous species. This story points to the importance of the discovery process (Lin, 2019).

There is also a famous painting from the reign of the Qing Dynasty, dating from 1600 AD, which shows a Chinese priest looking at *Ganoderma*. The specificity of this representation can be determined from the convolutions on the surface of the mushroom, which are considered to symbolize the cumulus clouds on which the immortals lived. Therefore, the belief was that those who eat this mushroom become immortal themselves (Wasson & Wasson, 1957). An extremely valuable silk painting by Emperor Qianlong of the Qing Dynasty in the collection at the Palace Museum in Taipei shows a vase containing pine branches, camellia and plum blossoms. Next to this vase, persimmons, lilies and *Ganoderma* are painted. The motifs in this particular painting symbolize the desire for happiness and fortune (Lin, 2019).

The worship of *Ganoderma* as a symbol of longevity, happiness, peace, well-being and prosperity lasted for centuries and is deeply engraved in the tradition and culture of the Chinese people to this day. Representations of this mushroom and "fortunate clouds" derived from its meaning are shown on temples, ancient buildings, palaces, paintings, sculptures, porcelain, embroideries, clothing and archaeological artifacts. They are visible in the Temple of Heaven on the ceiling of the Qinian Hall, on the pillar in front of the Tiananmen Square and in the Forbidden City, in the corridor through which the kings entered the Main Hall. Carvings representing *Ganoderma* bonsai are located on the fences of the Forbidden City, the Ancient Ministry of Education Building and the Confucius Temple and before the Sakyamuni statue in Yonghe Lamasery. A graphic representation of this sacred mushroom is carved in the Confucius Temple on the base of a stone slab (Lin, 2019).

Even today, after many centuries of respect and worship of the sacred mushroom, idioms are used in the Chinese language that vividly describe the specific relationship of people to *Ganoderma*; some of them are "ci fu jia ziang" (bestow blessing and happiness), "guo tai ming an" (progress is in the state and peace reigns among the people), "ji xiang ru yi" (good fortune and happiness) and "zeng tian shou kao" (longevity is a blessing) (Lin, 2019).

The approach to mushrooms in Western culture is completely different. In these countries, they were used primarily for lighting fires, as food and in ritual ceremonies, but there was no tradition of worship that was specific to the Far East. Therefore, there are no such significant artifacts. During the 20th century, works of art inspired by mushrooms appeared, primarily the drawing and engraving of the underside of the *G. applanatum* cap (known as artist's conk).

3.3.2 *Ganoderma* as Inspiration for Ancient Poetry

The influence of *Ganoderma* as a sacred mushroom with supernatural powers, which brings happiness, prosperity and longevity, even immortality, left its mark in Chinese poetry as well. Numerous poems inspired by this unusual and exceptional mushroom have been preserved. Apart from being a reflection of a time, they are even today a real treasure that testifies to the worship of the divine *Ganoderma* by people over the centuries.

Several examples of poems dedicated to *Ganoderma* are mentioned by Lin (2019). As a common motif in Chinese literature, *Ganoderma* is also a central figure in the poem "Nine Sons", written by the esteemed poet Ku Yuan of Chu. He wrote about the goddess, Mountain Spirit, who longs for love. In the poem, Sanxia represents *Ganoderma*, which can be found and harvested several times during the year. And the Mountain Spirit is the goddess who harvested *Ganoderma* on Mountain Wu.

The influence of this mushroom in various areas of human life is also reflected in the poem that describes the sacrificial ritual served by Emperor Hanwudi, during which 70 young girls and boys sing the song of *Ganoderma* accompanied by music. The purpose of the ritual is to invoke luck, well-being and longevity (Lin, 2019).

Another glorification of *Ganoderma* is described in the poetry of Li Heng (711–762, Tang Dynasty), in Second of Three Poems of Jade *Ganoderma* in Yanying Palace. He admires this divine mushroom that brings superhuman strength and enables the imperial power to justly rule China and bring peace.

A significant contribution to Chinese poetry and the worship of *Ganoderma* was made by the poet Cao Zhi, who often glorified this mushroom through his poetry during the period of the Three

Kingdoms. In the famous poem "Lingzhi Pian" ("On *Ganoderma*"), a chestnut-colored plant that grows along the bank of the Luo river is described as a symbol of the nation's prosperity and glory to God. Cao Zhi praises *Ganoderma* and attributes to it the power of creating heaven and earth. In the poem called "Luo Shen Fu" (which means "In Glory of the Goddess Luo"), Cao Zhi describes the beauty and loveliness of the *Ganoderma*-picking goddess Luo and expresses his great admiration for her. The poem "Fei Long Pian" ("Flying Dragon") is dedicated to the meeting between the poet and the Taoist monk on the misty mountain Tai. The monk was on a white deer, holding *Ganoderma* in his hand, and this meeting was very significant because the monk taught the poet about the magical influence of this mushroom on health.

A similar story, known as "Chang Ge Xing" ("Singing Trip"), belongs to a Han Dynasty Yuefu-style poem. The poet met a man riding a white deer; he had short hair and long ears, and he looked like God. The man took the poet to pick *Ganoderma* and then showed him in his home a red plant that has the power to improve health, darken hair and prolong life (Anonymous, 1997; Chen & Chen, 2003).

3.3.3 Myths about *Ganoderma*

The worship of *Ganoderma* by the Chinese people was so great that over the centuries, as a result of exaggeration, myths arose about the power of resurrection and immortality that this mushroom brings (Wang, 2020).

One of the more famous myths bears the name "The White Snake Stole the Celestial Herb". According to this myth, a white snake lived in the White Cloud Cave located on top of Mount Emei, in the Sichuan province of southwest China. Over time, with hard training, the snake gained extraordinary strength and transformed into a beautiful woman. The same thing happened with the green snake. The two became sisters and lived wonderfully together. Then, one day, they turned into young girls and traveled to Hangzhou West Lake. The white snake took the name Bai Suzhen, and the green one Qing'er, and they wore milky white and cyan dresses. However, the monk Fahai, the elder of Jinshan Temple, realizing that she was actually a white snake, forced Xu Xian to convince his wife to drink a cup of realgar wine during the Dragon Boat Festival. Because of that potion, Bai Suzhen appeared in her snake form, and Xu Xian died of fear. When Bai Suzhen discovered that her beloved was dead, even though she was pregnant, she embarked on a life-threatening journey to find and steal the celestial grass *Ganoderma*, the only medicine that could bring her husband back to life. Traveling through mountains and waters, she finally found *Ganoderma* and harvested it, but was discovered by the guardians of the fairy grass, Mei Tong and Lu Tong, who blocked her way and stabbed her with her sword. Tired, pregnant and wounded, Bai Suzhen was on the verge of death, but then an old man with a snowy head and a youthful complexion known as the Old Man of the South Pole appeared. He was moved by her strength and sincere desire to save her husband, so he asked the two immortal children to free her and give her *Ganoderma*. Deeply grateful to her savior, she took the almighty *Ganoderma* and hurried home, flying on clouds. Her sister Qing'er made *Ganoderma* soup and gave it to Xu Xian and he soon revived. Thus, thanks to *Ganoderma*, one great love survived. This love story with a happy ending was the inspiration for numerous dramas, films, novels and posters in China (Lin, 2019).

"Fairy Yunv and Lingzhi" is also a very important myth in Chinese history. It also talks about the supernatural powers of *Ganoderma*: Once upon a time, the Green Dragon, the youngest son of the Dragon King of the East Sea, was training in the stream of Dahong Mountain in Yingcheng. He had heard the story of the Fairy Junv of Heavenly Palace, who had loved a bath in a pool that was filled with water from a hot spring. Hoping to get some fairy air and gain some more skills, he also bathed in the same place as Fairy Yunv. Because he did not have deep magical power, he contaminated the hot water pool, and this caused the plague to appear. Upon learning of this thoughtless move by the Green Dragon, Fairy Yunv rushed to earth to help sick people. In order to detoxify the hot spring water, she pricked her finger with a hairpin and let out a drop of her own blood and then rushed to pick the *Ganoderma* on Dahong Mountain and heal the people. She stepped onto the clouds that carried her to the mountainside, where she saw a very large *Ganoderma* shining on the cliff. When she approached the mushroom, a large leopard-print python jumped out of the grass and tried to bite her. Realizing that she was in trouble, Fairy Yunv threw her pendant made of jade at the snake that was thus caught in the trap. The secret of this pendant is that it is a ruyi circle that can increase and decrease, so it closed around the python and he could no longer move. She looked resolutely at the snake, which instantly transformed into a hoe for gathering herbs. Rejoicing, she approached the *Ganoderma* with the hoe, but discovered a large brood of hornets flying around anxiously. Then Fairy Yunv waved her magic fan and the hornets flew into a basket for collecting herbs. The fairy, very happy to have removed the python and the hornets, took the basket and the *Ganoderma* and came back to make the medicine. On the way, the fairy encountered a magistrate of Yingcheng County named Diao, a very rapacious and concupiscent bureaucrat. Knowing that the use of *Ganoderma* not only contributes to the treatment of illnesses but also prolongs life, he tried to steal this precious mushroom from the fairy. Fascinated by her beauty, the judge confiscated the fairy, the hoe and the herb basket. He asked if she would agree to hand over the *Ganoderma* and share a great wealth with him, thus freeing herself from punishment. Fairy Yunv heard this, thought about it and then asked for three things to be done. First, to release the stolen women, to which the judge agreed because he considered that their beauty was not even a thousandth part of a fairy's. Next, the fairy demanded that the granary be opened and the victims released, which he did not initially agree to. But his evil fire grew with looking at the fairy, so he agreed to this request of hers. The fairy's third condition was to show the people Diao's severed head. Enraged, Diao ordered the servants to capture Yunv. Then the fairy waved her magic fan and the herb basket turned into a swarm of hornets that attacked the servants. Diao tried to escape, but Yunv waved her magic fan again, causing the basket to transform into a snake that wrapped tightly around Diao and he died. After the fairy waved her fan once more, the hornets flew into the beehive, which turned into a basket, and the snake turned into a hoe again. In the end, Fairy Yunv left the county government, said goodbye to the people and lowered Green Dragon to the foot of Wenfeng Tower with the help of magic powder.

There is also a well-known story from the Book of Shan Hai Jing (Mountains and Seas) which originates from the Warring States Period (476–221 BC), in which there is a story that Yaoji, the young daughter of Emperor Yan, transformed into the Yaocao plant (Grass of Yao). Song Yu, the poet from Chu, used this motif in a fairy tale love story. It is believed that Yaoji is actually *Ganoderma* (Lin, 2019).

Ge Hong, in The Legend of the Gods, describes the pretty goddess Magu who lived on Panlai Island and followed Taoism on Mount Guyu. As a special gift for the queen's birthday, she brewed *Ganoderma* wine. This myth had a significant influence on the creation of an image that sends messages of happiness and longevity in which there is Magu holding wine, a child holding a peach-shaped birthday cake, an old man with a goblet and a crane with *Ganoderma* in its beak. Even today, this motif is popular in folk art dedicated to birthday celebrations (Chen & Chen, 2003; Lin, 2009).

3.4 *GANODERMA* IN TRADITIONAL CHINESE MEDICINE

As far as is known today, the oldest pharmaceutical book dedicated to the descriptions of herbs and their medicinal values, among them *Ganoderma*, is *Shen Nong Ben Cao Jing* (*Shennong Materia Medica*). This work of inestimable importance was written around 100 BC as a summary of the knowledge and experience of ancient Chinese doctors; however, the real author of this remarkable book is unknown. Similar to the experience attributed to Shennong, the god of agriculture, the medical information about *Ganoderma* described in the *Shennong Materia Medica* is the result of centuries of practice. In this monumental work, 365 medical materials are listed in several categories, which are formed in line with the medical and toxic effects that these substances have on humans. The materials with the highest ratings are those that are nontoxic and have pronounced

TABLE 3.4
***Shennong Materia Medica* Categorization of *Ganoderma* according to Color, Taste, Medicinal Nature and Medicinal Effects, Which the Teachings of Traditional Chinese Medicine Are Based On (Lin, 2019)**

Color	Flavor	Medicinal Nature	Medicinal Effect
Red cizhi (or danzhi)	Bitter	Mild, nontoxic	Relieving chest congestion, improving memory
Black heizhi (or xuanzhi)	Salty	Mild nontoxic	Renal problems, increasing awareness
Blue qingzhi (or longzhi)	Acidic	Mild nontoxic	Improve eyesight, liver functions
White baizhi (or yuzhi)	Pungent	Mild nontoxic	Treatment of cough and lung disease
Yellow huangzhi (or jinzhi)	Sweet	Mild nontoxic	Treatment of heart, spleen and stomach diseases
Purple zizhi (or muzhi)	Sweet	Warm (mild) nontoxic	Treatment of difficulty hearing and arthritis

healing properties. Based on this categorization, and according to empirical data, all *Ganoderma* species (red cizhi, black heizhi, blue qingzhi, white baizhi, yellow huangzhi, and purple zizhi) were rated the highest (Lin, 2019).

As can be seen in Table 3.4, the *Shennong Materia Medica* reveals that cizhi relieves chest congestion and contributes to better memory, heizhi heals the kidneys and increases awareness, qingzhi improves eyesight and liver functions, baizhi enables the treatment of coughs and lung diseases, huangzhi helps treat diseases of the heart, spleen and stomach, while zizhi is effective in treating difficulty hearing and arthritis (Lin, 2019). It is very important to point out that all six types of *Ganoderma* can be used for a long time due to its mildness and nontoxicity; no harmful effects are known; and they contribute to maintaining good health, relaxation of the body, help in the treatment of numerous disorders, alleviate aging and ensure longevity (Wang et al., 2020).

The invaluable work *Ben Cao Gang Mu* (*Compendium of Materia Medica*), in addition to describing numerous medicines that come from nature and about which the Chinese have known for thousands of years also contains the statement of the respected naturalist and physician Li Shi Zhen (1518–1593) who said that *Ganoderma* contributes to the vital energy "qi" and helps those with chest problems. Even after long-term use of *Ganoderma*, the body will remain agile, and the years will approach the years of immortal fairies. The scientific qualities of *Ganoderma*, including its bionomics, herbal nature, classification, habitat and medicinal uses, were first explored in early Chinese literature, such as the *Shen Nong Ben Cao Jing* (*Shennong Materia Medica*) and countless other pieces of literature (Lin, 2019). At the same time, errors and unsubstantiated claims were corrected and removed. The *Shennong Materia Medica* is widely cited and analyzed in published works. This chapter covers the significant decade of traditional culture on *Ganoderma* (Lingzhi) from 2012 to 2022 with regard to its chemical profiles, folk medicinal usages, acceptance, major goods and country–based names. In other words, the Japanese word reishi is derived from the Chinese name for *G. lucidum*, a mushroom of the Ganodermataceae family. The native Japanese term is mannentake, while in English the species is referred to by the Chinese name lingzhi—meaning "miraculous mushroom"—or as bracket fungus or shelf fungus. The reishi is allocated to the first of these three categories as a substance that assists in the maintenance of health. It appears to have reached Japan too at an early date. Featured in both the *Nihon Shoki* (*Chronicles of Japan*) of 720 AD and the *Honzo Wamyo* of 918 AD, which is Japan's oldest book on medicinal herbs, it was known as a mushroom variety that promoted health.

3.5 CHEMICAL ASPECTS OF 20 PROMINENT *GANODERMA* SPECIES

Chinese culture has long associated lingzhi with superstitions and good fortune. Consequently, it is also well-known as xiancao, shenzhi, and Ruizhi, all of which have the connotation of mysterious strength and good fortune. Lingzhi's popularity, whether for curative or recreational purposes, can be attributed in large part to Taoism. Since ancient times, lingzhi has been the subject of several stories and poems in Chinese literature, where it has been the object of veneration and faith. This belief originated from the 20 most prominent species of *Ganoderma* (Gong et al., 2019) as depicted in Figure 3.2, which include *G. lucidum*, *G. lingzhi*, *G. orbiforme*, *G. amboinense*, *G. colossum*, *G. boninense*, *G. formosanum*, *G. atrum*, *G. cochlear*, *G. theaecolum*, *G. resinaceum*, *G. duropora*, *G. pfeifferi*, *G. tropicum*, *G. applanatum*, *G. tsugae*, *G. australe*, *G. capense*, *G. japonicum* and *G. sinense*. There are around 300 different types of triterpenes and 600 more bioactive chemicals that were produced by these species. These substances include polysaccharides, nucleobases, nucleosides, alkaloids, steroids, and meroterpenoids.

The key criterion in the use of *Ganoderma* in traditional medicine is its bioactive chemical capabilities, which are associated with each of the 20 most well-known species. According to the most reliable search engine, Web of Science (WoS), looking at the past decade (2012–2022), Table 3.5 provides the most utilized chemical from each *Ganoderma*. The most well-recognized component in these 20 *Ganoderma* species is polysaccharide, followed by terpenoid, alkaloid, steroid, nucleoside and nucleobase. In contrast to the use of dangerous chloroform in terpenoid extraction (Oluba, 2019), the fractionation steps of alkaloids (Liu et al., 2011), the time-consuming and costly separation of steroids (Yu et al., 2021) and the costly ZIC-HILIC separation of nucleosides and nucleobases (Chen et al., 2012), polysaccharides have been reported to be fast, safe, large and consistent extraction compounds that require only ethanol precipitation (Zhao et al., 2010).

The medicinal mushroom *G. lucidum* displayed the highest WoS hits, with 2,741 reported polysaccharide discoveries that focused on the effect of polysaccharides on human obesity,

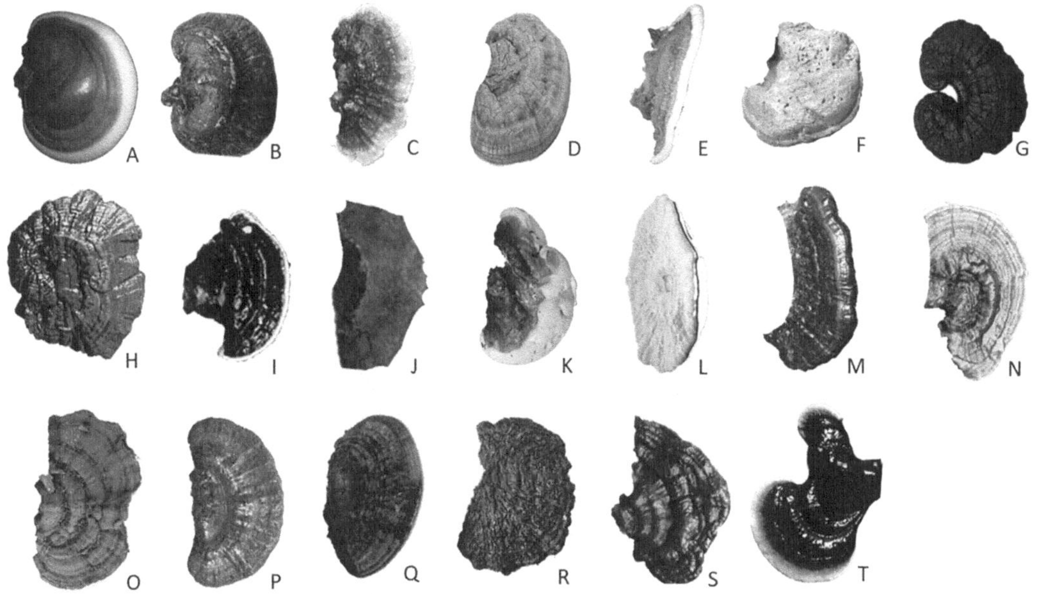

FIGURE 3.2 Top 20 prominent *Ganoderma* species. (A) *G. lucidum*, (B) *G. lingzhi*, (C) *G. orbiforme*, (D) *G. amboinense*, (E) *G. colossus*, (F) *G. boninense*, (G) *G. formosanum*, (H) *G. atrum*, (I) *G. cochlear*, (J) *G. theaecolum*, (K) *G. resinaceum*, (L) *G. duropora*, (M) *G. pfeifferi*, (N) *G. tropicum*, (O) *G. applanatum*, (P) *G. tsugae*, (Q) *G. australe*, (R) *G. capense*, (S) *G. japonicum* and (T) *G. sinense*.

TABLE 3.5
20 Most Bioactive Chemicals Found in *Ganoderma* Species in the Last 10 Years

No.	*Ganoderma*	WoS Hits	Keyword Combination	Reference	Prominent Use
1	*G. lucidum*	2741	AND Polysaccharides	(Chang et al., 2015)	Antiobesity
		164	AND Steroids		
		134	AND Terpenoids		
		104	AND Alkaloids		
		42	AND Nucleosides		
		16	AND Nucleobases		
2	*G. lingzhi*	179	AND Polysaccharides	(Lu et al., 2020)	Antioxidant
		18	AND Steroids		
		18	AND Terpenoids		
		12	AND Alkaloids		
		8	AND Nucleosides		
		6	AND Nucleobases		
3	*G. orbiforme*	5	AND Polysaccharides	(Sahebi et al., 2017)	Plant defense
		3	AND Steroids		
		5	AND Terpenoids		
		8	AND Alkaloids		
		2	AND Nucleosides		
		1	AND Nucleobases		
4	*G. amboinense*	6	AND Polysaccharides	(Ren et al., 2021)	Antialcoholic
		2	AND Steroids		
		1	AND Terpenoids		
		2	AND Alkaloids		
		1	AND Nucleosides		
		1	AND Nucleobases		
5	*G. cochlear*	3	AND Polysaccharides	(Wang et al., 2019)	Renoprotective
		2	AND Steroids		
		10	AND Terpenoids		
		5	AND Alkaloids		
		1	AND Nucleosides		
		1	AND Nucleobases		
6	*G. lingzhi*	179	AND Polysaccharides		*Not tested*
		18	AND Steroids		
		1	AND Terpenoids		
		2	AND Alkaloids		
		1	AND Nucleosides		
		1	AND Nucleobases		
7	*G. colossum*	2	AND Polysaccharides	(Yajima et al., 2014)	Ganomycin (Anti-HIV)
		2	AND Steroids		
		4	AND Terpenoids		
		2	AND Alkaloids		
		1	AND Nucleosides		
		1	AND Nucleobases		
8	*G. boninense*	5	AND Polysaccharides		*Not tested*
		3	AND Steroids		
		6	AND Terpenoids		
		9	AND Alkaloids		
		2	AND Nucleosides		
		1	AND Nucleobases		
9	*G. formosanum*	11	AND Polysaccharides	(Wang et al., 2012)	Immunity
		1	AND Steroids		
		0	AND Terpenoids		
		1	AND Alkaloids		
		1	AND Nucleosides		
		1	AND Nucleobases		

TABLE 3.5 *(Continued)*
20 Most Bioactive Chemicals Found in *Ganoderma* Species in the Last 10 Years

No.	*Ganoderma*	WoS Hits	Keyword Combination	Reference	Prominent Use
10	*G. atrum*	121	AND Polysaccharides	(Ferreira et al., 2015)	Antioxidant
		1	AND Steroids		
		0	AND Terpenoids		
		2	AND Alkaloids		
		3	AND Nucleosides		
		2	AND Nucleobases		
11	*G. theaecolum*	1	AND Polysaccharides	(Liu et al., 2014)	Hepatoprotective
		2	AND Steroids		
		1	AND Terpenoids		
		0	AND Alkaloids		
		0	AND Nucleosides		
		0	AND Nucleobases		
12	*G. resinaceum*	12	AND Polysaccharides	(Abdulghani et al., 2011)	Anti-staphylococcal
		5	AND Steroids		
		3	AND Terpenoids		
		2	AND Alkaloids		
		1	AND Nucleosides		
		1	AND Nucleobases		
13	*G. duropora*	1	AND Polysaccharides		*Not tested*
		1	AND Steroids		
		0	AND Terpenoids		
		1	AND Alkaloids		
		1	AND Nucleosides		
		1	AND Nucleobases		
14	*G. pfeifferi*	4	AND Polysaccharides		*Not tested*
		4	AND Steroids		
		1	AND Terpenoids		
		2	AND Alkaloids		
		1	AND Nucleosides		
		1	AND Nucleobases		
15	*G. tropicum*	1	AND Polysaccharides		*Not tested*
		1	AND Steroids		
		0	AND Terpenoids		
		1	AND Alkaloids		
		1	AND Nucleosides		
		1	AND Nucleobases		
16	*G. applanatum*	37	AND Polysaccharides	(Kozarski et al., 2012)	Antioxidant
		7	AND Steroids		
		5	AND Terpenoids		
		10	AND Alkaloids		
		4	AND Nucleosides		
		1	AND Nucleobases		
17	*G. tsugae*	27	AND Polysaccharides	(Chien et al., 2015)	Antiproliferation
		10	AND Steroids		
		0	AND Terpenoids		
		3	AND Alkaloids		
		1	AND Nucleosides		
		1	AND Nucleobases		
18	*G. australe*	8	AND Polysaccharides	(de Melo et al., 2016)	Phagocytic
		3	AND Steroids		
		2	AND Terpenoids		
		4	AND Alkaloids		
		1	AND Nucleosides		
		1	AND Nucleobases		

(*Continued*)

TABLE 3.5 ***(Continued)***
20 Most Bioactive Chemicals Found in *Ganoderma* Species in the Last 10 Years

No.	*Ganoderma*	WoS Hits	Keyword Combination	Reference	Prominent Use
19	*G. capense*	27	AND Polysaccharides	(Huang et al., 2015)	Antioxidant
		10	AND Steroids		
		0	AND Terpenoids		
		1	AND Alkaloids		
		1	AND Nucleosides		
		1	AND Nucleobases		
20	*G. japonicum*	9	AND Polysaccharides	(Subramaniam et al., 2017)	Antiglycemic
		2	AND Steroids		
		2	AND Terpenoids		
		2	AND Alkaloids		
		1	AND Nucleosides		
		1	AND Nucleobases		
21	*G. sinense*	43	AND Polysaccharides	(Han et al., 2014)	Immunity
		9	AND Steroids		
		5	AND Terpenoids		
		4	AND Alkaloids		
		3	AND Nucleosides		
		3	AND Nucleobases		

which is higher to its closest counterpart *G. lingzhi* at 179 (antioxidant). This was followed by medium trends of polysaccharides and their prevalent usage from *G. atrum* (antioxidant), *G. sinense* (immunity), *G. applanatum* (antioxidant), and *G. capense* (antioxidant). As anticipated, *G. lucidum* was the most popular species that has been used for at least a decade without exhibiting any signs of declining, whereas *G. tropicum* and *G. duropora* were entirely abandoned. Due to a number of unfavorable factors, four characterized species have not recently had their bioactive compounds tested. These species are *G. boninense* [pathogenicity to oil palm (Supramani et al., 2022)], *G. duropora* [mostly only available in the wild (Chen et al., 2007)], *G. pfeifferi* [new emerging European-like *Lingzhi* (Lindequist et al., 2015)] and the recently discovered *G. tropicum* (Luangharn et al., 2019).

These 20 well-known *Ganoderma* species have long been recognized for their ability to lengthen life, promote vitality and boost youthful vigor. By removing thrombi from the bloodstream, it also encourages healthy circulation. The person has fresh vigor as a result. The decline of the body and intellect is stopped. In fact, *Ganoderma* is a fungus with many uses. Animal investigations and clinical evaluations have been conducted recently to document *Ganoderma* structures. Results were occasionally reported, but no satisfying conclusion could be drawn. More study is being done, and the sheer volume of these studies demonstrates the intricacy and irrefutability of the subject matter. The following qualities are thought to be responsible for *Ganoderma*'s widespread use:

1) prevents thrombosis and phlebitis
2) prevents and treats disorders of the blood vessels and cardiovascular system
3) prevents hypo- and hypertension
4) effective in enhancing and restoring normal allergic reactions
5) effective against other ailments
6) stops the spread of cancer
7) as an adaptogen (Abate et al., 2020)

3.6 *GANODERMA* USAGE AS WORLD FUNCTIONAL FOOD

Ganoderma has been well documented as a folk medicine worldwide, especially in Asian culture which has changed the name from *G. lucidum* to *G. lingzhi* (Wang & Lin, 2019). Table 3.6 provides a clear justification of some of the world uses of *Ganoderma* as folk medicine in China, Malaysia, Argentina, Australia, Germany and Japan. However, China shows the highest interest in the use of *Ganoderma* species compared to European or American counterparts. The use of *Ganoderma* fruiting bodies are limited to its extracts, as it is non-culinary as compared to culinary mushrooms such as oyster and button mushrooms (Wan-Mohtar et al., 2020). Chinese folks utilize the mushroom as herbal tea and sometimes for worship due to ancient beliefs.

Japan (reishi and mannentake), which is the closest user to China, typically transforms the mushroom into a bioactive table essence. In Malaysia, instant coffee is used as a traditional treatment, whilst *G. oerstedii* (Fr.) Torrend is used in Argentina as a serum. The Australians turned *G. adspersum* (Schulzer) Donk into capsules of beta-glucan. German scientists confirmed the 2015 discovery of the novel European lingzhi, *G. pfeifferi*, as the new antibiotic ganomycin. Lingzhi's status as a therapeutic "drug" for medical purposes or as a food supplement for maintaining good health is still up for debate. There have been no reports of human research using lingzhi as a direct anticancer agent thus far, despite some data suggesting that it may be utilized as a potential supplement for cancer patients (Yuen & Gohel, 2005). Lingzhi improved the cellular immunological responses and mitogenic reactivity of cancer patients, according to two randomized and one nonrandomized trial, and in one research study, the quality of life of 65% of patients with lung cancer was enhanced. The direct cytotoxic and anti-angiogenesis mechanisms of lingzhi have been proven in in vitro studies, but clinical research should not be disregarded when determining the proper treatment in vivo. There is evidence that lingzhi may have direct in vivo anticancer advantages in addition to its current use as a dietary supplement to aid cancer patients. The classification of lingzhi or its compounds as an anticancer agent will depend on the availability of more recent and specific scientific data (Wasser, 2010).

3.7 CHALLENGES, CURRENT PRACTICES AND FUTURE SUSTAINABLE DEVELOPMENT GOALS OF *GANODERMA*

Modern biological research has demonstrated the wide range of advantageous properties of *Ganoderma*, including its capacity to shield the liver from chemical damage; reduce blood sugar; increase immunity; and possess anti-aging, antiradiation, anticancer, antiinflammation, and anti–blood clotting properties (Kozarski et al., 2023; Wan-Mohtar et al., 2021a, 2017). As a result, product development and applied research have been done across the board in the *Ganoderma* industrial franchise, from fundamental research to products, from new cultivar breeding to large–scale standardized processing, manufacturing plantations, and quality control. Over 150 teams or groups of researchers are working on scientific and technological studies of *Ganoderma* in China, while approximately 300 businesses are engaged in practical product development, research, and/or product sales. The creation and promotion of 3300 cosmetic goods, 180 medications, and 1200 functional foods, including herbal pharmaceuticals, have all been combined (Lai et al., 2004).

There have been 6260 *Ganoderma* discoveries reported to date, according to the most recent trends in Figure 3.3, which covers the years 1970 to 2022. The highest fields that reported more than 1000 discoveries during this 52-year period were biochemistry, molecular biology, and pharmacy, which are the most widely used medicinal mushrooms in pharmaceutical enterprises. With only 5.1%, microbiology had the lowest percentage, followed by chemistry applications (5–7%). Due to the difficult process of isolating and identifying new *Ganoderma* species from various places, mycology was still low, at roughly 9.58%. Food science technology, plant sciences and biotechnology applied microbiology are three significant areas of study that have begun to grow (11–13%).

TABLE 3.6
Some *Ganoderma* Products from Different World Cultures

Country	Another Name	Myths	Available Product	Reference
China	*Bai-zhi*, *Chi-zhi*, *Hei-zhi*, *Huang-zhi*, *Qing-zhi* and *Zi-zhi*	• God worshipping • Ancient poems	 Tea	(Du et al., 2019; Thyagarajan et al., 2007)
Malaysia	*Lingzhi*	• Traditional remedy	 3-in-1 coffee	(Tay et al., 2022)
Argentina	*G. oerstedii* (Fr.) Torrend	• Traditional remedy	 Serum	(Moncalvo et al., 1995)
Australia	*G. adspersum* (Schulzer) Donk	• Traditional remedy	 Beta-glucan capsules	(Badalyan et al., 2019)
Germany	*G. pfeifferi:* A European relative of *Ganoderma lucidum*	• New antibiotic: Ganomycin	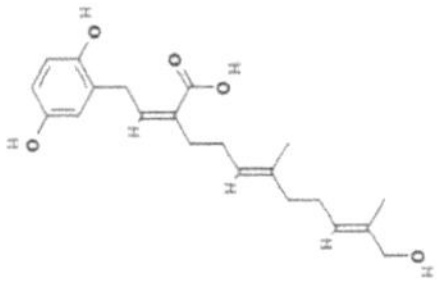 Antibiotic	(Lindequist et al., 2015)
Japan	*Reishi* *Mannentake*	• Bioactive essence	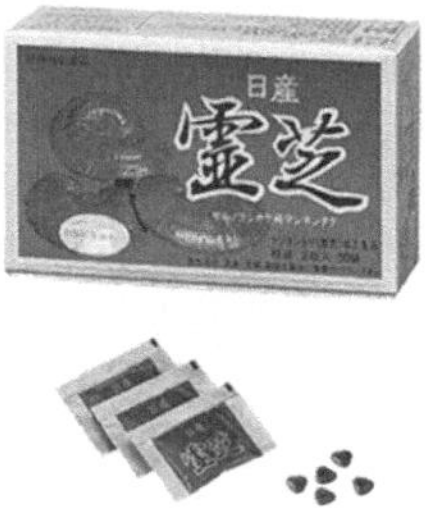 Tablet	(Mizuno, 1999)

Based on the 52-year Web of Science report, we have identified some key challenges in *Ganoderma* research.

3.8 CHAOTIC TAXONOMY

- The taxonomy of *Ganoderma* species is disorganized, and *G. lucidum* has been used to refer to the vast majority of glossy *Ganoderma* species. However, it is now understood that *G. lucidum* sensu stricto has a limited native range in Europe and some parts of China. It's likely that various *Ganoderma* species have varied quantities and types of medicinally significant substances (Loyd et al., 2018).
- *G. lingzhi*, which differs from *G. lucidum* in terms of morphology and genetics, is the medicinal species that is most frequently utilized (Hennicke et al., 2016).
- Several Chinese species that were once classified as *G. lucidum*, such as *G. lingzhi*, *G. tropicum*, *G. sichuanense*, *G. sinense*, *G. flexipes* and *G, multipileum*, are now thought to belong to other species (Raja et al., 2017).

3.9 HARD AND UNATTRACTIVE KIDNEY-SHAPED TEXTURE

- The woody texture, hard surface and occasionally shiny exterior of the *Ganoderma* species make them unsuitable for use as food-grade mushrooms (Tchotet Tchoumi et al., 2019).
- This mushroom's distinctive kidney-shaped cap and fan-like look are not well liked by customers (de Mattos-Shipley et al., 2016).

3.10 BITTERNESS

- Terpenoids are bitter in *Ganoderma* species, which prevents them from being used in food (Nishitoba et al., 1985).
- Ganoderic acid A and B provide bitterness for *G. lucidum* (Fr.) Karst (Kubota et al., 1982).

3.11 NEGATIVE ACCEPTANCE DUE TO *GANODERMA BONINENSE*

- The increasing epidemic of basal-stem rot on major palm oil–producing countries due to pathogenic *G. boninense* has painted an ugly picture of its goodness (Supramani et al., 2022).
- The attacks by *G. boninense* on the palm oil industries, particularly in Malaysia, Indonesia, and Vietnam, have confused stakeholders about the safety of all *Ganoderma* species, including the important medicinal *G. lingzhi* and *G. lucidum* (Parveez et al., 2020).
- In England and the UK woodlands, which are more badly infested with *Ganoderma* species than the other sites under investigation, airborne *Ganoderma boninense* spores are emerging from woods to agricultural and urban areas (Grinn-Gofroń et al., 2021).

3.12 PURCHASING THE RIGHT *GANODERMA*

- Since this mushroom does include a wide range of medicinal substances, research on the miraculous properties of *Ganoderma* will be substantially enhanced in the future. It is the combination of these components that makes *Ganoderma* so powerful at curing human illnesses (Wińska et al., 2019). It will take a great deal of time and effort to do additional study in order to be able to extract the several therapeutic components and use them for different medical applications.
- It has been determined that the *Ganoderma* essence obtained from the crown is superior to that from the stalk. *Ganoderma* typically comes in six different colors. For those with

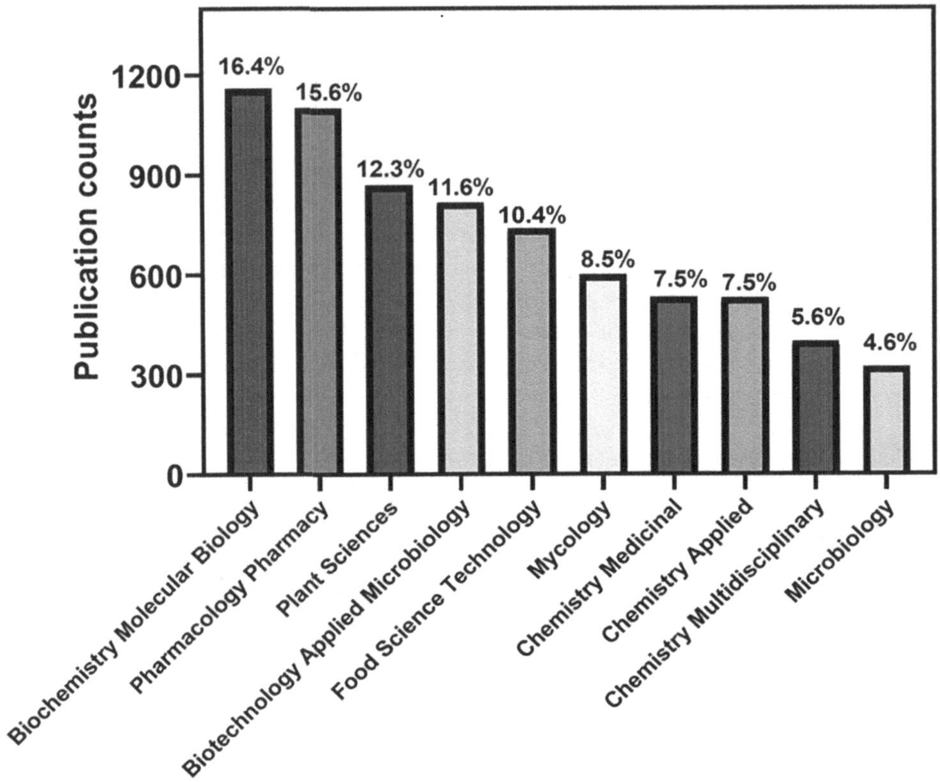

FIGURE 3.3 Total *Ganoderma* discoveries from 1970 to 2022 in various fields of interest.

high blood pressure and excessive cholesterol, the essence from the stalk is particularly effective. This has led us to the conclusion that the red variety of *Ganoderma* with a short stalk and a huge, thick crown is the most potent (Wasser, 2005).

- Some varieties of *Ganoderma* that have not fully developed will eventually form a structure that resembles a branch. About one-fifth, or 20%, of the full-grown thick crown fruiting body's potency will be represented by its essence (Hapuarachchi et al., 2016). Purchasers must also watch out for the unethical behavior of some businesspeople who sell low-quality *Ganoderma* products. Customers are frequently misled by the low pricing and end up with subpar goods. Worst of all, because of the manufacturer's lack of resources and lack of business ethics, the products may not have the required bacteria eradication control.
- Without employing the hot-water-boil extraction method, the user cannot properly pulverize the hard fruiting body of *Ganoderma*, even with the use of very fine grinders. This is a result of *Ganoderma*'s texture. Also, because *Ganoderma* is a fungus, there will always be a variety of microorganisms on the plant as it grows. The *Ganoderma* still includes a lot more bacteria than is required to control germs in food, which is a limit of 3000 bacteria per cm^3, even after being ground up and put in capsule form by these merchants. The Chinese used to always boil the *Ganoderma* and drink the essence rather than consuming the dry, crushed *Ganoderma*, and this is the sole reason for it (Chung & Youn, 2005).

3.12.1 Current Practices of *Ganoderma*

The consumer use of *Ganoderma* fruiting bodies in the form of either reishi or lingzhi is commonly utilized via a brewing or soup recipe as illustrated in Figure 3.4. Using a garden cutter, first slice or

FIGURE 3.4 The steps for the common traditional "brewing of *Ganoderma*" and "*Ganoderma* soup recipe".

chop the dried *Ganoderma* into small pieces or slices. Add three bowls of 600 mL water for each person's daily use of 3–5 gm of slices, and then boil the mixture for 30 minutes at a low temperature in a nonmetallic container like glassware or clay pots. Once the bitter taste has diminished, usually after three to four boils, the boiled *Ganoderma* can be used once more. The brew's bitter flavor can be eliminated by adding glucose and pure honey, mixing in Chinese wine and then storing it in a dry area for three to four months before serving. The residual *Ganoderma* pieces can be fed to wastewater as decolorizer (Mooralitharan et al., 2023) or bio-oil (Mohamad Jahis et al., 2022).

G. lucidum, which has a long history in Far Eastern medicine, may be a very intriguing raw material in the brewing industry as a supplement for producing a variety of products with specific physiological activity. The East Asian–native edible *Shiitake* (*Lentinula edodes*) fungus is used to enhance the Belgian beers "Shiimake" and "Epic Ales Terra — Saurus", which have long been regarded as both a delicacy and a healthy mushroom. An institution in Munich has created "Xanthohumol beer," which has a similar alcohol and calorie content to regular beer but has 10 times as much xanthohumol (Wunderlich et al., 2005). An alcohol extract of *Ganoderma* with a particular more or less bitter taste is acceptable, even though the sensory features of the finished product depend on added doses and a mix of medicinal herbs that may be used for flavoring and additional therapeutic improvement (Leskosek-Cukalovic et al., 2010). Such products could achieve a number of goals, including the launch of new beer products on the market, the development of creative items that satisfy consumer needs and the attraction of new beer drinkers who are concerned about their health benefits. According to certain consumer target groups and their needs, distinct phytopharmaceuticals may modify and adapt their composition, possible therapeutic activities and sensory characteristics (Jiménez-Colmenero et al., 2010).

In the meantime, the sliced *Ganoderma* will be soaked in 1.5 liters of water for 4 hours in the reishi soup recipe. After being withdrawn from the boiling water for 5 seconds, the quails are then put in a Chinese steamer and combined with sliced *Ganoderma*. Next, dry scallops and lean pig meat will be added in the boiling liquid along with a dash of cooking wine. The mixture will next be seasoned with salt and pepper in accordance with local customs and cooked for an additional 3 hours. Traditionally, the mixture will be sweetened with dried long-gang beef or dried red dates. Lingzhi is typically combined with ginseng, which is used to energize qi, treat asthma, calm the nerves and boost the immune system, according to a book on dietary Chinese herbs (Dong & Han, 2015).

3.12.2 Future of *Ganoderma* in Current Sustainable Development Goals (SDGs)

The use of *Ganoderma* is crucial, especially in a post-pandemic or second wave of COVID-19. Figure 3.5 provides nine important prospects of emerging *Ganoderma* discoveries during the post–COVID-19 era in fulfilling the 17 SDG goals which started in 25 September 2015. The use of *Ganoderma* in solar is quite new, as it just started in 2022 with 3 discoveries in fulfilling climate action (Ahmad et al., 2022), followed increasingly with fish-feed (8) (Wan-Mohtar et al., 2021b), building material (8), meat (12), COVID-19 (15) (Wan-Mohtar et al., 2017) and wastewater (29) (Mohd Hanafiah et al., 2019). The top two highest uses of *Ganoderma* in SDGs is starting with soil application (108), which is focused more on soil conditioning (Rashad et al., 2019), soil bioremediation (Rigas et al., 2007) and biochar (Chang et al., 2020). Meanwhile, more than half of the discoveries (412) are related to food utilization such as alternative cooking flour (Wan-Mohtar et al., 2018), functional food (El Sheikha, 2022) and dietary supplements (Tseng et al., 2008).

3.13 CONCLUSION

According to various archaeological records, epic historical collections, works of art, ancient poetry, myths and steadfast Chinese traditions, *Ganoderma* has influenced human traditions for at least 6000 years and will continue to do so until the year 2023. Twenty well-known *Ganoderma* species with their biochemical substances are the reasons why it hasn't been absorbed by time as a result of

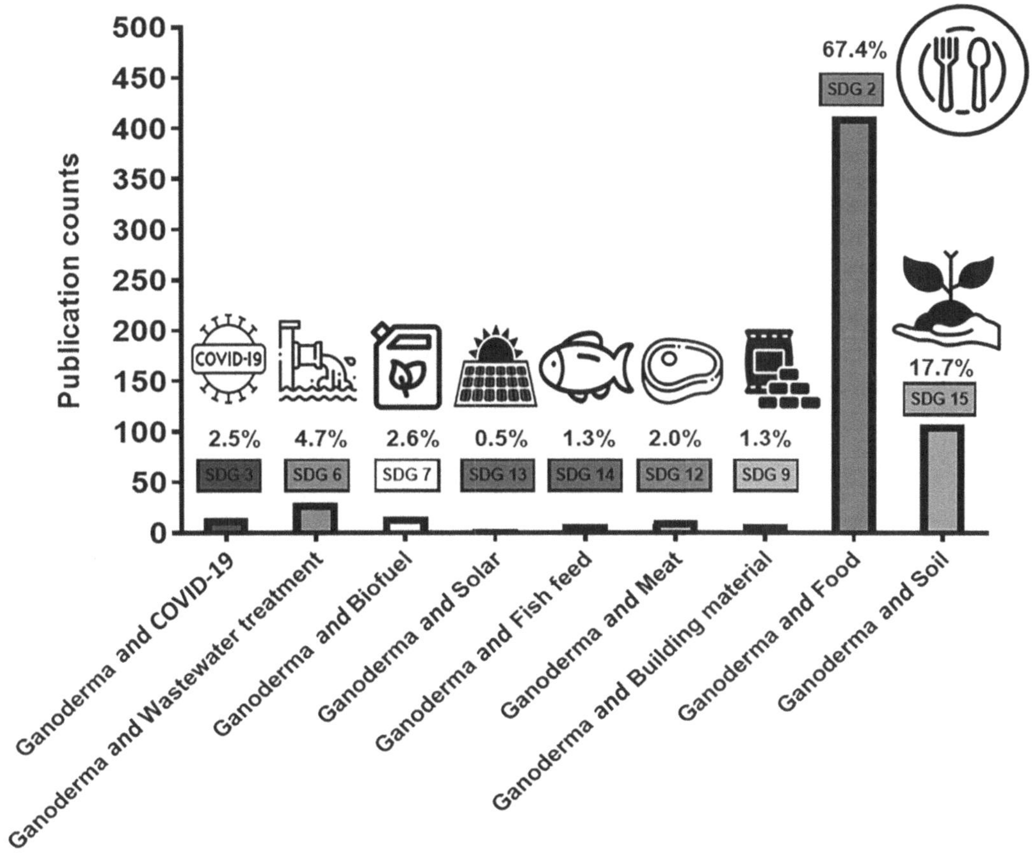

FIGURE 3.5 Nine prospects of *Ganoderma* in SDG goals from 2015 to 2022.

these findings. These have been used by all nations on all continents, and they are based on molecular biology and include things like tea, coffee and serum. With food serving as the primary motivating element, *Ganoderma* convincingly secured nine SDG targets in view of the epidemic that conflicts with them. It is thought that *Ganoderma* is a "godsend" and has the power to end all human suffering.

REFERENCES

Abate, M., Pepe, G., Randino, R., Pisanti, S., Basilicata, M. G., Covelli, V., Bifulco, M., Cabri, W., D'Ursi, A. M., Campiglia, P., & Rodriquez, M. (2020). *Ganoderma lucidum* ethanol extracts enhance re-epithelialization and prevent keratinocytes from free-radical injury. *Pharmaceuticals*, *13*(9), 224.

Abdel-Azeem, M. A., Blanchette, R. A., Mohesien, M. T., Salem, F. M., & Abdel-Azeem, M. A. (2016). *The conservation of mushroom in ancient Egypt through the present.* The First International Conference on Fungal Conservation in the Middle East and North of Africa (ICFC), Ismailia, 18–20 October, pp. 1–2.

Abdulghani, A., Jitendra, G. V., & Subhash, S. D. (2011). In vitro evaluation of anti-staphylococcal activity of *Ganoderma lucidum*, *Ganoderma praelongum* and *Ganoderma resinaceum* from Pune, India. *African Journal of Microbiology Research*, *5*(3), 328–333.

Ahmad, N., Vunduk, J., Klaus, A., Dahlan, N. Y., Ghosh, S., Muhammad-Sukki, F., Dufossé, L., Bani, N. A., & Wan-Mohtar, W. A. A. Q. I. (2022). Roles of medicinal mushrooms as natural food dyes and dye-sensitised solar cells (DSSC): Synergy of zero hunger and affordable energy for sustainable development. *Sustainability*, *14*(21), 13894.

Akers, B. P., Ruiz, J. F., Piper, A., Carl, A. P., & Ruck, C. A. P. (2011). A prehistoric mural in Spain depicting neurotropic psilocybe mushrooms. *Economic Botany*, *65*(2), 121–128.

Angelov Uzunov, B., & Petrova Stoyneva-Gärtner, M. (2015). Mushrooms and lichens in Bulgarian ethnomycology. *Journal of Mycology*, 1–7. https://doi.org/10.1155/2015/361053

Anonymous. (1997). *Shen Nong Ben Cao Jing (Shennong Materia Medica)*, 1st ed. Liaoning Science and Technology Press, Shenyang, p. 9.

Badalyan, S. M., Gharibyan, N. G., Iotti, M., & Zambonelli, A. (2019). Morphological and ecological screening of different collections of medicinal white-rot bracket fungus *Ganoderma adspersum* (Schulzer) Donk (Agaricomycetes, Polyporales). *Italian Journal of Mycology*, *48*, 1–15.

Berihuete-Azorín, M., Girbal, J., Piqué, R., Palomo, A., & Terradas, X. (2018). Punk's not dead: Fungi for tinder at the neolithic site of La Draga (NEIberia). *PLOS One*, *13*(4), e0195846.

Bernicchia, A., Maria, A. Fugazzola, M. A., Gemelli, V., Mantovani, B., Lucchetti, A., Cesari, M., & Speroni, E. (2006). DNA recovered and sequenced from an almost 7000 y-old neolithic polypore *Daedaleopsis tricolor*. *Mycological Research*, *110*, 14–17.

Brown, R. W. (1940). A bracket fungus from the late tertiary of Southwestern Idaho. *Journal of the Washington Academy of Sciences*, *30*(10), 422–424.

Chaney, R. W., & Mason, H. L. (1936). *Pleistocene Flora from Fairbanks, Alaska* (American Museum Novitates, 887). American Museum of Natural History, New York.

Chang, C. J., Lin, C. S., Lu, C. C., Martel, J., Ko, Y. F., Ojcius, D. M., Tseng, S. F., Wu, T. R., Chen, Y. Y. M., & Young, J. D. (2015). *Ganoderma lucidum* reduces obesity in mice by modulating the composition of the gut microbiota. *Nature Communications*, *6*(1), 1–19.

Chang, J., Zhang, H., Cheng, H., Yan, Y., Chang, M., Cao, Y., Huang, F., Zhang, G., & Yan, M. (2020). Spent *Ganoderma lucidum* substrate derived biochar as a new bio-adsorbent for Pb2+/Cd2+ removal in water. *Chemosphere*, *241*, 125121.

Chen, S. Y., & Chen, Q. W. (2003). Fungal anthropology and *Ganoderma* culture. *Journal of Hubei Agricultural College*, *23*(6), 426–433.

Chen, T. Q., Zhong, L. Y., Wu, J. Z., Li, Y., Lin, X. S., Zhu, P. G., Xu, J., & Qiu, F. P. (2007). Exploitation and utilization of wild Fujian *Ganoderma duropora* (Sect. Phaeonema) resource. *International Journal of Medicinal Mushrooms*, *9*(3–4), 238.

Chen, Y., Bicker, W., Wu, J., Xie, M., & Lindner, W. (2012). Simultaneous determination of 16 nucleosides and nucleobases by hydrophilic interaction chromatography and its application to the quality evaluation of *Ganoderma*. *Journal of Agricultural and Food Chemistry*, *60*(17), 4243–4252.

Chien, R. C., Yen, M. T., Tseng, Y. H., & Mau, J. L. (2015). Chemical characteristics and anti-proliferation activities of *Ganoderma tsugae* polysaccharides. *Carbohydrate Polymers*, *128*, 90–98.

Chung, H. S., & Youn, K. S. (2005). Comparison of pretreatment methods for extraction of selected components from *Ganoderma lucidum*. *Korean Journal of Food Preservation*, *12*(2), 130–134.

de Mattos-Shipley, K. M., Ford, K. L., Alberti, F., Banks, A., Bailey, A. M., & Foster, G. (2016). The good, the bad and the tasty: The many roles of mushrooms. *Studies in Mycology*, *85*(1), 125–157.

de Melo, R. H., do Amaral, A. E., Menolli, R. A., Ayala, T. S., Simao, R. D. C. G., de Santana-Filho, A. P., Sassaki, G. L., Kadowaki, M. K., & da Conceicao Silva, J. L. 2016. β-(1→3) glucan of the Southern bracket mushroom, *Ganoderma australe* (Agaricomycetes), stimulates phagocytosis and interleukin—6 production in mouse peritoneal macrophages. *International Journal of Medicinal Mushrooms*, *18*(4), 313–320.

Dickson, J., Oeggl, K., Holden, T. G., Handley, L. L., O'Connell, T. C., & Preston, T. (2000). The omnivorous Tyrolean Iceman: Colon contents (meat, cereals, pollen, moss, whipworm) and stable isotope analysis. *Philosophical Transactions of the Royal Society B*, *355*, 1843–1849.

Dikov, N. N. (1971). *Naskal'nyie zagadki drevnei Chukotki: Petroglify Pegtymelia (Rock art puzzles of ancient Chukotka: Pegtymel petroglyphs)*. Nauka, Moscow.

Dong, C., & Han, Q. (2015). *Ganoderma lucidum* 灵芝 (Lingzhi, *Ganoderma*). In: *Dietary Chinese herbs*. Springer, Singapore, pp. 759–766.

Du, Z., Dong, C. H., Wang, K., & Yao, Y. J. (2019). Classification, biological characteristics and cultivations of *Ganoderma*. In: Z. Lin & B. Yang (Eds.), *Ganoderma and health: Advances in experimental medicine and biology*. Springer, Singapore, vol. 1181, pp. 15–58.

El Sheikha, A. F. (2022). Nutritional profile and health benefits of *Ganoderma lucidum* "Lingzhi, Reishi, or Mannentake" as functional foods: Current scenario and future perspectives. *Foods*, *11*(7), 1030.

Ferreira, I. C., Heleno, S. A., Reis, F. S., Stojkovic, D., Queiroz, M. J. R., Vasconcelos, M. H., & Sokovic, M. (2015). Chemical features of *Ganoderma* polysaccharides with antioxidant, antitumor and antimicrobial activities. *Phytochemistry*, *114*, 38–55.

Gargano, M. L., van Griensven, L. J. L. D., Isikhuemhen, O. S., Lindequist, U., Venturella, G., Wasser, S. P., & Zervakis, G. I. (2017). Medicinal mushrooms: Valuable biological resources of high exploitation potential. *Plant Biosystems*, *151*(3), 548–565.

Gong, T., Yan, R., Kang, J., & Chen, R. (2019). Chemical components of *Ganoderma*. *Advances in Experimental Medicine and Biology*, *1181*, 59–106.

Grinn-Gofroń, A., Bogawski, P., Bosiacka, B., Nowosad, J., Camacho, I., Sadyś, M., Skjøth, C. A., Pashley, C. H., Rodinkova, V., & Çeter, T. (2021). Abundance of *Ganoderma* sp. in Europe and SW Asia: Modelling the pathogen infection levels in local trees using the proxy of airborne fungal spore concentrations. *Science of the Total Environment*, *793*, 148509.

Han, X. Q., Yue, G. L., Yue, R. Q., Dong, C. X., Chan, C. L., Ko, C. H., Cheung, W. S., Luo, K. W., Dai, H., & Wong, C. K. (2014). Structure elucidation and immunomodulatory activity of a beta glucan from the fruiting bodies of *Ganoderma sinense*. *PLOS One*, *9*(7), e100380.

Hapuarachchi, K., Wen, T., Jeewon, R., Wu, X., Kang, J., & Hyde, K. (2016). Mycosphere essays 7: *Ganoderma lucidum*—are the beneficial anti-cancer properties substantiated. *Mycosphere*, *7*(3), 305–332.

Hennicke, F., Cheikh-Ali, Z., Liebisch, T., Maciá-Vicente, J. G., Bode, H. B., & Piepenbring, M. (2016). Distinguishing commercially grown *Ganoderma lucidum* from *Ganoderma lingzhi* from Europe and East Asia on the basis of morphology, molecular phylogeny, and triterpenic acid profiles. *Phytochemistry*, *127*, 29–37.

Huang, Y., Li, N., Wan, J. B., Zhang, D., & Yan, C. (2015). Structural characterization and antioxidant activity of a novel heteropolysaccharide from the submerged fermentation mycelia of *Ganoderma capense*. *Carbohydrate Polymers*, *134*, 752–760.

Jiménez-Colmenero, F., Sánchez-Muniz, F. J., & Olmedilla-Alonso, B. (2010). Design and development of meat-based functional foods with walnut: Technological, nutritional and health impact. *Food Chemistry*, *123*(4), 959–967.

Kozarski, M., Klaus, A., Nikšić, M., Vrvić, M. M., Todorović, N., Jakovljević, D., & Van Griensven, L. J. (2012). Antioxidative activities and chemical characterization of polysaccharide extracts from the widely used mushrooms *Ganoderma applanatum*, *Ganoderma lucidum*, *Lentinus edodes* and *Trametes versicolor. Journal of Food Composition and Analysis*, *26*(1–2), 144–153.

Kozarski, M., Klaus, A., van Griensven, L., Jakovljevic, D., Todorovic, N., & Wan-Mohtar, W. A. A. Q. I., & Vunduk, J. (2023). Mushroom β-glucan and polyphenol formulations as natural immunity boosters and balancers: Nature of the application. *Food Science and Human Wellness*, *12*(2), 378–396.

Kubota, T., Asaka, Y., Miura, I., & Mori, H. (1982). Structures of Ganoderic acid A and B, two new lanostane type bitter triterpenes from *Ganoderma lucidum* (FR.) Karst. *Helvetica Chimica Acta*, *65*(2), 611–619.

Lai, T., Gao, Y., & Zhou, S. (2004). Global marketing of medicinal Lingzhi mushroom *Ganoderma* lucidum (W. Curt.: Fr.) Lloyd (Aphyllophoromycetideae) products and safety concerns. *International Journal of Medicinal Mushrooms*, *6*(2).

Leskosek-Cukalovic, I., Despotovic, S., Lakic, N., Niksic, M., Nedovic, V., & Tesevic, V. (2010). *Ganoderma lucidum*—Medical mushroom as a raw material for beer with enhanced functional properties. *Food Research International*, *43*(9), 2262–2269.

Lin, Z. B. (2009). *Lingzhi from mystery to science*, 1st ed. Peking University Medical Press, Beijing, pp. 1–19.

Lin, Z. B. (2019). *Ganoderma* (Lingzhi) in traditional Chinese medicine and Chinese culture. In: Z. Lin & B. Yang (Eds.), *Ganoderma and health: Advances in experimental medicine and biology*. Springer, Singapore, vol. 1181, pp. 1–13.

Lindequist, U., Jülich, W. D., & Witt, S. (2015). *Ganoderma pfeifferi*—a European relative of *Ganoderma lucidum*. *Phytochemistry*, *114*, 102–108.

Liu, J. Q., Wang, C. F., Peng, X. R., & Qiu, M. H. (2011). New alkaloids from the fruiting bodies of *Ganoderma sinense*. *Natural Products and Bioprospecting*, *1*(2), 93–96.

Liu, L. Y., Chen, H., Liu, C., Wang, H. Q., Kang, J., Li, Y., & Chen, R. Y. (2014). Triterpenoids of *Ganoderma theaecolum* and their hepatoprotective activities. *Fitoterapia*, *98*, 254–259.

Loyd, A. L., Richter, B. S., Jusino, M. A., Truong, C., Smith, M. E., Blanchette, R. A., & Smith, J. A. (2018). Identifying the "mushroom of immortality": Assessing the *Ganoderma* species composition in commercial Reishi products. *Frontiers in Microbiology*, *9*, 1557.

Lu, J., He, R., Sun, P., Zhang, F., Linhardt, R. J., & Zhang, A. (2020). Molecular mechanisms of bioactive polysaccharides from *Ganoderma lucidum* (Lingzhi), a review. *International Journal of Biological Macromolecules*, *150*, 765–774.

Luangharn, T., Karunarathna, S. C., Mortimer, P. E., Hyde, K. D., Thongklang, N., & Xu, J. (2019). A new record of *Ganoderma tropicum* (Basidiomycota, Polyporales) for Thailand and first assessment of optimum conditions for mycelia production. *MycoKeys*, *51*, 65.

Mason, H. L. (1934). Contribution to paleontology, studies of the Pleistocene paleobotany of California. IV. Pleistocene flora of the Tomales formation. *Carneige Institute of Washington Publications*, *415*, 81–179.

Mizuno, T. (1999). The extraction and development of antitumor—active polysaccharides from medicinal mushrooms in Japan. *International Journal of Medicinal Mushrooms*, *1*(1), 9–29.
Mohamad Jahis, B. M., Ilham, Z., Supramani, S., Sohedein, M. N. A., Ibrahim, M. F., Abd-Aziz, S., Rowan, N., & Wan-Mohtar, W. A. A. Q. I. (2022). Ganodiesel: A new biodiesel feedstock from biomass of the mushroom *Ganoderma lucidum*. *Sustainability*, *14*(17), 10764.
Mohd Hanafiah, Z., Wan Mohtar, W. H. M., Abu Hasan, H., Jensen, H. S., Klaus, A., & Wan-Mohtar, W. A. A. Q. I. (2019). Performance of wild-Serbian *Ganoderma lucidum* mycelium in treating synthetic sewage loading using batch bioreactor. *Scientific Reports*, *9*(1), 1–12.
Molodin, I. V., & Cheremisin, D. V. (1999). Pétroglyphes de l'âge du bronze du plateus d'Ukok. A propos des répresentations de personnages avec coiffure fongiforme (Petroglyphs of the Bronze Age of the Ukok Plateau). *Arts Asiatiques*, *54*, 148–152.
Moncalvo, J. M., Wang, H. F., & Hseu, R. S. (1995). Gene phylogeny of the *Ganoderma lucidum* complex based on ribosomal DNA sequences: Comparison with traditional taxonomic characters. *Mycological Research*, *99*(12), 1489–1499.
Money, N. P. (2016). Are mushrooms medicinal? *Fungal Biology*, *120*, 449–453.
Montroux, O., & Lundstrom-Baudais, K. (1979). Polyporacées des sites néolithiques de Chairvaux et Charavines (France). *Candollea*, *34*, 153–166.
Mooralitharan, S., Mohd Hanafiah, Z., Abd Manan, T. S. B., Muhammad-Sukki, F., Wan-Mohtar, W. A. A. Q. I., & Wan-Mohtar, W. H. M. (2023). Vital conditions to remove pollutants from synthetic wastewater using Malaysian *Ganoderma lucidum*. *Sustainability*, *15*(4), 3819.
Nishitoba, T., Sato, S., & Sakamura, S. (1985). New terpenoids from *Ganoderma lucidum* and their bitterness. *Agricultural and Biological Chemistry*, *49*(5), 1547–1549.
Oluba, O. M. (2019). *Ganoderma* terpenoid extract exhibited anti-plasmodial activity by a mechanism involving reduction in erythrocyte and hepatic lipids in *Plasmodium berghei* infected mice. *Lipids in Health and Disease*, *18*(1), 1–9.
O'Regan, H. J., Lamb, A. L., & Wilkinson, D. M. (2016). The missing mushrooms: Searching for fungi in ancient human dietary analysis. *Journal of Archaeological Science*, *75*, 139–143.
Parveez, G. K. A., Hishamuddin, E., Loh, S. K., Ong-Abdullah, M., Salleh, K. M., Bidin, M., Sundram, S., Hasan, Z. A. A., & Idris, Z. (2020). Oil palm economic performance in Malaysia and R&D progress in 2019. *Journal of Oil Palm Research*, *32*(2), 159–190.
Pegler, D. N. (2002). Useful fungi of the world: The Lingzhi—the mushroom of immortality. *Mycologist*, *16*, 100–101.
Peinter, U., Pöder, R., & Pumpel, T. (1998). The Iceman's fungi. *Mycological Research*, *102*, 1153–1162.
Pleszczyńska, M., Lemieszek, M., Siwulski, M., Wiater, A., Rzeski, W., & Szczodrak, J. (2017). *Fomitopsis betulina* (formerly *Piptoporus betulinus*): The Iceman's polypore fungus with modern biotechnological potential. *World Journal of Microbiology and Biotechnology*, *33*(5), 83.
Power, R. C., Salazar-García, D. C., Straus, L. G., González Morales, M. R., & Henry, A. G. (2015). Microremains from El Mirón Cave human dental calculus suggest a mixed plant-animal subsistence economy during the Magdalenian in Northern Iberia. *Journal of Archaeological Science*, *60*, 39–46.
Purdy, L. H., & Purdy, B. A. (1982). Ancient polypores from an archaeological wet site in Florida. *Botanical Gazette*, *143*, 551–553.
Raja, H. A., Baker, T. R., Little, J. G., & Oberlies, N. H. (2017). DNA barcoding for identification of consumer-relevant mushrooms: A partial solution for product certification? *Food Chemistry*, *214*, 383–392.
Rashad, F. M., El Kattan, M., Fathy, H. M., Abd El-Fattah, D. A., El Tohamy, M., & Farahat, A. (2019). Recycling of agro-wastes for *Ganoderma lucidum* mushroom production and *Ganoderma* post mushroom substrate as soil amendment. *Waste Management*, *88*, 147–159.
Ren, F., Chen, Q., Meng, C., Chen, H., Zhou, Y., Zhang, H., & Chen, W. (2021). Serum metabonomics revealed the mechanism of *Ganoderma amboinense* polysaccharides in preventing non-alcoholic fatty liver disease (NAFLD) induced by high-fat diet. *Journal of Functional Foods*, *82*, 104496.
Rigas, F., Papadopoulou, K., Dritsa, V., & Doulia, D. (2007). Bioremediation of a soil contaminated by lindane utilizing the fungus *Ganoderma* Australe via response surface methodology. *Journal of Hazardous Materials*, *140*(1–2), 325–332.
Robson, H. K. (2018). The star Carr fungi. In: N. Milner, C. Conneller, & B. Taylor (Eds.), *Star Carr, vol. 2: Studies in technology, subsistence and environment*. White Rose University Press, New York, pp. 437–445.
Roussel, B. (2005). *La production du feu par percussion de la pierre: Préhistoire, ethnographie, expérimentation*. Éditions Monique Mergoil, Montgnac.
Sahebi, M., Hanafi, M. M., van Wijnen, A. J., Akmar, A. S. N., Azizi, P., Idris, A. S., Taheri, S., & Foroughi, M. (2017). Profiling secondary metabolites of plant defence mechanisms and oil palm in response to *Ganoderma boninense* attack. *International Biodeterioration & Biodegradation*, *122*, 151–164.

Samorini, G. (1992). The oldest representations of hallucinogenic mushrooms in the world (Sahara Desert, 9000–7000 B.P.). *Integration, 2*(3), 69–78.
Singer, R., & Archangelsky, S. (1958). A petrified basidiomycete from Patagonia. *American Journal of Botany, 45*, 194–198.
Subramaniam, S., Raman, J., Sabaratnam, V., Heng, C. K., & Kuppusamy, U. R. (2017). Functional properties of partially characterized polysaccharide from the medicinal mushroom *Ganoderma neo-japonicum* (Agaricomycetes). *International Journal of Medicinal Mushrooms, 19*(10), 849–859.
Sun, Y. F., Costa-Rezende, D. H., Xing, J. H., Zhou, J. L., Zhang, B., Gibertoni, T. B., Gates, G., Glen, M., Dai, Y. C., & Cui, B. K. (2020). Multi-gene phylogeny and taxonomy of *Amauroderma* s.lat. (*Ganodermataceae*). *Persoonia, 44*, 206–239.
Supramani, S., Rejab, N. A., Ilham, Z., Wan-Mohtar, W. A. A. Q. I., & Ghosh, S. (2022). Basal stem rot of oil palm incited by *Ganoderma* species: A review. *European Journal of Plant Pathology, 164*, 1–20.
Tan, W. C. (2015). *Ganoderma neo-japonicum* Imazeki revisited: Domestication study and antioxidant properties of its basidiocarps and mycelia. *Scientific Reports, 5*, 12515.
Tay, Y. Z., Pan, A. F., & Chiam, K. H. (2022). Cholestatic liver injury from *Ganoderma lucidum* coffee extract—a case report. *AME Medical Journal, 7*, 29.
Tchotet Tchoumi, J. M., Coetzee, M. P. A., Rajchenberg, M., & Roux, J. (2019). Taxonomy and species diversity of *Ganoderma* species in the garden route national park of South Africa inferred from morphology and multilocus phylogenies. *Mycologia, 111*(5), 730–747.
Thyagarajan, A., Zhu, J., & Sliva, D. (2007). Combined effect of green tea and *Ganoderma lucidum* on invasive behavior of breast cancer cells. *International Journal of Oncology, 30*(4), 963–969.
Tseng, Y. H., Yang, J. H., & Mau, J. L. (2008). Antioxidant properties of polysaccharides from *Ganoderma tsugae*. *Food Chemistry, 107*(2), 732–738.
Wan-Mohtar, W. A. A. Q. I., Halim-Lim, S. A., Kamarudin, N. Z., Rukayadi, Y., Abd Rahim, M. H., Jamaludin, A. A., & Ilham, Z. (2020). Fruiting-body-base flour from an oyster mushroom waste in the development of antioxidative chicken patty. *Journal of Food Science, 85*(10), 3124–3133.
Wan-Mohtar, W. A. A. Q. I., Ilham, Z., Jamaludin, A. A., & Rowan, N. (2021a). Use of zebrafish embryo assay to evaluate toxicity and safety of bioreactor-grown exopolysaccharides and endopolysaccharides from European *Ganoderma applanatum* mycelium for future aquaculture applications. *International Journal of Molecular Sciences, 22*(4), 1675.
Wan-Mohtar, W. A. A. Q. I., Mahmud, N., Supramani, S., Ahmad, R., Zain, N. A. M., Hassan, N. A., Peryasamy, J., & Halim-Lim, S. A. (2018). Fruiting-body-base flour from an oyster mushroom—a waste source of antioxidative flour for developing potential functional cookies and steamed-bun. *AIMS Agriculture and Food, 3*(4), 481–492.
Wan-Mohtar, W. A. A. Q. I., Taufek, N. M., Yerima, G., Rahman, J., Thiran, J. P., Subramaniam, K., & Sabaratnam, V. (2021b). Effect of bioreactor-grown biomass from *Ganoderma lucidum* mycelium on growth performance and physiological response of red hybrid tilapia (*Oreochromis* sp.) for sustainable aquaculture. *Organic Agriculture, 11*(2), 327–335.
Wan-Mohtar, W. A. A. Q. I., Viegelmann, C., Klaus, A., & Lim, S. A. H. (2017). Antifungal-demelanizing properties and RAW264.7 macrophages stimulation of glucan sulfate from the mycelium of the mushroom *Ganoderma lucidum*. *Food Science and Biotechnology, 26*(1), 159–165.
Wang, C.-L., Lu, C.-Y., Pi, C.-C., Zhuang, Y.-J., Chu, C.-L., Liu, W.-H., & Chen, C.-J. (2012). Extracellular polysaccharides produced by *Ganoderma* formosanum stimulate macrophage activation via multiple pattern-recognition receptors. *BMC Complementary and Alternative Medicine, 12*(1), 1–10.
Wang, L., Li, J. Q., Zhang, J., Li, Z. M., Liu, H. G., & Wang, Y. Z. (2020). Traditional uses, chemical components and pharmacological activities of the genus *Ganoderma* P. Karst.: A review. *RSC Advances, 10*(69), 42084–42097.
Wang, X., & Lin, Z. (2019). Immunomodulating effect of *Ganoderma* (Lingzhi) and possible mechanism. In: Lin, Z., Yang, B. (eds) *Ganoderma* and Health. *Advances in Experimental Medicine and Biology*, vol 1182, 1–37.
Wang, X.-L., Wu, Z.-H., Di, L., Zhou, F.-J., Yan, Y.-M., & Cheng, Y.-X. (2019). Renoprotective phenolic meroterpenoids from the mushroom *Ganoderma* cochlear. *Phytochemistry, 162*, 199–206.
Wasser, S. P. (2005). Reishi or ling zhi (*Ganoderma lucidum*). *Encyclopedia of Dietary Supplements, 1*, 603–622.
Wasser, S. P. (2010). Medicinal mushroom science: History, current status, future trends, and unsolved problems. *International Journal of Medicinal Mushrooms, 12*(1), 1–16.
Wasson, V. P., & Wasson, R. G. (1957). *Mushrooms, Russia and history*, Pantheon Books, New York.
Watling, R., & Seaward, M. R. D. (1976). Some observations on puff-balls from British archaeological sites. *Journal of Archaeological Science, 3*, 165–172.
Wilford, J. N. (1998). Lessons in Iceman's prehistoric medicine kit. *The New York Times*.

Wińska, K., Mączka, W., Gabryelska, K., & Grabarczyk, M. (2019). Mushrooms of the genus *Ganoderma* used to treat diabetes and insulin resistance. *Molecules*, *24*(22), 4075.

Wunderlich, S., Zürcher, A., & Back, W. (2005). Enrichment of xanthohumol in the brewing process. *Molecular Nutrition & Food Research*, *49*(9), 874–881.

Yajima, A., Urao, S., Katsuta, R., & Nukada, T. (2014). Concise syntheses and biological activities of ganomycin I and fornicin A. *European Journal of Organic Chemistry*, *2014*(4), 731–738.

Yu, J.-H., Yu, S.-J., Liu, K.-L., Wang, C., Liu, C., Sun, J.-y., & Zhang, H. (2021). Cytotoxic ergostane-type steroids from *Ganoderma* lingzhi. *Steroids*, *165*, 108767.

Yuan, Y., Wang, Y. J., Sun, G. P., Wang, Y. R., Cao, L. J., Shen, Y. M., Yuan, B., Han, D., Huang, L. Q. (2018). Archaeological evidence suggests earlier use of *Ganoderma* in Neolithic China. *Chinese Science Bulletin*, *63*, 1180–1188.

Yuen, J. W., & Gohel, M. D. I. (2005). Anticancer effects of *Ganoderma lucidum*: A review of scientific evidence. *Nutrition and Cancer*, *53*(1), 11–17.

Zhao, L., Dong, Y., Chen, G., & Hu, Q. (2010). Extraction, purification, characterization and antitumor activity of polysaccharides from *Ganoderma lucidum*. *Carbohydrate Polymers*, *80*(3), 783–789.

4 It Is Said That Antioxidants Are Our Answer to Immortality

An Insight into the Antioxidant Activity of Ganoderma

Maja Kozarski[1] *and Jovana Vunduk*[2]

1 Institute for Food Technology and Biochemistry, Faculty of Agriculture, University of Belgrade, Serbia

2 Institute of General and Physical Chemistry, Belgrade, Serbia

4.1 INTRODUCTION: THE IMMORTALITY QUEST

A peek into the register of deaths of Scotland reveals that Queen Elizabeth, the Queen Mother, died of "old age" at the age of 96 (Aberdeenshire Council, 2022). It appears that she is not a rare case. As a matter of fact, about 65% of all people who die, especially in developed countries, die of old age (Verburgh, 2018). And we took it for granted. But what is the precise scientific explanation hiding behind this "old age"? Why do we age, and is there such a thing as dying of old age? Is it really as natural as we are used to thinking of it? And finally, why do we have to die? Whether it is a philosopher's stone, the Holy Grail, ambrosia, soma the elixir of eternal life, or teonanakatl, humankind has always been obsessed with immortality. Millennia of progress transferred these once-legitimate means of achieving eternal life into myths and folklore; however, the lure of immortality is as strong as ever. The industry, business, and research of "forever young" is blooming. Numerous companies and research centers are founded by rich people with the goal of beating death. Genetic manipulations, including the research on lengthening telomeres, give the promise of at least longer if not eternal life (Moore, 2019). Recently, scientists proved that some organisms, like the sea anemone, manage to maintain lifelong homeostasis of neuroglandular cell lineages, unlike vertebrates in which it happens only in the embryo stage (Steger et al., 2022). This implies that immortality *can* be obtained. Our skin is among the first to expose our age and senescence (Ganceviciene et al., 2012). No wonder that the majority of immediate attempts to counteract aging are directed toward our largest organ, the skin. Moreover, there is a sociological pressure to stay and appear young, heavily supported by mass media, fashion, plastic surgery, science, food, and the health industry, to mention a few. In all these cases, the main concern is to achieve "physical immortality". On the other hand, there is still no consensus about why we age and whether aging is a natural process or rather a lethal disease. Caplan (2005) even hypothesized about the ethical justification of research on aging and prolonging life and concluded that all existing theories failed to prove the naturalness of aging. According to the same author, there is no intrinsic ethical reason not to try to get closer to immortality, putting aside the fact that most probably only the very rich people will be able to afford it. There are numerous aging theories, as summarized by Miyazawa et al. (2022), proving that our interest in the topic of immortality has not diminished. On the contrary, it expanded using current technology and science. So far, the most up-to-date and comprehensive project, The Immortality Project, worth 5.1 million US$, headed by the University of California, summarized the state of progress in different research fields concerning immortality, explaining that the existential terror (the fright of no longer existing after death) will ensure the future fascination with this topic (Boissoneault, 2022). But until, and if, we get to the point when death is just another disease followed by myriad senescence conditions,

DOI: 10.1201/9781003354789-4

the most affordable and nongenetic-dependent way to slow our aging is in the hands of the food, pharmaceutical, and wellness industries. Several diets were analyzed through cohort studies and proved to positively impact aging-related diseases like the low-carbohydrate diet, the Mediterranean diet (MedDiet), the DASH diet, the SEAD diet, the Baltic Sea diet, and some others (Leitão et al., 2022). Their main effect seems to be obtained through preventing oxidative and DNA damage. On the other hand, the theory of oxidative damage postulates that senescence and death come as a consequence of oxidative imbalance in a human organism, so the lifespan can be extended by the intake of external antioxidants, mostly supplements. Their beneficial effect can be differently accessed, but the most often measured parameter is the influence on lifespan. Sadowska-Bartosz and Bartosz (2014) summarized the observed effects of around 50 synthetic and natural antioxidants, from vitamins to polyphenols, showing that most of them increased the average lifespan. Among natural products, supplements based on mushrooms, and especially *Ganoderma* species, are gaining more attention lately. As reported by Shaher et al. (2020) the PubMed (2022) database had 33 research papers dealing with the antioxidant effect of just one *Ganoderma* species, *Ganoderma lucidum*, in a period of 30 years. Its exceptional antioxidant and antiaging properties are recognized by industry and consumers as well, which propelled its use as the chief ingredient in cosmetics and personal care products (Bhavana and Roshan, 2022). Besides being the most common and popular commercial mushroom, *G. lucidum* is also among the most researched medicinal mushroom species, mainly as an adjuvant cancer treatment and immune response modifier, so there are more than 80 preclinical and clinical studies (Zeng et al., 2018; Zizak et al., 2014). The same mushroom has a rising popularity in the food sector, especially as the ingredient of functional food, beverages and spirits, primarily for its antioxidant activity (Vunduk and Veljović, 2021; Kozarski and van Griensven, 2022).

This chapter discusses the role of *Ganoderma* and its bioactive compounds/metabolites in preventing oxidative stress and its contribution to vivacity and longevity. It provides information on the compounds' chemical structure, the molecular mechanism of action, bioavailability, safety profile, and preclinical and human clinical research. Finally, attention is drawn to the perspectives of *Ganoderma* nutraceutical and cosmeceutical formulations aimed to strengthen or support cellular redox homeostasis. The evidence provided in this chapter should also promote *Ganoderma*'s clinical application as an antiaging medicine of natural origin.

4.2 THE EARLY ROOTS OF *GANODERMA* AS THE SECRET OF IMMORTALITY

Examples from human history show that achieving immortality was always approached through taking some special substance, often called an elixir, to be ingested or applied to the human body. Mostly, it was a gift from gods for those who were fortunate enough. One such substance was lingzhi or *G. lucidum*. Most of the literature evidence traces the origin of belief in the divine mushroom to China, like the term *zhi* (chih) meaning "mushroom" but also referring to good health, longevity and immortality when ingested (Wasson, 1968; Needham, 1974). The semantics of *zhi* can be traced to as early as the 4th to 3rd century BC. The first records of *zhi* in the Chinese classics are the "Records of the Grand Historian" known as *Shiji* from the 1st century BC (Sima, 1996). In this monumental book, *zhi* refers to the *mushroom of immortality*. The mushroom was considered magical and precious, and only the alchemists, known as *fangshi*, knew the secret location where it grew. Wasson (1968) mentions that the expression *ling chih*, meaning a supernatural mushroom, appeared first in the Ch'in Dynasty (221–207 BC) in the talks of the First Emperor (Shih-huang) and his necromancers, who were looking for this mushroom with miraculous properties. Next, Emperor Wu (around 109 BC) found *ling chih* growing in an inner pavilion of his palace. It was such a glorious event that the emperor served food to hundreds of families, gave amnesty to prisoners, and even composed an ode in this mushroom's honor. This poem was the first time that the expression *chi* was used in combination with *ling*.

Another later work, the *Baopuzi* (320 CE), enlists *zhi* as one of the medicines of immortality and *of the highest type* (Ware, 1981). In one of the later works, *Laozi yuxia zhongzhi jing shenxian bishi*

("The Secret of Divine Immortals", 7th to 8th century CE), *zhi* mushrooms are divine and instantaneously bring immortality to the one who consumes them, but only if they grow above specific minerals (Steavu, 2018). This unity incorporates them with the breath of Heaven and Earth, as well as of yin and yang. Moreover, the consumption of *zhi* as intrinsically perfect provides a bridge, a direct link, to the dao or divine. As described by the same author, the Queen Mother of the West is traditionally associated with lingzhi and immortality cults. Some Chinese idioms like *zeng tian shou kao* also point to lingzhi's connection with immortality, meaning "blessed with longevity" (Lin, 2009; Wachtel-Galor et al., 2011). In addition, the famous Chinese legend of the white snake tells a love story of a snake-woman deity who went through many hardships to find the herb of immortality (lingzhi) in order to save the man she loved. The Chinese culture has a strong-rooted belief that immortality can be achieved, as explained by Lin (2009), and the cause should be looked for in Taoism. This religion declares that people can learn to be immortal, and taking magical herbs, like *G. lucidum*, is one of the means. In pursuing immortality, the Taoist religion suggests that a person should consume lingzhi. That person will never grow old or die. The fact that *G. lucidum* bestows a person with eternal youth brings one essential quality to contemporary efforts: besides living forever, it guarantees a certain quality, which is to never grow old. On the other hand, the father of ethnomycology, Gordon Wasson, claimed that *ling chih* was not indigenous to China. He suggested that it originated from the Indian Rigveda and that it was a "literary" reflection of soma, which became an integral part of Chinese culture through a phenomenon he named *idea diffusion* (Wasson, 1968).

Contemporary research indeed tried to focus on examining each promising substance one by one, very often herbs, foods, vitamins, minerals, or a specific diet like the Nordic diet (Miyazawa et al., 2022). Although some general conclusions might be made, a definite answer has not been achieved. Among specific groups of natural compounds, the most promising antiaging effect is reported from polyphenols (Meccariello and D'Angelo, 2021). They are also reputed to be the most potent natural antioxidants (Pandey and Rizvi, 2009). It has been proposed that dietary antioxidants enhance endogenous enzymatic defense capacity and suppress free radical (FR) formation in the fruit fly model (Peng et al., 2014). However, mushrooms and *G. lucidum* have never been examined in this context. Keeping in mind that the *Ganoderma* genus has the most diverse dietary antioxidant profile among macromycetes, future studies might bring more definite conclusions concerning its overall antiaging effect (Ahmad et al., 2021; Mwangi et al., 2022; Sułkowska-Ziaja et al., 2022).

4.3 AGING AND OXIDATIVE DAMAGE THEORY

Aging is a natural process. However, its biological nature and "inevitability" do not make it simple. On the contrary, aging involves more than a few factors. Based on the latest research, the World Health Organization (WHO) points out that aging is a consequence of the accumulation of molecular damage, providing a basis for a theory of aging (Peng et al., 2014; Sadowska-Bartosz and Bartosz, 2014). The main processes contributing to this damage are reactions of FR species and oxidative stress and an imbalance between pro-oxidant and antioxidant species, apart from reactions of metabolites and spontaneous errors in biochemical processes, as supported by several studies (Leyane et al., 2022). The free radical theory of aging (FRTA), also termed the oxidative damage theory of aging, speculates that aging is not inevitable, but comes because of the failure of several defensive mechanisms. They fail to respond to FR-induced damage, especially in the mitochondria (Tan et al., 2018). Age-related diseases happen due to structural changes in mitochondria. They are accompanied by alterations in the biophysical properties of the membrane, decreased fluidity, and changes in the electron transport chain complexes' activities, resulting in energy imbalance followed by mitochondrial failure (Tan et al., 2018). These perturbations aggravate cellular homeostasis and mitochondrial function, leading to vulnerability to oxidative stress, as shown in Figure 4.1. Moreover, elderly people are susceptible to oxidative stress due to a decline in the efficiency of

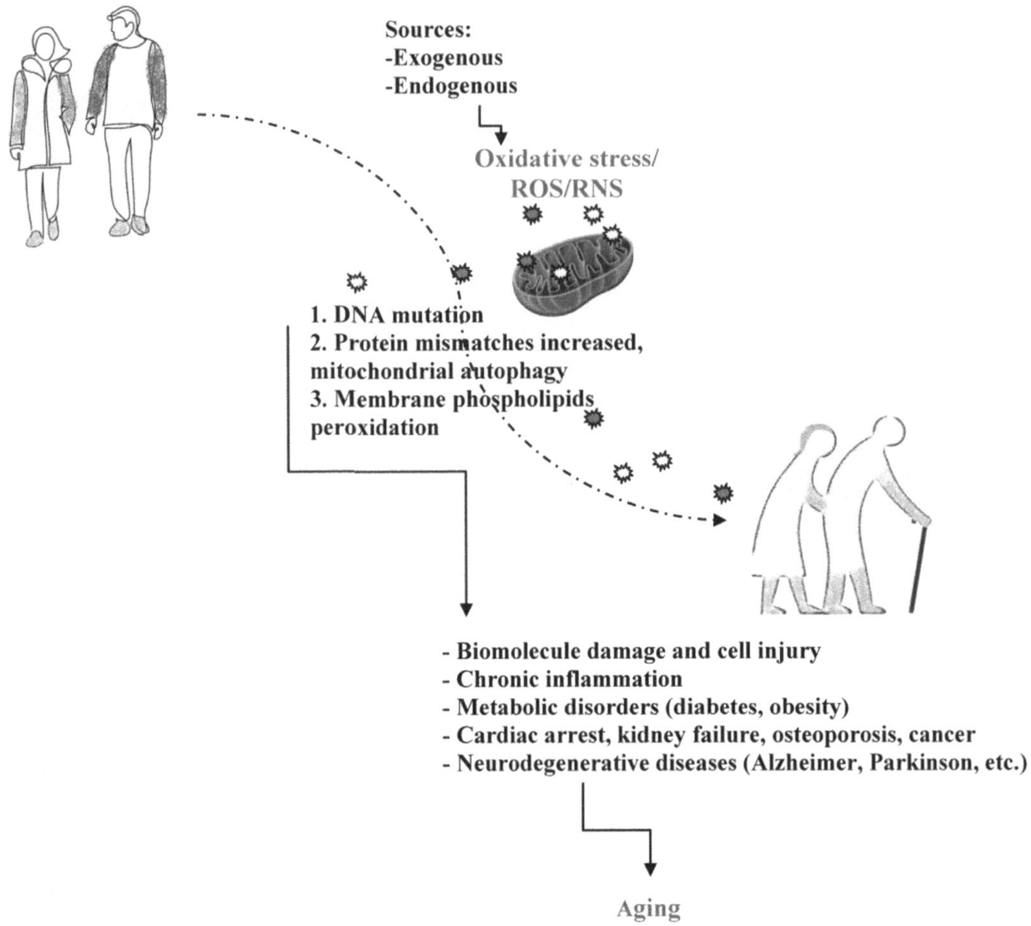

FIGURE 4.1 Oxidative stress leads to cellular senescence through mitochondrial dysfunction, which is mostly accompanied by age-related diseases.

their endogenous antioxidant systems. Organs with high rates of oxygen consumption and limited respiration levels (the brain and heart) are the most susceptible to this phenomenon, which partially explains the high prevalence of cardiovascular diseases (CVDs).

In the pharmaceutical and nutraceutical industries, antioxidants are the current hype. They are intended as prophylactic and therapeutic agents commercialized as antiaging wonder compounds. A PubMed (on 10 December 2022) search with "antioxidants and aging" as keywords resulted in 26,471 publications; this underlines the immense interest of the scientific community, especially considering the philosophical, ethical, personal, and commercial basis of the topic.

Supplements for boosting internal antioxidant defenses are focused on enhancing nonenzymatic and enzymatic protection systems. For example, vitamin C provides direct protection against oxidative damage and synergistically scavenges the reactive FR, resulting in the enhanced function of endogenous enzymatic antioxidants (Kozarski and van Griensven, 2022). Moreover, exogenous antioxidants can interact with various redox signaling pathways. This interaction happens via modulation of the activity of redox enzymes and the formation of secondary metabolites with bioactive properties (Hunyadi, 2019). Additionally, antioxidants may influence significant immunomodulatory mechanisms to bestow a better oxidant or antioxidant profile (Kozarski and van Griensven, 2022; Kozarski et al., 2023).

4.4 *GANODERMA* AS A RECOGNIZED AND COMMERCIAL SUPPORTER OF BODY REDOX HOMEOSTASIS

For over two millennia *Ganoderma* has been known and used in Asian countries as a folk remedy and vivacity and longevity elixir among many traditional medicines (Hapuarachchi et al., 2018; El Mansy, 2019). In modern times *G. lucidum* (lingzhi) and *Ganoderma sinense* (zizhi) are legal medicinal agents listed in the *Pharmacopoeia of the People's Republic of China* (State Pharmacopoeia Commission of the PRC, 2005). Besides, the Chinese government authorized *G. lucidum*, *G. sinense*, and *Ganoderma tsugae* and included them in "the list of fungal species that can be used in health food" (Zhang et al., 2019). Nowadays *Ganoderma* is included in the *American Herbal Pharmacopoeia and Therapeutic Compendium* and is gaining more and more popularity in European countries (Upton, 2000).

As seen in Figure 4.2, the chemistry for *Ganoderma* spp. reveals numerous active compounds having the potential to support and enhance antioxidant defense, including polyphenols, carbohydrates, sterols, triterpenoids, meroterpenoids, nucleosides and nucleotides, protein and nonprotein amino acids e.g. ergothioneine, fatty acids (e.g., polyunsaturated fatty acids known as PUFAs), vitamins, minerals, and trace elements (Kozarski et al., 2011, 2012; Zengin et al., 2015; El Mansy, 2019; Zhang et al., 2019; Ahmad et al., 2021; Fraile-Fabero et al., 2021; El Sheikha, 2022; Kolniak-Ostek et al., 2022; Sułkowska-Ziaja et al., 2022). In accordance with these findings, in the global market of mushroom antioxidants, *Ganoderma* nutraceuticals formulated to balance and support antioxidant

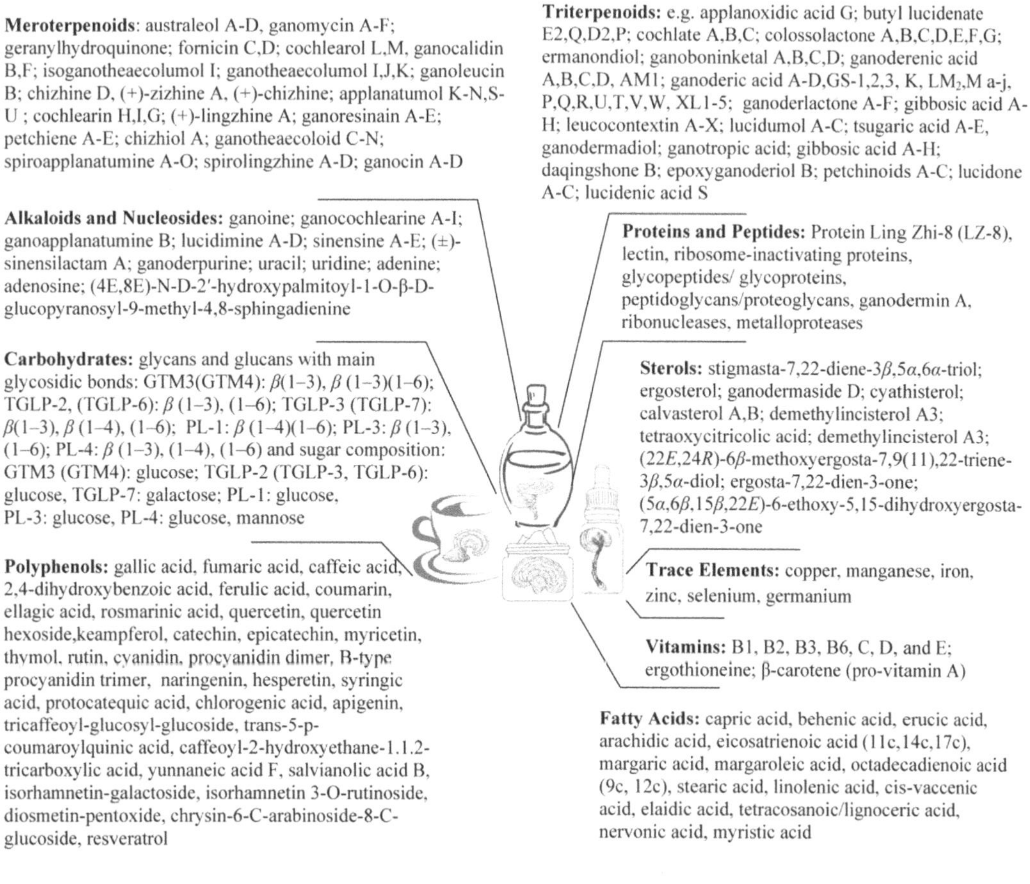

FIGURE 4.2 Biologically active compounds of *Ganoderma* spp. with antioxidant activity and potential to regulate redox homeostasis.

and immune defense systems are well-positioned and among the most sought after (Bhavana and Roshan, 2022; Kozarski et al., 2023).

Ganoderma spp. has been recognized as a traditional optimal source of natural bioactive components. Its bioactive compounds are extracted from different parts of mycelia, fruit body, or spores and marketed in the form of popular beverages like coffee and tea; powder; dietary supplements including spore-derived products, syrups, and drinks; and personal hygiene products like toothpaste, soaps, lotions, facial masks, and serums (Kozarski et al., 2023). *Ganoderma* spp. has been globally commercialized as a drug supplement able to support redox balance, enhance the body's immune system, and improve metabolic function (Ahmad et al., 2021). It is also considered a constituent of functional food. Cosmetics is the latest niche of *Ganoderma* product diversification and includes skin-lightening preparations, which are produced mainly in China, Korea, and lately in the United States, followed to a lesser extent by some other Asian countries. (Ahmad et al., 2021). For example, many facial masks containing *Ganoderma* extracts are commercialized as skin-whitening products (Ahmad et al., 2021; Kozarski et al., 2023). *G. lucidum*, the best-known *Ganoderma* spp. on the global market, was valued at 3,096.9 million USD in 2019. It is projected to reach more than 5 million USD by 2027, and its compound annual growth rate (CAGR) is estimated to reach 8.04% for the period 2021–2027 (Bhavana and Roshan, 2022).

Meanwhile, a huge presence on the market and intensive scientific research on oxidative stress prevention activities of *Ganoderma*, both in vitro and in vivo (in Table 4.1), demonstrating its antiaging effect and with it connected extension of the average lifespan, are only the beginning. Moreover, it is debatable whether *Ganoderma* is an antioxidant supplement for functional food to maintain health or a therapeutic "drug" under the official medicine palette.

TABLE 4.1
Common Therapeutic Effects of Different *Ganoderma* spp. to Strengthen and Support Body Redox Homeostasis

Therapeutic Effects	*Ganoderma* spp.	Major Bioactive Compounds	References
Radical scavenging activity	*G. lucidum*	Extract rich in triterpenoids (ganoderic acids as the most abundant) and polyphenols from the phenolic acids, flavonoids, and stilbenes families	Kolniak-Ostek et al., 2022
Radical scavenging activity; increases the activity of SOD, CAT, and GPx	*G. lucidum*	Aqueous extracts	Hasnat et al., 2013
Improves liver function in diabetes via antioxidant activity and short-chain fatty acid excretion	*G. atrum*	Polysaccharides (PSG-1)	Zhu et al., 2016
Radical scavenging activity, inhibition of LPO, chelating ability	*G. lucidum* *G. applanatum* *G. resinaceum*	Aqueous extracts; Polysaccharide extracts	Kozarski et al., 2011, 2012, 2020
Antiradical potential and inhibition on hyperpigmentation and skin extracellular matrix enzymes	*G. lucidum*	Polysaccharide and phenol extracts	Kozarski et al., 2019
Regulation of bromodomain proteins (acetylation of histones, remodeling of chromatin, recruiting other factors necessary for transcription)	*G. cochlear*	Cochlearols A and B, polycyclic meroterpenoids	Dou et al., 2014
Radical scavenging activity	*G. lucidum*	Polysaccharides	Kana et al., 2015
Skin wound healing, mitigating postburn infection, and preventing skin flap ischemia	*G. lucidum*	Extracts rich in polysaccharides and ganoderic acids	Yin et al., 2019
Down-regulating prostate-specific antigen (PSA) concentrations	*G. lucidum*	Extracts of fruiting bodies	Grammatikopoulou et al., 2020

TABLE 4.1 *(Continued)*
Common Therapeutic Effects of Different *Ganoderma* spp. to Strengthen and Support Body Redox Homeostasis

Therapeutic Effects	*Ganoderma* spp.	Major Bioactive Compounds	References
Inhibition of LPO and ROS; increase SOD, CAT and liposomal GSH activity; protects against neural damage after hypoxia-ischemia/ reperfusion in diabetes	*G. lucidum*	Aqueous extracts	Iwata et al., 2008; Okazaki et al., 2008b, 2008a
Protection against oxidative stress; expression of Skn7 regulator and increase production of mitochondrial inner membrane protein UTH1	*G. lucidum*	Ganodermasides A, B, C and D isolated from spores	Weng et al., 2010, 2011
Antioxidant activity in the promotion of neuronal differentiation; antiglycation activity	*G. capense*	Polysaccharide from submerged fermentation culturing mycelium powder	Yan et al., 2013; Huang et al., 2015; Jiang et al., 2016
Inhibition of LPO and ROS production; activation of the Nrf2/ARE, antioxidant, and cytoprotective effects on H_2O_2-induced human umbilical vein endothelial cells	*G. lingzhi*	Pentaeptide (GLP4) from mycelium	Huang et al., 2022
Radical scavenging activity, chelating ability	*G. adspersum*, *G. applanatum*, *G. carnosum*, *G. lucidum*, *G. pfeifferi*, *G. resinaceum*	Methanol extracts of mycelial cultures with phenolic acids, non-hallucinogenic indoles, ergosterol and ergosterol peroxide, kojic acid	Sułkowska-Ziaja et al., 2022
Radical scavenging activity	*G. lucidum*	Hydrochloric acid extract of 18 amino acids, with the most abundant being leucine	Zhang et al., 2018
Radical scavenging activity	*G. neojaponicum*, *G. applanatum*		Lee et al., 2009, Nguyen et al., 2013
Resistance to the oxidative stress induced by paraquat and Cr^{6+}, influence on TOR signaling pathways	*G. lucidum*	Aqueous extract	Cuong et al., 2019
Radical scavenging activity, inhibition of LPO, restoring antioxidant protection of kidneys	*G. lucidum*	Chloroform and methanol extracts; triterpenoids (lingzhin E and F); polysaccharides	Dong et al., 2019; Zheng et al., 2020; Joseph et al., 2009; Sheena et al., 2003
Action against H_2O_2-induced apoptosis; inhibition of sphingomyelinases	*G. lucidum*	β-1,3-glucan	Kao et al., 2012
Gradually increased GSH levels in HepG2 cells via HO-1 expression; upregulation of Nrf2 levels	*G. lucidum* *G. tsugae*	Triterpene-ganodermanondiol	Li et al., 2013
Extension of the lifespan by regulating the expression of UTH1 oxidative stress-responsive genes	*G. lucidum*	Spores methanol extract rich in ergosterols, and ganodermasldase A-B	Weng et al., 2010
Restoration the level of antioxidant enzymes, TNF-α and IL-1β, and activation of NF-κB	*G. lucidum*	Selenium-enriched protein	Chang et al., 2014
Skin wound healing/antioxidant activity	*G. lucidum*	Aqueous extract	Krupodorova et al., 2015
Radical scavenging activity, inhibition of LPO	*G. adspersum*	Methanol, hexane, ethyl acetate, and water extracts	Tel-Cayan et al., 2015
Radical scavenging activity	*G. lucidum*	Supercritical carbon dioxide extract rich in ganoderic acids	Karimi et al., 2022

4.4.1 Stages and Mechanisms in the Prevention of Oxidative Stress

Ganoderma metabolites can express protective properties at different stages of the oxidation process and by different mechanisms in an organism:

1. As primary antioxidants/scavengers involved directly in the neutralization of radical species (•RS);
2. As secondary antioxidants, they are involved in the regeneration of primary antioxidants inhibition or breakdown of lipid hydroperoxides, deactivation of metals, singlet oxygen ($^1O_2$) quenching, etc.;
3. With the potential to interact with different redox signaling pathways by modulating the activity of redox enzymes and the generation of bioactive secondary metabolites. Mechanisms include the transduction of signals through specific pathways and the upregulation of target genes to enhance the level of their products, as well as •RS sensing (Kozarski et al., 2015). Our organism possesses a complex system of adaptive responses to •RS exposure. It consists of several components, as shown in Table 4.2. When the oxidative stress is of low intensity, the Kelch-like ECH-associated protein 1/NF-E2-related factor 2 (Keapl/Nrf2) system up-regulates genes that further encode antioxidant enzymes. It is very sensitive and activates by minute amounts of •RS. If nuclear factor kappa-light-chain-enhancer of activated B cells (NF-κB), activator protein-1 (AP1), and mitogen-activated protein kinases (MAPKs) are employed, we are talking about upregulation of antioxidant enzymes and inflammation protein induction caused by intermediate-intensity oxidative stress (Kozarski and van Griensven, 2022). The third option is high-intensity oxidative stress, which causes disturbance of mitochondrial permeability transition pores, activation of the apoptosis cascade, and destruction of electron transporters, As a consequence, apoptosis and/or necrosis might happen (Kozarski et al., 2015). Active metabolites of *Ganoderma* up-regulate antioxidant systems and transduction of signals through specific pathways and regulation of target genes, making them more capable of eliminating •RS. In this way, an autoregulated negative feedback control loop is created.
4. Having the potential for immune regulatory activity. The disruption of the oxidative balance often may be connected to the immune system, since oxidative stress and inflammatory damage are multistage processes. Antioxidant molecules may trigger notable immunomodulatory mechanisms to award a better oxidant/antioxidant profile. The ability of antioxidant molecules to strategically modulate T helper type 1/T helper type 2 (Thl/Th2) immune responses may be exploited in both therapeutic and prophylactic management of diseases (Ahmad et al., 2021; Vunduk and Veljović, 2021; Kozarski and van Griensven, 2022).

TABLE 4.2
Systems of Adaptive Responses to •RS Exposure in Humans and Animals That Can Be Influenced by *Ganoderma* Bioactive Metabolites by Creating an Autoregulated Negative Feedback Control Loop

Oxidative Stress Intensity		
Low	**Intermediate**	**High**
Type of redox signaling pathways that can be regulated by *Ganoderma* metabolites		
– Keapl/Nrf2 system: regulation of genes encoding antioxidant enzymes	– Regulation of antioxidant enzymes and inflammation via NF-κB, AP1 and MAPKs	– Regulation of perturbations of mitochondrial permeability transition pore and destruction of electron transporters – Regulation of apoptosis cascade

A variety of methods are used to measure the antioxidative properties and to examine the mechanisms of action of *Ganoderma* bioactive ingredients in the form of extracts or tonics, semi-purified and highly purified fractions, as well as different compound classes, isolated compounds, nanoparticles, etc. Methods are commonly based on:

- Electron (ET) e.g. 1,1-diphenyl-2-picrylhydrazyl (DPPH) free radical scavenging, trolox equivalent antioxidant capacity (TEAC) decolorization, Folin-Ciocalteu to determine total phenolics, cupric ion–reducing antioxidant capacity (CUPRAC);
- Hydrogen atom transfer (HAT) e.g. lipid peroxidation inhibition capacity (LPIC), oxygen radical absorbance capacity (ORAC), scavenging of peroxide (ROO$^•$) radicals, 2,2′-azino-bis(3-ethylbenzothiazoline-6-sulfonic acid) (ABTS) radical scavenging method, scavenging of super oxide ($^•O_2$–) radical formation by alkaline (SASA), etc.;
- Electron spin resonance (ESR) method;
- measurement of thermodynamic parameters, e.g. ionization potential, bond dissociation enthalpy, proton affinity and electron-transfer enthalpy, proton dissociation enthalpy;
- Ability to chelate transition metal ions;
- Erythrocyte hemolysis;
- monitoring of enzymes activity e.g., catalase (CAT), superoxide dismutase (SOD), and glutathione peroxidase (GPx);
- Quantitative and qualitative measurement of biomarkers (Kozarski et al., 2014; Dhama et al., 2019; Klaus et al., 2021; Kozarski and van Griensven, 2022).

Furthermore, every individual part (stipe, caps, spores, etc.) of *Ganoderma* may be investigated for the exact amount of antioxidant mycochemicals for the purpose of standardization of isolation, purification, antioxidant, and antiaging activities.

4.5 *GANODERMA* PRIMARY AND SECONDARY ANTIOXIDANTS

4.5.1 Low-Molecular-Weight Compounds

The most common primary and secondary antioxidants of *Ganoderma* spp. involved in mycoceutical formulations are different classes of low-molecular-weight compounds e.g., phenolic acids, flavonoids, vitamins, nonprotein amino acids e.g. ergothioneine, and terpenoids (Zengin et al., 2015; Dong et al., 2019; El Mansy, 2019; Zhang et al., 2019; Ahmad et al., 2021; Fraile-Fabero et al., 2021; El Sheikha, 2022; Kolniak-Ostek et al., 2022; Kozarski and van Griensven, 2022; Mwangi et al., 2022; Sułkowska-Ziaja et al., 2022; Vunduk et al., 2022).

4.5.1.1 Phenolic Acids and Flavonoids

A diversity of phenolic acids and flavonoids was observed in fruiting bodies, spores, and mycelium of *Ganoderma*. Some of the main phenolic acids found in *Ganoderma* spp. are reported to be hydroxybenzoic acid (HBA) and hydroxycinnamic acid (HCA) derivatives: *p*-hydroxybenzoic-, vanillic-, protocatechuic-, gallic-, 5-sulphosalicylic-, chlorogenic-, caffeic-, ferulic-, ellagic-, trans-cinnamic-, benzoic-, rosmarinic-, *p*-coumaric-, *o*-coumaric-, ferulic-, vanillic-acids, and vanillin, as shown in Figure 4.2. Among flavonoids were detected catechin, epicatechin, rutin, apigenin, kaempferol, pyrogallol, quercetin, hesperetin, myricetin, biochanin, formononetin, coumarin, and resveratrol (Kim et al., 2008; Ferreira et al., 2009; Tel-Cayana et al., 2015; Zengin et al., 2015; Dong et al., 2019; Dong et al., 2019a; Fraile-Fabero et al., 2021; Kolniak-Ostek et al., 2022; Kozarski et al., 2023).

As primary antioxidants, phenolic acids and flavonoids react on several mechanisms of the scavenging actions of $^•$R: via HAT, ET, proton and electron transfers (PETs), and radical adduct (RA) formation mechanisms. These mechanisms may play different roles in varied proportions, depending on the corresponding physiological conditions of the reaction, e.g. pH (Kozarski and van Griensven, 2022). According to these mechanisms, antioxidants scavenge harmful •R and form a new radical (•R'), which is more stable and less reactive than the previous one. This inhibits the initial phase of harmful oxidation and interrupts the propagation phase by capturing •R before they reach target cells. HAT is often assumed to be the predominant mechanism regarding *Ganoderma* phenolic acids (Badhani and Kakkar, 2018; Kozarski and van Griensven, 2022). The O-H bond of the hydroxyl group (-OH) directly attached to the benzene ring is the preferred place of the •R attack, e.g. gallic acid, commonly detected in *Ganoderma* spp. (Veljović et al., 2017; Dong et al., 2019); see Figure 4.3. As an example, the methanol extract of wild-growing *G. lucidum* from Serbia had only one phenolic compound detected, gallic acid (1103 μg/g). However, its content was significantly higher than reported by other authors (Karaman et al., 2010). In addition, gallic acid is considered among the best peroxyl radical (ROO•) scavengers identified so far in nonpolar (lipid) media. It is capable of scavenging hydroperoxyl radicals (HOO•) with rate constants on the order of 10^5 M^{-1} s^{-1} and hydroxyl radicals (HO•) at diffusion-limited rates (Marino et al., 2014). It has been predicted that when in an aqueous solution, the deprotonation of gallic acid increases its protective action against oxidative stress.

At physiological pH, •R that are most common in biological systems, as well as their reduced forms: a ROO•/ROOH, alkoxyl radical (RO•/ROH), HO•/H_2O, are uncharged, so the reduction of •R is formally a HAT. In terms of mechanism, it can be either a transfer of a single hydrogen atom or separate electron transfer and proton equilibration steps. In general, the HAT is an energy-favored mechanism (Njus and Kelley, 1991, Kozarski and van Griensven, 2022). Among various cellular and tissue systems, red blood cells are particularly sensitive to oxidative stress because they lack a nucleus and mitochondria, are unable to synthesize fresh protein along with the degradation of detoxification enzymes, etc. When the alterations in the redox status of the body are happening, red blood cells are the first to be affected (Dhama et al., 2019). The mechanism to directly and primarily act with RS permits *Ganoderma* formulations rich in phenolic acids to react with •R in plasma and to protect erythrocytes, on the first line of defense, from the harmful effects of oxidative stress (Njus and Kelley, 1991; Kozarski and van Griensven, 2022). Besides phenolic acids, the double bond, presence, and position of the multiple -OH groups in the flavonoid structure of *Ganoderma* spp. formulations allow their scavenging activity by yielding a more stable flavonoid radical via conjugation and electron delocalization (Zhong and Xiao, 2009; Kozarski et al., 2015Gonzalez-Paramas et al., 2019).

A common secondary antioxidant mechanism of action of polyphenolic compounds present in *Ganoderma* spp. includes the deactivation of metals (Kozarski and van Griensven, 2022). Among these compounds, flavonoids are the most potent chelators of transition metals in edible mushrooms (Ferreira et al., 2009). Carbonyl (>C=) and -OH groups present in flavonoids enable them to coordinate metal ions and form complexes. Chelating complexes with divalent cations may be formed between the 3-OH or 5- OH and 4-oxo group or between the 5- and 7-OH in the A ring or 3'- and 4'-OH in the B ring (Gonzalez-Paramas et al., 2019). Quercetin, commonly detected in *Ganoderma*

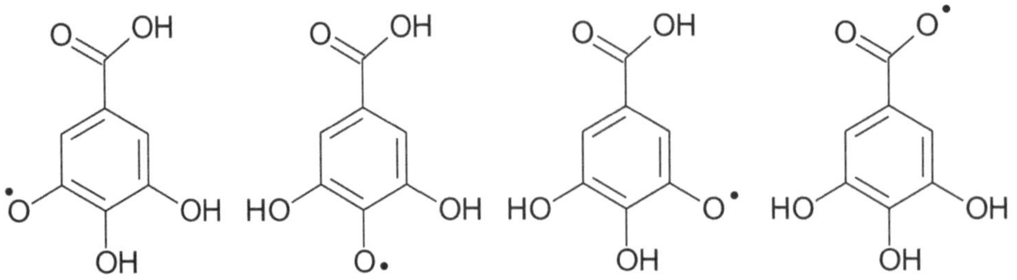

FIGURE 4.3 Possible sites for the donation of hydrogen atoms of gallic acid, a common phenolic compound of the *Ganoderma* spp. (Based on Badhani and Kakkar, 2018.)

(Ferreira et al., 2009), has iron-chelating and iron-stabilizing abilities due to its potential binding sites via the 3-hydroxyl and 4-carbonyl group of the C ring, the 4-carbonyl-5-hydroxyl site of the A and C rings, or the catechol moiety of the B ring. This suggests a little difference between aglycones and glycosides in the metal-complexing ability, as shown in Figure 4.4. The factors determining the preferred binding site are the flavonoid type, the ion, and the pH value (Kozarski and van Griensven, 2022).

It is important to mention that quercetin can also regulate the cell's oxidative status by inhibiting oxidative enzymes responsible for $^{\bullet}O_2^-$ production, e.g., lipoxygenase and nicotinamide adenine dinucleotide phosphate (NADPH) oxidase (Gonzalez-Paramas et al., 2019). For example, quercetin is a flavonoid with a benzopyrone ring system with free hydroxyl substituents at the 3′, 4′, and 7 positions, which can inhibit NADPH via the inhibition of protein kinase C (Gonzalez-Paramas et al., 2019; Kozarski and van Griensven, 2022). Polyphenols affect many biochemical processes, from central metabolism to signaling events, and this has been demonstrated in several in vitro studies. A system view has been provided through protein (amino acids) binding to polyphenols (Lacroix et al., 2018).

4.5.1.1.1 Enhancing Biosynthesis of the Polyphenols

Ganoderma formulations rich in polyphenols are popular as nutraceuticals, and various techniques and methodologies have been proposed to improve and stimulate their biosynthesis. Among the latest research, Dong et al. (2019a) pointed out that high oxygen enhances the content of phenolic compounds. In their study, air or 60% and 80% oxygen was used for 6 days to treat freshly harvested *G. lingzhi* fruiting bodies It was observed that high oxygen treatment significantly increased the scavenging with DPPH$^{\bullet}$, ABTS$^{\bullet}$, HO$^{\bullet}$, and $^{\bullet}O_2^-$ and DNA damage protective activity. Based on these results, a high oxygen concentration treatment presents a promising method for induction of

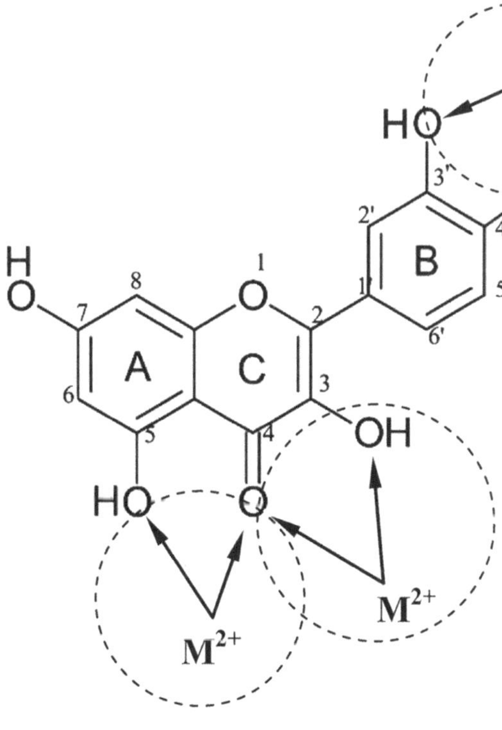

FIGURE 4.4 Structural requirements associated with the metal chelating activity of quercetin. (Based on Gonzalez-Paramas et al., 2019.)

the bioactive compound synthesis, resulting in improved bioactivity and technological quality of *Ganoderma* spp. fruit bodies.

4.5.1.2 Nonprotein Amino Acids

In addition to mushroom polyphenols, ergothioneine (ET), a trimethyl-betaine derivative of histidine, is an effective chelator of divalent metal ions and a potent antioxidant (Cheah and Halliwell, 2021). Mushrooms are a primary source of this compound. Lee et al. (2009) reported that ET makes up 0.06 mg/g of *Ganoderma applanatum* to 0.08 mg/g dry weight (DW) of *G. lucidum* fruit bodies. Besides, it was observed that mycelia accumulated higher levels of ET (approximately 10 times) in comparison with fruit bodies of mushroom species such as *Ganoderma neo-japonicum* and *G. applanatum*, which are considered economically important Moreover, additional enhancement of ET accumulation in mycelia cultures of *G. neo-japonicum* (2.3 mg/g DW) was reported by supplementation with methionine (Lee et al., 2009).

Diet can cause the accumulation of high levels of ET, meaning that it plays an important physiological role in human health and development, and possibly in disease prevention or treatment. Tissues predisposed to oxidative stress like blood cells, brain, bone marrow, and eye tissues have high levels of ET, although other tissues may also accumulate significant levels of ET with continuous administration (Cheah and Halliwell, 2021). Aging and various diseases cause a decrease in ET levels in the blood.

ET is an effective chelator, and the most stable ET complex is with Cu^{2+}, with the highest complex formation constant (Cheah and Halliwell, 2021). In contrast to the generation of RS by glutathione (GSH) in the presence of Cu^{2+} via the formation of a redox-active Cu(I)-$[GSH]_2$ complex, the ET- Cu^{2+} complex is quite stable and does not decompose to generate $^{\bullet}R$ (Zhu et al., 2011). It is stated that ET is accumulated in mitochondria. indicating its specific role in protecting mitochondrial components from oxidative damage. Hence, it is associated with the mitochondrial generation of $^{\bullet}O_2^-$ (Martinez-Medina et al., 2021).

4.5.1.3 Vitamins

Ganoderma spp. are a rich source of antioxidant vitamins: C, D, and E (Stamets, 2005; Hossain et al., 2007; Pehlivan, 2017; Amara, 2017; Fraile-Fabero et al., 2021). Vitamin C is easily transported across cell membranes and acts as a primary antioxidant with scavenging potential by the HAT mechanism (Njus and Kelley, 1991). In opposition to polyphenols, which may also react by ET and PET mechanisms and sometimes may include the production of unfavorable intermediates, either a protonated radical, e.g. $ROOH^{+\bullet}$ or reduced anion, e.g., ROO^-, a favored HAT mechanism enables vitamin C to react efficiently with $^{\bullet}R$ without adverse pro-oxidant effects (Njus and Kelley, 1991; Kozarski and van Griensven, 2022). Moreover, in opposition to flavonoids, which act at the membrane/water interface, vitamin C can work both inside and outside the cells because it is a water-soluble compound. Accordingly, it is involved in the first line of antioxidant defense, protecting proteins and lipid membranes from oxidative damage (Pehlivan, 2017). On one hand, it is a cofactor for enzymes involved in regulating hormone biosynthesis and regeneration of other antioxidants also included in cell division and growth regulation, and on the other hand, vitamin C is involved in signal transduction and has roles in immune stimulation; iron absorption; and synthesis of collagen, hormones, and neurotransmitters, as well as in detoxifying the body from heavy metals (Pehlivan, 2017). Its content in *G. lucidum* is found to be 32.2 mg/100 g. By way of comparison, the reported average amount of vitamin C found in strawberries is the highest (60 mg/100 g) and a bit lower in citrus fruits (30–50 mg/100 g), while apples, pears, and plums have very small amounts of ascorbic acid (3–5 mg/100 g) (Kozarski et al., 2015; El Sheikha, 2022). Although the main function of vitamin C is as an antioxidant, under certain circumstances, it can also express a pro-oxidant effect by maintaining the Fe^{3+} and Cu^{2+} in their reduced forms (Kozarski et al., 2015). Next, these metal ions react with H_2O_2 and form the highly reactive $HO^{\bullet}$ in a process known as the Fenton reaction. However, clear evidence of these reactions' significance in vivo has not been provided:

$$Fe^{3+} + \text{Vitamin C} \rightarrow Fe^{2+} + {}^{\bullet}\text{Vitamin C} \quad (6.1)$$

$$Fe^{2+} + H_2O_2 \rightarrow HO^{\bullet} + HO^{-} + Fe^{3+} \quad (6.2)$$

The term "vitamin E" is used to designate a family of chemically related compounds, namely tocopherols and tocotrienols, which share a common structure. α, β, γ, and δ tocopherols were confirmed and quantified in *Ganoderma* spp. (Pehlivan, 2017; Amara, 2017). Furthermore, Amara (2017) reported that the antioxidant activity and tocopherol content of the *G. lucidum* mycelium were significantly higher than for fruiting bodies. The mycelium of *G. lucidum* showed good results in DPPH scavenging activity (EC_{50} = 10.4 mg/mL) and in the reducing power assay (EC_{50} = 0.32 mg/mL). *G. lucidum* mycelium was especially rich in δ-tocopherol (362 μg/g extract) and β-tocopherol (272 μg/g extract), followed by γ-tocopherol (68 μg/g extract) and α-tocopherol (15 μg/g extract). Significant differences in tocopherol amounts were reported in the fruiting body of the same species originating from the northeastern region of Portugal (α-, β-, and δ-tocopherol: 1.93, 26.7, and 15.4 μg/g extract, respectively), Serbia and China (Stojković et al., 2013). Considering the decently high detected contents, the mycelium of *G. lucidum* (717 μg tocopherols/g extract) might be considered as a potential source of these lipophilic antioxidants, compared with food sources known for their high vitamin E content, e.g., almonds (262 μg/g), roasted sunflower seeds (363 μg/g), avocados (21 μg/g), tofu (53 μg/g), shrimp (22 μg/100 g), etc. (Kozarski et al., 2015).

Although a primary and secondary antioxidant, vitamin E's main role is the protection of cell membranes from lipid peroxidation (LPO). LPO has been a culprit in several diseases together with aging, including atherosclerosis, rheumatoid arthritis, cataracts, and neurodegenerative disorders (Dhama et al., 2019). Tocopherols terminate the activity of LPO by scavenging lipid peroxyl radicals ($LOO^{\bullet}$). However, it converts itself into a reactive radical during this reaction:

$$LH + \text{Oxidant initiator} \rightarrow L^{\bullet} \quad (6.3)$$

$$L^{\bullet} + O_2 \rightarrow LOO^{\bullet} \quad (6.4)$$

$$LOO^{\bullet} + \text{Tocopherols} \rightarrow LOOH + \text{Tocopherols}^{\bullet} \quad (6.5)$$

Vitamin C has the potential to recycle vitamin E by reducing the generated radicals and re-establishing its antioxidant activity (Fabre et al., 2015). Vitamin $C^{\bullet}$ (semidehydroascorbate) is reduced to vitamin C by NADH-dependent semidehydroascorbate reductase by the rapid reaction of vitamin C with tocopherols$^{\bullet}$ (Kozarski et al., 2015):

$$\text{Vitamin C} + \text{Tocopherols}^{\bullet} \rightarrow \text{Vitamin } C^{\bullet} + \text{Tocopherols} \quad (6.6)$$

$$\text{Vitamin } \mathbf{C}^{\bullet} + NADH \rightarrow \text{Vitamin C} + NAD^{\bullet} \quad (6.7)$$

Ganoderma spp. are a rich source of ergocalciferol/vitamin D_2. They contain significant levels of ergosterol, a terpenoid compound that is present in large amounts in the cell wall, and which is converted to ergocalciferol/vitamin D_2 by exposure to ultraviolet (UV) light (Stamets, 2005; Cardwell et al., 2018). Stamets (2005) reported that 6–9 hours of sunlight exposure stimulated the production of vitamin D in *G. lucidum* fruiting bodies after they had been harvested and dried indoors. At normal growth conditions and filtered light and no sun exposure after harvesting and drying, vitamin D content was reported to be 66 IU/100 g. However, after exposure to 6–8 hours of sun, vitamin D content in dry fruiting bodies increased to 2,760 IU/100 g (Stamets, 2005). Vitamin D_2 and its active metabolite 1,25-dihydroxycholecalciferol are membrane antioxidants and inhibit Fe-dependent liposomal LPO (Cardwell et al., 2018). These lipophilic compounds may accumulate in membranes and decrease membrane fluidity, which is suspected of being responsible for the inhibition of Fe-dependent liposomal LPO (Outila et al., 1999). It has already been shown that ergocalciferol from lyophilized and homogenized wild edible mushrooms is well absorbed in humans (Cardwell et al., 2018).

4.5.1.4 Carotenoids

Since carotenoids have been found in significant amounts in the *Ganoderma* species, they became important on par with fruits and vegetables (Smina et al., 2011; Celık et al., 2014; Rajoriya et al., 2015). Rajoriya et al. (2015) reported a total carotenoid content of 4.47 to 7.43 mg/g of *G. lucidum*, *G. tsugae*, and *G. applanatum* acetone extracts. Among these species, the highest content of β-carotene (3.63 mg/g) and lycopene (0.224 mg/g) was observed for *G. lucidum*. All these extracts expressed excellent scavenging potential against DPPH radicals (92–96%). In addition, Smina et al. (2011) confirmed that 100 μg/mL of the purified triterpenes from *G. lucidum* showed 81.81% DPPH scavenging activity. The IC_{50} value in the DPPH activity of triterpene extract was 41.45 μg/mL. Carotenoids have the potential to act as chain-breaking antioxidants in a lipid environment, especially under low oxygen partial pressure. Carotenoids are particularly susceptible to attack $LOO^{\bullet}$ due to the systems of double bonds. The result is the formation of inactive products and carotenoid radicals, which are very short-lived species. Factors affecting their reactivity are the characteristics of the end groups and the conjugated double bond chain (Kozarski et al., 2015). Furthermore, carotenoids are very potent natural $^1O_2$ quenchers, both in vitro and in vivo. Their energy levels lie close to that of $^1O_2$, and the process of quenching is very efficient, especially for carotenoids having 11 conjugated double bonds as well as lycopene and β-carotene common carotenoids in *Ganoderma* (Smina et al., 2011; Celık et al., 2014; Rajoriya et al., 2015; Kozarski and van Griensven, 2022). Furthermore, lycopene was shown to be present in the central nervous system in low concentrations as a potential $^1O_2$ quencher due to its ability to cross the blood-brain barrier (Rao and Rao, 2007).

4.5.2 High-Molecular Weight Compounds

Several research studies have shown that compounds of higher molecular masses of *Ganoderma* spp. e.g., peptides and carbohydrates, possess significant scavenging activity with little or no side effects. Sun et al. (2004) screened the antioxidant activity of *G. lucidum* peptide (GLP) using several oxidation systems. In comparison with butylated hydroxytoluene, GLP had higher antioxidant activity in the soybean oil system. Soybean lipoxygenase activity was blocked by GLP in a dose-dependent manner with an IC_{50} value of 27.1 μg/mL. The IC_{50} value of scavenging activity toward $HO^{\bullet}$ produced in a deoxyribose system was 25 μg/mL. A dose-dependent manner of quenching of $^{\bullet}O_2^-$ produced by pyrogallol autoxidation has been reported (Sun et al., 2004). Moreover, GLP showed prominent antioxidant activity in rat liver tissue homogenates and mitochondrial membrane peroxidation systems. GLP also blocked the auto-hemolysis of rat red blood cells in a dose-dependent manner. The authors suggested that GLP could participate in the inhibition of lipid peroxidation in biological systems.

Polysaccharides obtained from *Ganoderma* spp. show a wide range of pharmacological antioxidant properties (Zhonghui et al., 2014; Seweryn et al., 2021). The presence of H from specific monosaccharide units and the type of their binding on side branches of the main chain condition the ability of polysaccharide molecules to scavenge $^{\bullet}R$ (Kozarski et al., 2015). Direct $^{\bullet}R$ scavenging with carbohydrates may occur in vivo in higher aerobic organisms, e.g., in gastrointestinal (GI) tract cells (Van den Ende et al., 2011). The high antioxidant capacity of polysaccharides can prevent LPO and the pathogenesis of various GI diseases like GI cancers, peptic ulcers, and inflammatory bowel disease, partly due to oxidative stress (Kozarski et al., 2014). Likewise, *G. lucidum* polysaccharides (GL-PS) may attenuate exercise-induced oxidative stress in skeletal muscle. In the study of Zhonghui et al. (2014), GL-PS in three different concentrations, 50, 100, and 200 mg/kg body weight of mice, were administered per day. The mice were subjected to a series of analyses: an exhaustive swimming exercise, along with the determination of SOD, CAT, and GP_X activities and malondialdehyde (MDA) levels in the skeletal muscle after 28 days. The results showed that GL-PS could increase antioxidant enzyme activities and decrease the MDA levels in the skeletal muscle of mice. The evidence provided by this study pointed out that GL-PS supplementation can protect against oxidative stress.

4.6 POTENTIAL TO EXPRESS ANTIOXIDANT ACTIVITY BY INTERACTING WITH VARIOUS SIGNALING PATHWAYS

Besides the chemical structure of *Ganoderma* antioxidants that enables them to directly scavenge RS, their antioxidant action demonstrated in vivo can also be linked to their ability to interact with various redox signaling pathways by modulating the activity of redox enzymes and/or molecules. Cells respond to antioxidants mainly through direct interactions with receptors or enzymes involved in signal transduction. As a consequence, a modification of the cell's redox status may trigger a series of redox-dependent reactions (Lacroix et al., 2018; Cateni et al., 2021). By that, *Ganoderma* ingredients may modulate different signaling pathways such as nuclear Nrf2/ARE, NF-κB, mammalian target of rapamycin (mTOR), forkhead-box O (FoxO), and NADPH oxidase (NOX) (Li et al., 2013; Elsayed et al., 2014; Kim et al., 2020; Barros Camara and Brandao, 2021).

4.6.1 Nrf2/ARE Pathway Upregulation

Nfr2 is a stress-sensitive transcription factor that responds to the presence of RS through the activation of several genes that encode antioxidant enzymes such as CAT, glucose-6-phosphate dehydrogenase (G-6-PDH), GP_X, GSH, glutathione-S-transferase (GST), gamma-glutamyl cysteine synthetase (γ-GCS), paraoxonases (PONs), gamma-glutamyl cysteine synthetase (γ-GCS), SOD, and detoxifying enzymes such as alcohol and carbonyl dehydrogenase, cytochrome P450 (CYP), and prostaglandin reductase (Lee et al., 2019; Barros Camara and Brandao, 2021). Among the *Ganoderma* compounds that activate the Nrf2/ARE pathway are ET (Hseu et al., 2015; Hseu et al., 2020), polyphenols, e.g., coumaric and ferulic acid (Catino et al., 2016; Sabitha et al., 2020), vitamin D_2 (Kim et al., 2020), and other biologically actively triterpenes e.g., ganodermanondiol isolated from *G. lucidum* and *G. tsugae* (Li et al., 2013). Ganodermanondiol treatment may gradually increase GSH levels in the human hepatoma cell line (HepG2) through the expression of the antioxidant enzyme heme oxygenase-1 (HO-1). Ganodermanondiol significantly increased Nrf2 levels by realizing it from a cytosolic complex with kelch-like enoyl-CoA hydratase-associated protein 1 (Keapl) and successfully promoted the translocation of phosphorylated Nrf2 into the nucleus in HepG2 cells, inducing HO-1 expression (Li et al., 2013). Activation of Nrf2-mediated antioxidant genes with ET is also important in the protection of skin from photodamage. Hseu et al. (2020) reported that submicromolar concentrations of ET, ranging from 0.125 to 0.5 μM, mitigate ultraviolet A (UVA) radiation–induced reactive oxygen species (ROS) generation tested in human dermal fibroblasts. An increase in the number of antioxidant genes: HO-1, NAD(P) H: quinone oxidoreductase 1 (NQO-1), and gamma-glutamylcysteine synthetase (γ-GCLC) after treatment with ET was observed. Also, the GSH level significantly increased. This was associated with the upregulation of Nrf2 expressions in a dose- and/or time-dependent manner. Moreover, it was shown that signaling cascades mediated by an extracellular signal-regulated kinase (ERK), c-jun N-terminal kinase (JNK), and protein kinase C (PKC) induced Nrf2 translocation (Hseu et al., 2020). Vitamin D has been demonstrated to exert an antioxidant role by vitamin D receptor (VDR)-mediated transcriptional upregulation of Nrf2 and downregulation of NOX2, a major isoform of NADPH oxidase (Kim et al., 2020). Such action is in line with the regulation of critical biomarkers of oxidative stress, including 8-hydroxy-2'-deoxyguanosine tested in VDR knockout mice (Kallay et al., 2001). Antioxidants derived from Nrf2 are known as indirect antioxidants, meaning that their physiological effects last longer than those exhibited by direct antioxidants, so they are efficient in relatively small doses. Moreover, Nrf2 is an important mediator of neuroprotection. Therefore, it represents a promising target for the development of novel therapies against amyotrophic lateral sclerosis, Alzheimer's disease, Parkinson's disease, and Huntington's disease (Zgorzynska et al., 2021).

4.6.2 NF-κB Pathway Upregulation

Ganoderma antioxidants have the potential to exclude NF-*κ*B activation and to inhibit the increase in intracellular effects of oxidative stress conditions (Elsayed et al., 2014). In support of this, *G. lucidum* ethanol extract (EGL) inhibited the excessive production of nitric oxide (NO), prostaglandin E2 (PGE2) that is generated by cyclooxygenase 2 (COX2) conversion of arachidonic acid, and proinflammatory cytokines, including tumor necrosis factor-α (TNF-*α*) and interleukin (IL)-1β, in a concentration-dependent manner without cytotoxic effects (Yoon et al., 2013). In addition, EGL suppressed NF-κB translocation and transcriptional activity by blocking IκB degradation and inhibiting toll-like receptor 4 (TLR4) and myeloid differentiation primary response 88 (MyD88) expression in lipopolysaccharide (LPS)-stimulated inflammatory responses in BV2 microglial cells (Yoon et al., 2013). Therefore, EGL inhibits mediator responses in activated microglia, which makes it promising in the treatment of neurodegenerative diseases. It is also important to mention that TNF-*α* is a main proinflammatory cytokine in inflammatory diseases such as Crohn's disease, septic shock, or rheumatoid arthritis (Chen et al., 2008). TNF-α is secreted by endothelial cells and induces intracellular ROS formation, providing a potential mechanism by which this cytokine can activate and injure endothelial cells, and the outcome is endothelial dysfunction, involved in coronary artery disease, atherosclerosis, chronic renal failure, hypertension, vascular complications of diabetes, insulin resistance, and hypercholesterolemia (Chen et al., 2008). Furthermore, it was observed that ganoderic acid, lucidenic acid, and vitamin D_2–derived ergosterol have the potential to exclude NF-κB activation and to inhibit the increase in intracellular ROS (Elsayed et al., 2014). As an example, vitamin D_2 has been reported to stimulate sirtuin 1 (SIRT1), also known as NAD-dependent deacetylase sirtuin-1, which exerts cardioprotective effects. The mechanism of its action includes the increase of autophagy and mitochondrial function by inhibition of the mTOR pathway and reduction of oxidative stress and inflammatory responses, activation of FoxO-dependent antioxidant pathways, and inhibition of NF-κB signaling, respectively (Kim et al., 2020). Vitamin D acts as a guardian of cellular homeostasis. It also protects from oxidative stress in various cell types, including human endothelial cells, due to its ability to regulate crosstalk between redox signaling and autophagy. This system is an essential cellular antioxidant system based on the removal of damaged or dysfunctional proteins and organelles (Kim et al., 2020; Kozarski and van Griensven, 2022).

4.7 UPREGULATION OF PERTURBATIONS OF MITOCHONDRIAL PERMEABILITY TRANSITION PORE AND DESTRUCTION OF ELECTRON TRANSPORTERS

Among neurological diseases, epilepsy is among the most common and it is not age dependent. (Jiang et al., 2018). Based on the experimental data originating from animal and human studies, certain seizures can cause damage to the hippocampal neurons, associated with a myriad of apoptotic cells. LPO induced by ROS may cause damage to mitochondria in the epileptiform hippocampal neurons, and it is usually controlled by antiepileptic drugs taken daily. Unfortunately, available antiepileptic drugs are not effective in more than one-third of patients, which may exhibit side effects, like other nervous system problems (Jiang et al., 2018). Ganoderic acids have the potential to regulate mitochondrial LPO and stabilize the mitochondrial membrane potential difference ($\Delta\psi m$) to maintain the normal structure of mitochondria. Jiang et al. (2018) reported that ganoderic acid A can significantly improve SOD activity. On the other hand, ganoderic acids A and B are able to stabilize $\Delta\psi m$ in hippocampal neurons, thereby protecting the hippocampal neurons by apoptosis inhibition. In this study, it was found that the SOD activity (118.84 U/mg proteins) and $\Delta\psi m$ (244.08 mV) of the epileptic hippocampal neurons were significantly lower than control ones

(135.95U/mg protein and 409.81 mV), associated with the obvious increase of cell apoptosis (31.88% vs. 8.84%). The treatment with ganoderic acid A/ganoderic acid B improved the results for SOD, 127.15 ± 3.82/120.52 ± 4.30 U/mg protein; for $\Delta\psi m$, 372.35/347.28 mV; and for cell apoptosis (%), 14.93/20.52.

4.8 ANTIOXIDANT ACTIVITY OF *G. LUCIDUM* AND CLINICAL STUDIES

A search of the official database for clinical studies, clinicaltrials.gov, with the keyword "*Ganoderma* lucidum" gave a list of 28 clinical studies and their prospect statuses. Among them, only a few were focused on the antioxidant effect. Ironically, the study that directly connects antioxidant activity and condition improvement goes in line with the issues deriving from prolonged life expectancy. At the moment the number-one cause of blindness in the developed world is age-related macular degeneration (AMD), with the prognosis of an increase due to the exponential aging of the global population (Wong et al., 2014). AMD affects the elderly population; it is very progressive and irreversible, meaning it leads to irreparable blindness (Cabral de Guimaraes et al., 2022). As Cabral de Guimaraes et al. (2022) summarized, the pathogenesis of AMD is influenced by factors like genetic, environmental, inflammatory, and ischemic, as well as oxidative stress. Datta et al. (2017) reported that factors like smoking and a high-fat diet are the main modifiable factors contributing to the already high oxidative stress environment of the macula. Furthermore, oxidative stress induces inflammation, which even further worsens the stage of AMD. Thus, one of the proposed treatments is supplementation with antioxidants like vitamin C, vitamin E, etc., and the results of several clinical trials were reviewed by Banerjee et al. (2021). As the authors warned, some degree of beneficial effects might be expected only in patients with moderate to advanced AMD. In the interventional, monocentric, prospective, randomized, and double-blind clinical study, a complex oral supplement, Macuprev, consisting of 14 natural compounds, vitamins, and minerals, including *G. lucidum*, was administered to 30 patients (15 received the supplement, two tablets per day, while the other 15 were given two tablets of placebo daily) over a 6-month period. The subjects eligible for the study were 50–80 years old and with age-related eye degeneration. As explained, patients with this problem are suffering from the deterioration of the antioxidant mechanisms of the retina. Due to the oxidative imbalance, the increased production of FR damages the eye tissue. As reported by Parravano et al. (2019) the patients receiving Macuprev showed a significant increase in the function of the macular preganglionic elements. They successfully showed that antioxidant supplementation can reduce disease progression while also reducing the loss of visual acuity.

In another double-blind, randomized, placebo-controlled clinical study, 65 patients with rheumatoid arthritis were supplemented with a combination of *G. lucidum* (4g/day) and San Miao San (2.4 g/day) (a traditional Chinese medicine herb) in the form of capsules (three capsules, two times per day) for 24 weeks (Li et al., 2007). Monitored parameters were a clinical response as defined by the American College of Rheumatology (ACR) criteria: changes in the number of swollen and tender joint counts, the levels of the inflammatory markers, cytokine levels, and oxidative stress. A 20% ACR score was achieved in 15% of the patients from the group receiving the supplement in comparison with 9.1% in the placebo group. A significant improvement in pain score and patient's global score was noticed only in the group taking *G. lucidum* and San Miao San herb. However, the levels of inflammatory markers, cytokines, and oxidative stress were the same in both test groups. In conclusion, the combination of two TCM herbs' analgesic effect was reported, and the product was generally safe and well tolerated, while antioxidant, anti-inflammatory, and immunomodulating effects were not detected. In one more randomized crossover clinical trial, Chiu et al. (2017) examined whether *G. lucidum* (GL-TP) enriched with polysaccharides and triterpenoids influence antioxidation and hepatoprotective efficacy and if the possible mechanism of its action is the suppression of oxidative stress. Of healthy subjects, 42 of them (22 male and 20 female) were recruited and segregated into two groups as experimental or placebo. They were taking *G. lucidum* preparation in a dose of 225 mg after lunch or dinner for 6 consecutive months with 1 month washout period in between. As reported, each capsule (225 mg) contained 7% triterpenoid present as a set of ganoderic acids (A-E, and G) and 6% polysaccharide peptides. Essential amino acids and trace elements were also present. On the other hand, the placebo capsule contained starch (90%) and GL-TP residues (10%) and appeared similar to GL-TP capsules. As a result, the total antioxidant capacity (TEAC; from 79.33% to 84.04%), total thiols (0.19–0.28 mM/mL), and GSH content (6–8.05 μM/L) in plasma and the activities of antioxidant enzymes, SOD (1155.98–1385.63 IU/gHb), CAT (246.26–279.21 IU/gHb), G-6-PDH (11.99–13.56 IU/g Hb), and GPx (13.16–15.44 IU/g Hb) were substantially improved in the group consuming GL-TP tablets. However, the levels of thiobarbituric acid reactive substances (TBARSs; 3.37–2.47%), 8-hydroxy-deoxy-guanosine (8-OH-dG; 15.99–11.98%), and hepatic marker enzymes glutamic-pyruvic transaminase (GPT) and glutamic-oxaloacetic transaminase (GOT) were concomitantly reduced (42% and 27%) after the same treatment with GL-TP. Furthermore, an alteration in the hepatic condition (mild fatty liver condition [initial] reversed to normal condition) was observed after the abdominal ultrasound examination of patients who consumed GL-TP. Also, in clinical comparative research by Collado et al. (2015) of *G. lucidum* and *Ceratonia siliqua* on physical fitness in women suffering from fibromyalgia, the antioxidant potential of *G. lucidum* was confirmed. *C. siliqua* was used for comparison as an already confirmed natural therapy with high levels of antioxidants with potential benefits on health. Sixty-four women with fibromyalgia were administered 6 g of *G. lucidum* or *C. siliqua* per day for 6 weeks. The company Mundo Reishi provided the substances. To evaluate the functional effect on upper and lower body muscular strength, upper and lower body flexibility, velocity, balance, agility, aerobic endurance, balance, and trunk endurance were measured. After the treatment period, improved aerobic endurance, lower body flexibility, and velocity were reported and attributed to the *G. lucidum* antioxidant effect. In the group administered *C. siliqua* no significant improvement in any of the enlisted physical tests was observed. These results point out that *G. lucidum* might be an effective natural-origin dietary supplement for patients suffering from fibromyalgia due to its effect on their physical performance.

4.9 THE NANOPARTICLES AND *GANODERMA* ANTIOXIDANT COMPOUNDS

There are several drawbacks to herbal natural medicine: adverse effects, toxicities, the need for repeated administration, and low to medium levels of therapeutic value, which all can be reduced or eliminated with nanoparticles (Ahmad et al., 2021). The nanoparticles such as polymeric nanoparticles, nano gels, nanoemulsions, solid lipid nanoparticles, and liposomes can enhance the product's solubility, bioavailability, and targeted treatment approaches, thus converting its primary poor therapeutic properties into satisfactory or excellent ones. When it comes to the use of nanoparticles as delivery systems for *Ganoderma*, the available data is negligible. Extensive studies are needed for all types of *Ganoderma* preparations (dry powdered fruit bodies, extracts, capsules, tinctures, syrups, or isolated pure compounds) to make them drug candidates. This kind of research should enhance its targeted properties and decrease the doses necessary for therapeutic effects as well as the possible toxicity (Ahmad et al., 2021). Among rare studies is the research of Karimi et al. (2022) on the *G. lucidum* extract nanoparticle production process. This study was based on expanding the supercritical fluid solution (ESS) method. Supercritical carbon dioxide extraction was used to produce *G. lucidum* extract, which included ganoderic acid B, ganoderic acid C_2, ganoderic acid D, ganoderic acid H, and ganolucidic acids. The problem with ganoderic acids is that they have poor water solubility. Accordingly, low solubility lowers biological activity, which further limits their use in industry and medicine. Karimi et al. (2022) also showed, using the DPPH method, that the antioxidant properties improve when the active compound is in the form of a nanoparticle. In this case, the IC_{50} values for the produced nanoparticles and the extract were 580 and 724 ppm, respectively. The ESS process proved to be a beneficial approach to producing nanoparticles. In this way, the bioavailability of insoluble or poorly soluble phytochemical compounds of *Ganoderma* spp. can be enhanced.

4.10 CONCLUSIONS AND FUTURE PERSPECTIVES

Ganoderma spp. possess unique primary and secondary metabolites that may prevent oxidative stress damage and improve cellular longevity and lifespan. The efficiency of oxidative stress prevention of its metabolites is connected to their ability to interact with various RS and redox signaling pathways by generating bioactive secondary metabolites and modulating the activity of molecular redox systems. Traditional texts, millennia of use, and current research provide a sufficient platform to support *Ganoderma* use as a therapeutic compound with a strong antioxidant activity, which should be implemented in the modern health care system. There is intensive research to determine which antioxidant components are effective, less costly, and have a higher safety profile. Moreover, in accordance with the literature and research, in the modern global market of mushroom antioxidants, *Ganoderma* nutraceuticals designed to balance and support antioxidant systems are among the most sought after.

It is important to point out that a gap in antioxidant potential research exists for *Ganoderma* even today. One of the main problems is that high-quality progressive clinical trials with large sample sizes, proper randomization, multicenter designs, and trials in the special population are still required. Without this, no prospects of establishing an effective and safe dose can be provided, not for highly purified therapeutic or supplement *Ganoderma* products. Likewise, *Ganoderma* mutagenicity and genotoxicity have to be assessed in adequately designed clinical trials. If *Ganoderma* antioxidant compounds are expected to become drug candidates, in-depth studies should include an examination of changes at the receptor and genetic levels. Another unsolved problem of *Ganoderma* antioxidant formulations is the lack of literature regarding quality variation and standardization. The cultivation process, quality of the raw material, and storage conditions all affect the sample/product quality, which further reflects on the *Ganoderma* antioxidant compounds and the therapeutic quality of the final product. Accordingly, it is essential to evaluate uniformity concerning the quality of *Ganoderma* samples based on their geographical origin. A very important issue is the lack of pharmaceutical control for marketed *Ganoderma* antioxidant products. The proper guidelines, including legislation, are lacking, while the health controlling agencies are of utmost importance to ensure the approval, registration, and availability of pharmaceutical *Ganoderma* antioxidant products with valid clinical data in terms of claims and longevity. Overall, *Ganoderma* might not provide actual immortality, but it might bring us closer to the elusive goal of dying of old age; at least at an affordable price.

ACKNOWLEDGMENTS

This work was supported by the "Agreement on the implementation and financing of scientific research work in 2022 and 2023 between the Faculty of Agriculture in Belgrade, Institute of General and Physical Chemistry and the Ministry of Education, Science and Technological Development of the Republic of Serbia", contract record numbers: 451-03-68/2022-14/200116, 451-03-47/2023-01/200116, 451-03-68/2022-14/200051, and 451-03-47/2023-01/200051, and by the Science Fund of the Republic of Serbia, #Grant No: 7748088, "Composite clays as advanced materials in animal nutrition and biomedicine-AniNutBiomedCLAYs".

REFERENCES

Aberdeenshire Council. 2022. *Registrar General Releases Extract of Death Entry for HM the Queen, National Records of Scotland*. www.nrscotland.gov.uk/files//images/entry-in-the-register-of-deaths-hm-the-queen.jpg (accessed December 20, 2022).

Ahmad, R., M. Riaz, A. Khan, et al. 2021. *Ganoderma lucidum* (Reishi) an edible mushroom; a comprehensive and critical review of its nutritional, cosmeceutical, mycochemical, pharmacological, clinical, and toxicological properties. *Phytother Res* 35:6030–62.

Amara, K. 2017. *In Vitro Production of Ganoderma Lucidum Mycelium from Northeast Portugal: The Antioxidant Potential of Tocopherols Extract in the Preservation of Natural Yogurt.* MSc thesis, Polytechnic Institute of Braganca, Braganca.

Badhani, B., and R. Kakkar. 2018. Influence of intrinsic and extrinsic factors on the antiradical activity of gallic acid: A theoretical study. *Struct Chem* 29:359–73.

Banerjee, M., R. Chawla, and A. Kumar. 2021. Antioxidant supplements in age-related macular degeneration: Are they actually beneficial? *Ther Adv Ophthalmol* 13 (June):1–14. www.ncbi.nlm.nih.gov/pmc/articles/PMC8404659/pdf/10.1177_25158414211030418.pdf

Barros Camara, A., and I. A. Brandao. 2021. The relationship between vitamin D deficiency and oxidative stress can be independent of age and gender. *Int J Vitam Nutr Res* 91:108–23.

Bhavana, T., and D. Roshan. 2022. *Reishi Mushroom Market.* Allied Market Research. www.alliedmarketresearch.com/reishi-mushroom-market-A10352 (accessed November 15, 2022).

Boissoneault, L. 2022. *Pre-Life, Afterlife, and the Drive for Immortality: The Science of Immortality.* John Templeton Foundation. www.templeton.org/news/pre-life-afterlife-and-the-drive-for-immortality (accessed November 12, 2022).

Cabral de Guimaraes, T. A., M. Daich Varela, M. Georgiou, and M. Michaelides. 2022. Treatments for dry age-related macular degeneration: Therapeutic avenues, clinical trials and future directions. *Br J Ophthalmol* 106:297–304.

Caplan, A. L. 2005. Death as an unnatural process. *EMBO Reports* 6:S72–5.

Cardwell, G., J. F. Bornman, A. P. James, and L. J. Black. 2018. A review of mushrooms as a potential source of dietary vitamin D. *Nutrients* 10 (October):1–11. www.ncbi.nlm.nih.gov/pmc/articles/PMC6213178/pdf/nutrients-10-01498.pdf

Cateni, F., M. L. Gargano, G. Procida, G. Venturella, and F. Cirlincione. 2021. Mycochemicals in wild and cultivated mushrooms: Nutrition and health. *Phytochem Rev* 21:339–83.

Catino, S., F. Paciello, F. Miceli, et al. 2016. Ferulic acid regulates the Nrf2/heme oxygenase-1 system and counteracts trimethyltin-induced neuronal damage in the human neuroblastoma cell line SH-SY5Y. *Front Pharmacol* 6, no. 305 (January):1–12. www.ncbi.nlm.nih.gov/pmc/articles/PMC4705308/pdf/fphar-06-00305.pdf

Celık, G. Y., D. Onbasl, B. Altınsoy, and H. All. 2014. *In vitro* antimicrobial and antioxidant properties of *Ganoderma lucidum* extracts grown in Turkey. *European J Med Plants* 4:709–22.

Chang, G. M., T. W. Hong, X. Zhen, and S. Jie. 2014. Effects of selenium-enriched protein from *Ganoderma lucidum* on the levels of IL-1β and TNF-α, oxidative stress, and NF-κB activation in ovalbumin-induced asthmatic mice. *Evid Based Complement Alternat Med* 2014, no. 182817 (February):1–6. www.ncbi.nlm.nih.gov/pmc/articles/PMC3934624/pdf/ECAM2014-182817.pdf

Cheah, I. K., and B. Halliwell. 2021. Ergothioneine, recent developments. *Redox Biol* 42 (June):1–10. www.ncbi.nlm.nih.gov/pmc/articles/PMC8113028/pdf/main.pdf

Chen, X., B. T. Andresen, M. Hill, J. Zhang, F. Booth, and C. Zhang. 2008. Role of reactive oxygen species in tumor necrosis factor-alpha induced endothelial dysfunction. *Curr Hypertens Rev* 4:245–55.

Chiu, H. F., H. Y. Fu, Y. Y. Lu, et al. 2017. Triterpenoids and polysaccharide peptides enriched *Ganoderma lucidum*: A randomized, double-blind placebo controlled crossover study of its antioxidation and hepatoprotective efficacy in healthy volunteers. *Pharm Biol* 55:1041–6.

Collado, M. D., F. Pazzi, F. J. Domínguez Munoz, et al. 2015. *Ganoderma lucidum* improves physical fitness in women with fibromyalgia. *Nutr Hosp* 32:2126–35.

Cuong, V. T., W. Chen, J. Shi, et al. 2019. The anti-oxidation and anti-aging effects of *Ganoderma lucidum* in *Caenorhabditis elegans. Exp Gerontol* 117:99–105.

Datta, S., M. Cano, K. Ebrahimi, L. Wang, and J. T. Handa. 2017. The impact of oxidative stress and inflammation on RPE degeneration in non-neovascular AMD. *Prog Retin Eye Res* 60:201–18.

Dhama, K., S. K. Latheef, M. Dadar, et al. 2019. Biomarkers in stress related diseases/disorders: Diagnostic, prognostic, and therapeutic values. *Front Mol Biosci* 6, no. 91 (October):1–50. www.ncbi.nlm.nih.gov/pmc/articles/PMC6843074/pdf/fmolb-06-00091.pdf

Dong, Q., D. He, X. Ni, H. Zhou, and H. Yang. 2019. Comparative study on phenolic compounds, triterpenoids, and antioxidant activity of *Ganoderma lucidum* affected by different drying methods. *Food Meas* 13:3198–205.

Dong, Q., Y. Li, G. Liu, Z. Zhang, H. Zhou, and H. Yang. 2019a. High oxygen treatments enhance the contents of phenolic compound and ganoderic acid, and the antioxidant and DNA damage protective activities of *Ganoderma lingzhi* fruiting body. *Front Microbiol* 10, no. 2363 (October):1–11. www.ncbi.nlm.nih.gov/pmc/articles/PMC6813255/pdf/fmicb-10-02363.pdf

Dou, M., L. Di, L. L. Zhou, et al. 2014. Cochlearols A and B, polycyclic meroterpenoids from the fungus *Ganoderma cochlear* that have renoprotective activities. *Org Lett* 16:6064–7.

El Mansy, S. M. 2019. *Ganoderma*: The mushroom of immortality. *Microb Biosyst* 4:45–57.
El Sheikha, A. F. 2022. Nutritional profile and health benefits of *Ganoderma lucidum* "Lingzhi, Reishi, or Mannentake" as functional foods: Current scenario and future perspectives. *Foods* 11:1030. https://doi.org/10.3390/foods1107103
Elsayed, A. E., H. E. Enshasy, M. A. M. Wadaan, and R. Aziz. 2014. Mushrooms: A potential natural source of anti-inflammatory compounds for medical applications. *Mediators Inflamm* 2014, no. 805841 (November):1–15. www.ncbi.nlm.nih.gov/pmc/articles/PMC4258329/pdf/MI2014-805841.pdf
Fabre, G., I. Bayach, K. Berka, et al. 2015. Synergism of antioxidant action of vitamins E, C and quercetin is related to formation of molecular associations in biomembranes. *Chem Comm* 51:7713–16.
Ferreira, I. C. F. R., L. Barros, and R. M. V. Abreu. 2009. Antioxidants in wild mushrooms. *Curr Med Chem* 16:1543–60.
Fraile-Fabero, R., M. V. Ozcariz-Fermoselle, J. A. Oria-de-Rueda-Salgueiro, et al. 2021. Differences in antioxidants, polyphenols, protein digestibility and nutritional profile between *Ganoderma lingzhi* from industrial crops in Asia and *Ganoderma lucidum* from cultivation and Iberian origin. *Foods* 10, no. 8 (July):1–13. www.ncbi.nlm.nih.gov/pmc/articles/PMC8394434/pdf/foods-10-01750.pdf
Ganceviciene, R., A. I. Liakou, A. Theodoridis, E. Makrantonaki, and C. C. Zouboulis. 2012. Skin anti-aging strategies. *Derm-Endocrinol* 4:308–19.
Gonzalez-Paramas, A. M., B. Ayuda-Duran, S. Martinez, S. Gonzalez-Manzano, and C. Santos-Buelga. 2019. The mechanisms behind the biological activity of flavonoids. *Curr Med Chem* 26:6976–90.
Grammatikopoulou, M. G., K. Gkiouras, S. T. Papageorgiou, et al. 2020. Dietary factors and supplements influencing prostate specific-antigen (PSA) concentrations in men with prostate cancer and increased cancer risk: An evidence analysis review based on randomized controlled trials. *Nutrients* 12, no. 10 (September):1–37. www.ncbi.nlm.nih.gov/pmc/articles/PMC7600271/pdf/nutrients-12-02985.pdf
Hapuarachchi, K. K., W. A. Elkhateeb, S. C. Karunarathna, et al. 2018. Current status of global *Ganoderma* cultivation, products, industry and market. *Mycosphere* 9:1025–52.
Hasnat, M. A., M. Pervin, and B. O. Lim. 2013. Acetylcholinesterase inhibition and in vitro and in vivo antioxidant activities of *Ganoderma lucidum* grown on germinated brown rice. *Molecules* 18:6663–78.
Hossain, M. S., N. Alam, S. M. R. Amin, M. A. Basunia, and A. Rahman. 2007. Essential fatty acid contents of *Pleurotus ostreatus, Ganoderma lucidum* and *Agaricus bisporus*. *Bangladesh J Mushroom* 1:1–7.
Hseu, Y. C., Y. V. Gowrisankar, X. Z. Chen, Y. C. Yang, and H. L. Yang. 2020. The antiaging activity of ergothioneine in uva-irradiated human dermal fibroblasts via the inhibition of the AP-1 pathway and the activation of Nrf2-mediated antioxidant genes. *Oxid Med Cell Longev* 2020, no. 576823 (February):1–13. www.ncbi.nlm.nih.gov/pmc/articles/PMC7038158/pdf/OMCL2020-2576823.pdf
Hseu, Y. C., H. W. Lo, M. Korivi, Y. C. Tsai, M. J. Tang, and H. L. Yang. 2015. Dermato-protective properties of ergothioneine through induction of Nrf2/ARE-mediated antioxidant genes in UVA-irradiated human keratinocytes. *Free Radic Biol Med* 86:102–17.
Huang, P., F.-J. Luo, Y.-C. Ma, et al. 2022. Dual antioxidant activity and the related mechanisms of a novel pentapeptide GLP4 from the fermented mycelia of *Ganoderma lingzhi*. *Food Funct* 13, no. 9032 (August):1–17. https://pubs.rsc.org/en/content/articlelanding/2022/FO/D2FO01572B
Huang, Y., N. Li, J. B. Wan, D. Zhang, and C. Yan. 2015. Structural characterization and antioxidant activity of a novel heteropolysaccharide from the submerged fermentation mycelia of *Ganoderma* capense. *Carbohydr Polym* 134:752–60.
Hunyadi, A. 2019. The mechanism(s) of action of antioxidants: From scavenging reactive oxygen/nitrogen species toredox signaling and the generation of bioactive secondary metabolites. *Med Res Rev* 39:2505–33.
Iwata, N., M. Okazaki, C. Kasahara, et al. 2008. Protective effects of a water-soluble extract from culture medium of *Ganoderma lucidum* mycelia against neuronal damage after cerebral ischemia/reperfusion in diabetic rats. *J Jpn Soc Nutr Food Sci* 61:119–27.
Jiang, J., F. Kong, N. Li, D. Zhang, C. Yan, and H. Lv. 2016. Purification, structural characterization and in vitro antioxidant activity of a novel polysaccharide from Boshuzhi. *Carbohydr Polym* 147:365–71.
Jiang, Z. M., H. B. Qiu, S. Q. Wang, J. Guo, Z. W. Yang, and S. B. Zhou. 2018. Ganoderic acid A potentiates the antioxidant effect and protection of mitochondrial membranes and reduces the apoptosis rate in primary hippocampal neurons in magnesium free medium. *Pharmazie* 73:87–91.
Joseph, S., B. Sabulal, V. George, T. P. Smina, and K. K. Janardhanan. 2009. Antioxidative and antiinflammatory activities of the chloroform extract of *Ganoderma lucidum* found in South India. *Sci Pharm* 77:111–21.
Kallay, E., P. Pietschmann, S. Toyokuni, et al. 2001. Characterization of a vitamin D receptor knockout mouse as a model of colorectal hyperproliferation and DNA damage. *Carcinogenesis* 22:1429–35.
Kana, Y., T. Chen, Y. Wu, and J. Wu. 2015. Antioxidant activity of polysaccharide extracted from *Ganoderma lucidum* using response surface methodology. *Int J Biol Macromol* 72:151–7.

Karaman, M., E. Jovin, R. Malbasa, M. Matavulj, and M. Popovic. 2010. Medicinal and edible lignicolous fungi as natural sources of antioxidative and antibacterial agents. *Phytother Res* 24:1473–81.

Karimi, M., F. Raofie, and M. Karimi. 2022. Production *Ganoderma lucidum* extract nanoparticles by expansion of supercritical fluid solution and evaluation of the antioxidant ability. *Sci Rep* 12, no. 9904 (June):1–12. www.ncbi.nlm.nih.gov/pmc/articles/PMC9198024/pdf/41598_2022_Article_13727.pdf

Kao, P.-F., S.-H. Wang, W.-T. Hung, Y.-H. Liao, C.-M. Lin, and W.-B. Yang. 2012. Structural characterization and antioxidative activity of low-molecular weights beta-1,3-glucan from the residue of extracted *Ganoderma lucidum* fruiting bodies. *J Biomed Biotechnol* 2012, no. 673764 (December):1–8. www.ncbi.nlm.nih.gov/pmc/articles/PMC3236510/pdf/JBB2012-673764.pdf

Kim, H. A., A. Perrelli, A. Ragni, F. Retta, T. M. De Silva, C. G. Sobey, and S. F. Retta. 2020. Vitamin D deficiency and the risk of cerebrovascular disease. *Antioxidants* 9, no. 4 (April):1–22. www.ncbi.nlm.nih.gov/pmc/articles/PMC7222411/pdf/antioxidants-09-00327.pdf

Kim, M. Y., P. Seguin, J. K. Ahn, et al. 2008. Phenolic compound concentration and antioxidant activities of edible and medicinal mushrooms from Korea. *J Agric Food Chem* 56:7265–70.

Klaus, A., W. A. A. Q. I. Wan-Mohtar, B. Nikolic, S. Cvetkovic, and Vunduk, J. 2021. Pink oyster mushroom *Pleurotus flabellatus* mycelium produced by an airlift bioreactor—the evidence of potent in vitro biological activities. *World J Microbiol Biotechnol* 37, no. 17 (January):1–11. https://link.springer.com/article/10.1007/s11274-020-02980-6

Kolniak-Ostek, J., J. Oszmianski, A. Szyjka, H. Moreira, and E. Barg. 2022. Anticancer and antioxidant activities in *Ganoderma lucidum* wild mushrooms in Poland, as well as their phenolic and triterpenoid compounds. *Int J Mol Sci* 23, no. 16 (August):1–15. www.ncbi.nlm.nih.gov/pmc/articles/PMC9408863/pdf/ijms-23-09359.pdf

Kozarski, M., A. Klaus, D. Jakovljevic, N. Todorović, W. A. A. Q. I. Wan-Mohtar, and M. Niksic, M. 2019. *Ganoderma lucidum* as a cosmeceutical: Study of anti-radical potential and inhibitory effect on hyperpigmentation and skin extracellular matrix degradation enzymes. *Arch. Biol Sci* 71:253–64.

Kozarski, M., A. S. Klaus, D. Jakovljevic, et al. 2015. Antioxidants of edible mushrooms. *Molecules* 20, no. 10 (October):19489–525. www.ncbi.nlm.nih.gov/pmc/articles/PMC6331815/pdf/molecules-20-19489.pdf

Kozarski, M., A. S. Klaus, M. P. Niksic, L. J. L. D. Jakovljevic, J. P. F. G. Helsper, and L. J. L. D. Van Griensven. 2011. Antioxidative and immunomodulating activities of polysaccharide extracts of the medicinal mushrooms *Agaricus bisporus, Agaricus brasiliensis, Ganoderma lucidum* and *Phellinus linteus. Food Chemistry* 129:1667–75.

Kozarski, M. S., A. S. Klaus, M. P. Niksic, L. J. L. D. Van Griensven, M. M. Vrvic, and D. M. Jakovljevic. 2014. Polysaccharides of higher fungi: Biological role, structure and antioxidative activity. *Hem Ind* 68:305–20.

Kozarski, M., A. S. Klaus, M. P. Niksic, et al. 2012. Antioxidative activities and chemical characterization of polysaccharide extracts from the widely used mushrooms *Ganoderma applanatum*, *Ganoderma lucidum*, *Lentinus edodes* and *Trametes versicolor*. *J Food Compos Anal* 26:144–53.

Kozarski, M., A. S. Klaus, L. J. L. D. van Griensven, et al. 2023. Mushroom β-glucan and polyphenol formulations as natural immunity boosters and balancers: Nature of the application. *Food Sci Hum Wellness* 12:378–96.

Kozarski, M., A. S. Klaus, J. Vunduk, D. Jakovljevic, M. Jadranin, and M. Niksic. 2020. Health impact of the commercially cultivated mushroom *Agaricus bisporus* and the wild-growing mushroom *Ganoderma resinaceum*—a comparative overview. *J Serbian Chem Soc* 85:721–35.

Kozarski, M., and L. J. L. D. van Griensven. 2022. Oxidative stress prevention by edible mushrooms and their role in cellular longevity. In *Wild Mushrooms Characteristics, Nutrition, and Processing*, ed. S. B. Dhull, A. Bains, P. Chawla, and P. K. Sadh, 319–48. London: Taylor & Francis Ltd.

Krupodorova, T. A., P. P. Klymenko, V. Y. Barshteyn, Y. I. Leonov, D. W. Shytikov, and T. N. Orlova. 2015. Effects of *Ganoderma lucidum* (Curtis) P. Karst. and *Crinipellis schevczenkovi* Buchalo aqueous extracts on skin wound healing. *J Phytopharmacol* 4:197–201.

Lacroix, S., J. Klicic Badoux, M. P. Scott-Boyer, et al. 2018. A computationally driven analysis of the polyphenol-protein interactome. *Sci Rep* 8, no. 1 (February):1–13. www.nature.com/articles/s41598-018-20625-5

Lee, M. T., W. C. Lin, and T. T. Lee. 2019. Potential crosstalk of oxidative stress and immune response in poultry through phytochemicals. *Asian-Australas J Anim Sci* 32:309–19.

Lee, W. Y., E.-J. Park, J. K. Ahn, and K.-H. Ka. 2009. Ergothioneine contents in fruiting bodies and their enhancement in mycelia cultures by the addition of methionine. *Mycobiology* 37:43–7.

Leitão, C., A. Mignano, M. Estrela, et al. 2022. The effect of nutrition on aging—a systematic review focusing on aging-related biomarkers. *Nutrients* 14, no. 3 (February):1–29. www.ncbi.nlm.nih.gov/pmc/articles/PMC8838212/pdf/nutrients-14-00554.pdf

Leyane, T. S., S. W. Jere, and N. N. Houreld. 2022. Oxidative stress in ageing and chronic degenerative pathologies: Molecular mechanisms involved in counteracting oxidative stress and chronic inflammation. *Int J Mol Sci* 23, no. 7272 (July):1–28. https://www.ncbi.nlm.nih.gov/pmc/articles/PMC9266760/pdf/ijms-23-07273.pdf

Li, B., D. S. Lee, Y. Kang, N. Q. Yao, R. B. An, and Y. C. Kim. 2013. Protective effect of *Ganoderma*nondiol isolated from the Lingzhi mushroom against tert-butyl hydroperoxide-induced hepatotoxicity through Nrf2-mediated antioxidant enzymes. *Food Chem Toxicol* 53:317–24.

Li, E. K., L.-S. Tam, C. K. Wong, et al. 2007. Safety and efficacy of *Ganoderma lucidum* (lingzhi) and San Miao San supplementation in patients with rheumatoid arthritis: A double-blind, randomized, placebo-controlled pilot trial. *Arthritis Rheum* 57, no. 7 (October):1143–50. https://onlinelibrary.wiley.com/doi/epdf/10.1002/art.22994

Lin, Z. B. 2009. *Lingzhi: From Mystery to Science*. Beijing, China: Peking University Medical Press.

Marino, T., A. Galano, and N. Russo. 2014. Radical scavenging ability of gallic acid toward OH and OOH radicals: Reaction mechanism and rate constants from the density functional theory. *J Phys Chem B* 118:10380–9.

Martinez-Medina, G. A., M. L. Chavez-Gonzalez, D. K. Verma, et al. 2021. Bio-functional components in mushrooms, a health opportunity: Ergothionine and huitlacohe as recent trends. *J Funct Foods* 77, no. 104326 (February):1–17. www.sciencedirect.com/science/article/pii/S1756464620305508

Meccariello, R., and S. D'Angelo. 2021. Impact of polyphenolic-food on longevity: An elixir of life. An overview. *Antioxidants* 10, no. 4 (April):1–26. www.ncbi.nlm.nih.gov/pmc/articles/PMC8064059/pdf/antioxidants-10-00507.pdf

Miyazawa, T., C. Abe, G. Carpentero Burdeos, A. Matsumoto, and M. Toda. 2022. Food antioxidants and aging: Theory, current evidence and perspectives. *Nutraceuticals* 2:181–204.

Moore, A. W. 2019. Is the quest for immortality worse than death? Silicon Valley entrepreneurs are obsessed with prolonging life—but they could be deluded in what they wish for. *The New Statesman*. www.newstatesman.com/world/2019/12/is-the-quest-for-immortality-worse-than-death-2 (accessed November 12, 2022).

Mwangi, R. W., J. M. Macharia, I. N. Wagara, and R. L. Bence. 2022. The antioxidant potential of different edible and medicinal mushrooms. *Biomed Pharmacother* 147, no. 112621 (January):1–16. www.sciencedirect.com/science/article/pii/S0753332222000099

Needham, J. 1974. *Science and Civilisation in China. Volume 5: Chemistry and Chemical Technology. Part 2: Spagyrical Discovery and Invention: Magisteries of Gold and Immortality*. New York and London: Cambridge University Press.

Nguyen, T. H., R. Nagasaka, and T. Ohshima. 2013. The natural antioxidant ergothioneine: Resources, chemical characterization, and applications. In *Lipid Oxidation, Challenges in Food Systems*, ed. A. Logan, U. Nienaber, and X. Pan, 381–415. Urbana, IL: Elsevier, AOCS Press.

Njus, D., and P. M. Kelley. 1991. Vitamins C and E donate single hydrogen atoms *in vivo*. *FEBS Lett* 284:147–51.

Okazaki, M., N. Iwata, S. Horiuchi, et al. 2008a. Protective effects of a water-soluble extract from culture medium of *Ganoderma lucidum* mycelia against neuronal damage after hypoxia-ischemia in mice. *Jpn J Compl Altern Med* 5:153–62.

Okazaki, M., A. Tanaka, Y. Hatta, et al. 2008b. Antioxidant properties of a water-soluble extract from culture medium of *Ganoderma lucidum* (Rei-shi) mycelia and antidiabetic effects in streptozotocin-treated mice. *Jpn J Compl Altern Med* 5:209–18.

Outila, T. A., P. H. Mattila, V. I. Piironen, and C. J. E. Lamberg-Allardt. 1999. Bioavailability of vitamin D from wild edible mushrooms (*Cantharellus tubaeformis*) as measured with a human. *Am J Clin Nutr* 69:95–8.

Pandey, K. B., and S. I. Rizvi. 2009. Plant polyphenols as dietary antioxidants in human health and disease. *Oxid Med Cellr Longev* 2:270–8.

Parravano, M., M. Tedeschi, D. Manca, et al. 2019. Correction to: Effects of Macuprev® supplementation in age-related macular degeneration: A double-blind randomized morpho-functional study along 6 months of follow-up. *Adv Ther* 36:2493–505.

Pehlivan, F. E. 2017. Vitamin C: An antioxidant agent. In *Vitamin C*, ed. A. Hamza. London: IntechOpen Limited.

Peng, C., X. Wang, J. Chen, et al. 2014. Biology of ageing and role of dietary antioxidants. *Biomed Res Int* 2014, no. 831841 (April):1–13. www.ncbi.nlm.nih.gov/pmc/articles/PMC3996317/pdf/BMRI2014-831841.pdf

PubMed[R]. 2022. *National Library of Medicine*. Rockville Pike, Bethesda: The National Center for Biotechnology Information. https://pubmed.ncbi.nlm.nih.gov (accessed December 10, 2022).

Rajoriya, A., S. S. Tripathy, and N. Gupta. 2015. *In vitro* antioxidant activity of selected *Ganoderma* species found in Odisha, India. *Trop Plant Res* 2:72–7.

Rao, A. V., and L. G. Rao. 2007. Carotenoids and human health. *Pharmacol Res* 55:207–16.
Sabitha, R., K. Nishi, V. P. Gunasekaran, et al. 2020. p-Coumaric acid attenuates alcohol exposed hepatic injury through MAPKs, apoptosis and Nrf2 signaling in experimental models. *Chem Biol Interact* 321, no. 109044 (April):1–11. www.sciencedirect.com/science/article/pii/S0009279719310968?via%3Dihub
Sadowska-Bartosz, I., and G. Bartosz. 2014. Effect of antioxidants supplementation on aging and longevity. *Biomed Res Int* 2014, no. 404680 (March):1–17. www.ncbi.nlm.nih.gov/pmc/articles/PMC3982418/pdf/BMRI2014-404680.pdf
Seweryn, E., A. Ziala, and A. Gamian. 2021. Health-promoting of polysaccharides extracted from *Ganoderma lucidum*. *Nutrients* 13, no. 8 (August):1–14. www.ncbi.nlm.nih.gov/pmc/articles/PMC8400705/pdf/nutrients-13-02725.pdf
Shaher, F., H. Qiu, S. Wang, et al. 2020. Associated targets of the antioxidant cardioprotection of *Ganoderma lucidum* in diabetic cardiomyopathy by using open targets platform: A systematic review. *BioMed Res Int* 2020, no. 7136075 (July):1–20. www.ncbi.nlm.nih.gov/pmc/articles/PMC7397440/pdf/BMRI2020-7136075.pdf
Sheena, N., T. A. Ajith, and K. K. Janardhanan. 2003. Prevention of nephrotoxicity induced by the anticancer drug cisplatin, using *Ganoderma lucidum*, a medicinal mushroom occurring in South India. *Curr Sci* 85:478–82.
Sima, Q. 1996. *Records of the Grand Historian: Qin Dynasty*, translated by Burton Watson. New York: Columbia University Press.
Smina, T. P., J. Mathewa, K. K. Janardhanana, and T. P. A. Devasagayam. 2011. Antioxidant activity and toxicity profile of total triterpenes isolated from *Ganoderma lucidum* (Fr.) P. Karst. occurring in South India. *Environ Toxicol Pharmacol* 32:438–46.
Stamets, P. 2005. Notes on nutritional properties of culinary-medicinal mushrooms. *Int J Med Mushrooms* 7:103–10.
State Pharmacopoeia Commission of the PRC. 2005. *Pharmacopoeia of the People's Republic of China: Volume I*. Beijing: People's Medical Publishing House, Co. Ltd.
Steavu, D. 2018. The marvelous fungus and the secret of divine immortals. *Micrologus* 26:353–83.
Steger, J., A. G. Cole, A. Denner, et al. 2022. Single-cell transcriptomics identifies conserved regulators of neuroglandular lineages. *Cell Rep* 40, no. 111370 (September):1–22. www.cell.com/cell-reports/pdf/S2211-1247(22)01202-5.pdf
Stojković, D. S., L. Barros, R. C. Calhelha, et al. 2013. A detailed comparative study between chemical and bioactive properties of *Ganoderma lucidum* from different origins. *Int J Food Sci Nutr* 65:42–7.
Sułkowska-Ziaja, K., G. Zengin, A. Gunia-Krzyzak, et al. 2022. Bioactivity and mycochemical profile of extracts from mycelial cultures of *Ganoderma* spp. *Molecules* 27, no. 1 (January):1–15. www.ncbi.nlm.nih.gov/pmc/articles/PMC8746335/pdf/molecules-27-00275.pdf
Sun, J., H. He, and B. J. Xie. 2004. Novel antioxidant peptides from fermented mushroom *Ganoderma lucidum*. *J Agric Food Chem* 52:6646–52.
Tan, B. L., M. E. Norhaizan, W.-P.-P. Liew, and H. S. Rahman. 2018. Antioxidant and oxidative stress: A mutual interplay in age-related diseases. *Front Pharmacol* 9, no. 1162 (October):1–28. www.ncbi.nlm.nih.gov/pmc/articles/PMC6204759/pdf/fphar-09-01162.pdf
Tel-Cayana, G., M. Ozturka, M. E. Duru, et al. 2015. Phytochemical investigation, antioxidant and anticholinesteraseactivities of *Ganoderma* adspersum. *Ind Crops Prod* 76:749–54.
Upton, R. 2000. Reishi mushroom: *Ganoderma lucidum*: Standards of analysis, quality control, and therapeutics. In *American Herbal Pharmacopoeia and Therapeutic Compendium*, ed. R. Upton and C. Petrone, 1–28. Scotts Valley: American Herbal Pharmacopoeia.
Van den Ende, W., D. Peshev, and L. De Gara. 2011. Disease prevention by natural antioxidants and pre-biotics acting as ROS scavengers in the gastrointestinal tract. *Trends Food Sci Technol* 22:689–97.
Veljović, S., M. Veljović, N. Nikićević, et al. 2017. Chemical composition, antiproliferative and antioxidant activity of differently processed *Ganoderma lucidum* ethanol extracts. *J Food Sci Technol* 54, no. 5:1312–20.
Verburgh, K. 2018. *The Longevity Code: Secrets to Living Well for Longer from the Front Lines of Science*, narrated by Pete Cross. New York: Dreamscape Media Audio.
Vunduk, J., D. Tura, and A. Y. Biketova. 2022. Medicinal mushroom nutraceutical commercialization: Two sides of a coin. In *Wild Mushrooms Characteristics, Nutrition, and Processing*, ed. S. B. Dhull, A. Bains, P. Chawla, and P. K. Sadh, 89–131. London: Taylor & Francis Ltd.
Vunduk, J., and S. Veljović. 2021. Macrofungi in the production of alcoholic beverages: Beer, wine, and spirits. In *Advances in Macrofungi: Industrial Avenues and Prospects*, ed. K. R. Sridhar and S. K. Deshmukh. Boca Raton: CRC Press.

Wachtel-Galor, S., J. Yuen, J. A. Buswell, and I. F. F. Benzie. 2011. *Ganoderma lucidum* (Lingzhi or Reishi): A medicinal mushroom. In *Herbal Medicine: Biomolecular and Clinical Aspects*, ed. I. F. F. Benzie and S. Wachtel-Galor. Boca Raton: CRC Press/Taylor & Francis.
Ware, J. R. 1981. *Alchemy, Medicine and Religion in the China of A.D. 320: The Nei Pien of Ko Hung*. New York: Dover Pubns.
Wasson, R. G. 1968. *Soma: Divine Mushroom of Immortality*. New York: Harcourt, Brace & World, Inc.
Weng, Y., J. L. L. Xiang, A. Matsuura, Y. Zhang, Q. Huang, and J. Qi. 2011. *Ganoderma*sides C and D, two new anti-aging ergosterols from spores of the medicinal mushroom *Ganoderma lucidum*. *Biosci Biotechnol Biochem* 75:800–3.
Weng, Y., L. Xiang, A. Matsuura, Y. Zhang, Q. Huang, and J. Qi. 2010. *Ganoderma*sides A and B, two novel anti-aging ergosterols from spores of a medicinal mushroom *Ganoderma lucidum* on yeast via UTH1 gene. *Bioorg Med Chem* 18:999–1002.
Wong, W. L., X. Su, X. Li, et al. 2014. Global prevalence of age-related macular degeneration and disease burden projection for 2020 and 2040: A systematic review and meta-analysis. *The Lancet Glob Health* 2, no. 2 (January):1–11. www.thelancet.com/journals/langlo/article/PIIS2214-109X(13)70145-1/fulltext
Yan, C., F. Kong, D. Zhang, and J. Cui. 2013. Anti-glycated and antiradical activities *in vitro* of polysaccharides from *Ganoderma capense*. *Pharmacogn Mag* 9:23–7.
Yin, Z., B. Yang, and H. Ren. 2019. Preventive and therapeutic effect of *Ganoderma* (Lingzhi) on skin diseases and care. *Adv Exp Med Biol* 1182:311–21.
Yoon, H. M., K. J. Jang, M. S. Han, et al. 2013. *Ganoderma lucidum* ethanol extract inhibits the inflammatory response by suppressing the NF-κB and toll-like receptor pathways in lipopolysaccharide-stimulated BV2 microglial cells. *Exp Ther Med* 5:957–63.
Zeng, P., Z. Guo, X. Zeng, et al. 2018. Chemical, biochemical, preclinical and clinical studies of *Ganoderma lucidum* polysaccharide as an approved drug for treating myopathy and other diseases in China. *J Cell Mol Med* 22:3278–97.
Zengin, G., C. Sarikurkcu, E. Gunes, et al. 2015. Two *Ganoderma* species: Profiling of phenolic compounds by HPLC-DAD, antioxidant, antimicrobial and inhibitory activities on key enzymes linked to diabetes mellitus, Alzheimer's disease and skin disorder. *Food Funct* 6:2794–802.
Zgorzynska, E., B. Dziedzic, and A. Walczewska. 2021. An overview of the Nrf2/ARE pathway and its role in neurodegenerative diseases. *Int J Mol Sci* 22, no. 17 (September):1–23. www.ncbi.nlm.nih.gov/pmc/articles/PMC8431732/pdf/ijms-22-09592.pdf
Zhang, H., H. Jiang, X. Zhang, and J. Yan. 2018. Amino acids from *Ganoderma lucidum*: Extraction optimization, composition analysis, hypoglycemic and antioxidant activities. *Currt Pharm Anal* 14:562–70.
Zhang, J., Y. Liu, Q. Tang, S. Zhou, J. Feng, and H. Chen. 2019. Polysaccharide of *Ganoderma* and its bioactivities. In *Ganoderma and Health Biology, Chemistry and Industry, Advances in Experimental Medicine and Biology*, ed. Z. Lin and B. Yang, 107–35. Singapore: Springer Nature Pvt. Ltd.
Zheng, S., W. Zhang, and S. Liu. 2020. Optimization of ultrasonic-assisted extraction of polysaccharides and triterpenoids from the medicinal mushroom *Ganoderma lucidum* and evaluation of their in vitro antioxidant capacities. *PLOS One* 15, no. 12 (December):1–16. www.ncbi.nlm.nih.gov/pmc/articles/PMC7774858/pdf/pone.0244749.pdf
Zhu, B. Z., L. Mao, R. M. Fan, et al. 2011. Ergothioneine prevents copper-induced oxidative damage to DNA and proteinby forming a redox-inactive ergothioneine-copper complex. *Chem Res Toxicol* 24:30–4.
Zhu, K. X., S. P. Nie, L. H. Tan, et al. 2016. A polysaccharide from *Ganoderma atrum* improves liver function in type 2 diabetic rats via antioxidant action and short-chain fatty acids excretion. *J Agri and Food Chem* 64:1938–44.
Zhong, J.-J., and J.-H. Xiao. 2009. Secondary metabolites from higher fungi: Discovery, bioactivity, and bioproduction. *Adv Biochem Eng Biotechnol* 113:79–150.
Zhonghui, Z., Z. Xiaowei, and F. Fang. 2014. *Ganoderma lucidum* polysaccharides supplementation attenuates exercise-induced oxidative stress in skeletal muscle of mice. *Saudi J Biol Sci* 21:119–23.
Zizak, Z., A. Klaus, M. Kozarski, J. Vunduk, M. Niksic, and Z. Juranic. 2014. Antitumor activities of some macrofungi extracts. *Eur J Cancer* 50:S30.

5 Hepatoprotective Effect of *Ganoderma lucidum* (Curt.:Fr.) P. Karst

Thekkuttuparambil A. Ajith[1] *and Kainoor K. Janardhanan*[2]
1 Amala Institute of Medical Sciences, Kerala, India
2 Amala Cancer Research Centre, Kerala, India

5.1 INTRODUCTION

Nonalcoholic fatty liver disease (NAFLD) comprises a progression of liver conditions with varying severity of liver injury, finally leading to fibrosis and cirrhosis. Among the liver conditions, hepatic steatosis (fatty liver) is an early condition, while nonalcoholic steatohepatitis (NASH) is described as a more fatal condition with inflammation resulting in damage to hepatocytes. When compared to NASH, the risk of adverse outcomes in patients with only NAFLD is very low (Singh et al., 2015). Pericellular fibrosis associated with NASH may be progressed to cirrhosis. Cirrhosis, hepatocellular carcinoma and liver failure were described as the NASH-related hepatic outcomes, whereas increased malignancy and cardiovascular disease were the non–liver-associated adverse outcomes (Rinella and Sanyal, 2016; Lindenmeyer, 2018). Seventy five percent of type 2 diabetes individuals have NAFLD. A higher prevalence of NASH with advanced fibrosis was reported among individuals with diabetes with NAFLD than subjects with nondiabetic disease and NAFLD (Bazick et al., 2015; Portillo-Sanchez et al., 2015). The risk of developing liver-related complications in subjects with NAFLD and diabetes was found to be high (Anstee et al., 2013). Insulin resistance was recognized as one of the major integral components of NAFLD pathogenesis, which worsens with progression of the disease (Choudhury and Sanyal, 2004). Furthermore, subjects with NAFLD were demonstrated to be at increased risk for developing diabetes (Ballestri et al., 2016). The risk for incidence of metabolic syndrome and type 2 diabetes among NAFLD subjects was increased by almost two-fold. Oxidative stress was demonstrated in the pathogenesis of fibrosis and cirrhosis (Reyes-Gordillo et al., 2008).

Ganoderma lucidum, or reishi, a polypore mushroom, is known as the 'king of herbs'. Among the 131 species found in the world, 20 kinds of *G. lucidum*, *G. lingzhi*, *G. sinense* and others have medicinal value, which is demonstrated in various experimental studies (Wang et al., 2020). Products of *G. lucidum* alone or in combination with various herbal medicines have long been used in Asian countries. The use of *G. lucidum* in treating chronic hepatitis and hepatopathy was found in the folk medicine of Japan and China (Jong and Birmingham, 1992). The pharmacological properties described such as immunomodulatory, anti-inflammatory, hypolipidemic, antifibrotic and anti-oxidant make it an effective hepatoprotective agent. Approximately 400 various bioactive compounds such as polysaccharides, triterpenoids, sterols, nucleotides, fatty acids, trace elements and protein/peptides were isolated from the spores, mycelia and fruiting body of *G. lucidum* (Kim and Kim, 1999). The quality of *G. lucidum* is found to be directly associated with its polysaccharide and triterpenoid contents (Li et al., 2016). The major triterpenoids reported were ganodermic acid (Figure 5.2) and lucidenic acid (Boh et al., 2007).

5.2 MECHANISM OF LIVER INJURY IN NONALCOHOLIC FATTY LIVER DISEASE

An important factor associated with the etiology of fatty liver in humans was increased calorie consumption rather than dietary fat composition (Green and Hodson, 2014). NAFLD, as a hepatic

 DOI: 10.1201/9781003354789-5

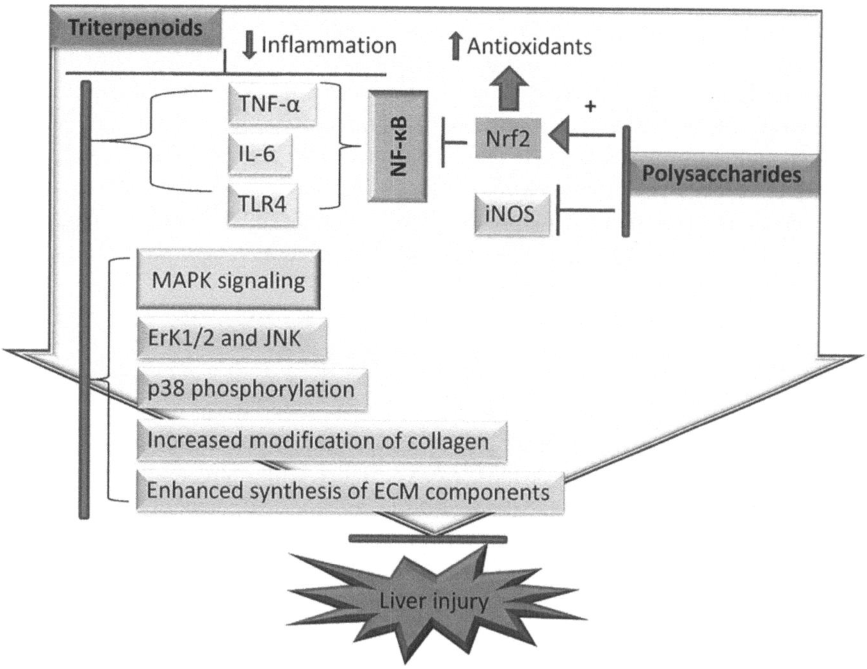

FIGURE 5.1 Various mechanisms for protecting the liver from injury by phytochemicals present in *G. lucidum*. Terpenoids can prevent the formation of proinflammatory cytokines and block various cell signaling pathways that result in modification of the extracellular matrix. Polysaccharides can activate the nuclear factor erythroid 2–related factor 2 (Nrf2) to block the activation of transcription factor, NF-kB, which can alleviate the inflammation as well as improve the antioxidant status in the hepatocytes. Antioxidant and anti-inflammation properties can prevent the stellate cell activation and progression of liver fibrosis. MAPK: Mitogen-activated protein kinase; TNF-α: Tumor necrosis factor alpha; TLR-4: Toll-like receptor 4; NF-κB: Nuclear factor kappa B; ErK 1/2: Extracellular signal–regulated protein kinase; JNK: c-Jun N-terminal kinase; ECM: Extracellular matrix; IL-6: Interleukin 6; iNOS: Inducible nitric oxide synthase.

constituent of metabolic syndrome, includes a wide spectrum of disease entities that begin from the fatty liver (steatosis) to more severe steatohepatitis, cirrhosis and hepatocellular carcinoma. Most of the stages of the disease are coupled with marked inflammation and fibrosis (Tariq et al., 2014). Dysfunctions of mitochondria have long been demonstrated in the progression of NAFLD. The pathophysiology of liver mitochondrial dysfunction is correlated to the generated reactive oxygen species (ROS) and pro-inflammatory molecules (Simões et al., 2018). Downregulation of mitochondrial oxidative phosphorylation and mitochondrial biogenesis were found to be associated with the inflammatory pathways (Malik et al., 2019).

Free fatty acids have a pivotal role in the pathogenesis of NASH. A defect in beta-oxidation of fatty acids or overwhelmed formation of triglycerides leads to accumulation of fatty acids, which can augment the development of lipotoxic species. Oxidant stress, stress to endoplasmic reticulum and activation of inflammasomes were responsible for NASH. Furthermore, stellate cell activation, inflammation, hepatocellular injury and increased accumulation of the extracellular matrix were found to be associated with NASH. Lifestyle modifications such as decreased calorie intake, regular exercise and healthy eating habits can reduce the overload of fat. Diversion of excess fat to effective metabolites can reverse or prevent the development of NASH. The sites of action of many drugs, based on their primary locus of activity within the liver, are currently in phase 2 or 3 clinical trials. Various drug targets in the management of NASH include regulation of lipid accumulation, mitochondrial oxidative stress, glucose homeostasis, stellate cell activation and inflammatory signals that converge to hepatocytes and fibrogenesis. While progressing NAFLD to NASH and cirrhosis, an increased nitro-oxidative stress may also be considered. The generated ROS and reactive nitrogen species (RNS) were the main contributors in the development of fibrosis. ROS and RNS increase the generation of interleukin-6 (IL-6), tumor necrosis factor-α (TNF-α) and transforming growth factor-β to induce the formation of fibrogenesis (Karkucinska-Wieckowska et al., 2022). As an increased hepatic lipogenesis with suppressed beta-oxidation of fatty acids is suggested to be an underlying cause in the development of NAFLD, agents with activities to regulate these processes will be promising prophylactic agents (Green et al., 2020).

5.3 HEPATOPROTECTIVE ROLE OF *G. LUCIDUM*

G. lucidum has long been used in the management of hepatopathy. It was demonstrated to be effective against a wide range of different liver diseases, including hepatocellular carcinoma, NAFLD, hepatitis B, alcoholic liver disease, toxin-induced liver injury, inflammation and fibrosis (Qiu et al., 2019). The possible hepatoprotective mechanism of *G. lucidum* is depicted in Figure 5.1. Extract from this red mushroom exhibits a protective effect against toxic chemical–induced liver injury. A previous study demonstrated that an alcoholic extract from *G. lucidum* could alleviate carbon tetrachloride– and alcohol-induced acute hepatotoxicity and fibrosis (Xu et al., 2020). The polysaccharide fractions and triterpenes triterpenoids have shown liver-protective effects in human studies. Ganoderic acids S and R, isolated from the *G. lucidum* cultured mycelia, showed a hepatoprotective effect against galactosamine-induced cytotoxicity in cultured rat hepatocytes. Extracts from reishi can protect against CCl_4-and galactosamine-induced liver damage in rats (Yanling et al., 2008) Triterpenoids isolated from *G. lucidum* showed significant protection against immunological liver damage in mice (Dudhgaonkar et al., 2009). *G. lucidum* at doses of 100 and 250 mg/kg body wt, when administered prior toCCl_4 once daily for 15 days, could prevent the deterioration of mitochondrial enzyme activities and thus maintain the function of electron transport chain complexes in a rat experimental model (Sudheesh et al., 2012a).

The liver-protective effect of polysaccharides isolated from *G. lucidum* was also evaluated in a mouse model. Treatment with polysaccharides diminished histological changes of liver injury and inflammatory infiltration of lymphocytes. The protective effect of polysaccharides was found to be mediated by repressing the inducible nitric oxide synthase and thus decreased the excess production of nitric oxide in liver cells. The protein-bound polysaccharides isolated from *G. lucidum* were found to reduce the activities of serum alkaline phosphatase, aspartate transaminase (AST), alanine

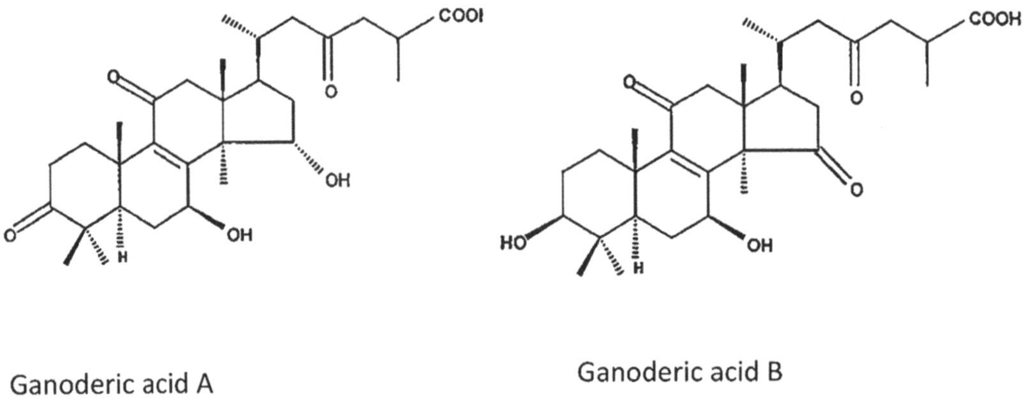

FIGURE 5.2 Structure of triterpenoids present in *G. lucidum*. A) Ganoderic acid B and B) ganoderic acid A.

transaminase (ALT) and total bilirubin level in rats with cirrhosis. An experimental study of liver injury demonstrated that triterpenes of *G. lucidum* reduces the liver IL-6 and TNF-alpha production by interfering with the signaling pathways mediated from mitogen-activated protein kinase and nuclear factor kappa B (NF-κB). The effect was found to be mediated through inhibiting the expressions and interactions of toll-like receptor (TLR), nuclear translocation of NF-κB and its binding to DNA, p38 phosphorylation, c-Jun N-terminal kinase and extracellular signal–regulated protein kinase 1/2 (Hu et al., 2020). Spores of *G. lucidum* also showed an anti-inflammatory effect in an experimental liver injury model. The protective effect was mediated through suppressing TLR and inhibiting NF-κB (Chen et al., 2022). *G. lucidum* intervention regulated the fatty acid metabolism and homeostasis of bile acids. The effect was mediated through regulating the messenger RNA (mRNA)expression from genes associated with the hepatic bile acids and fatty acid biosynthesis (Guo et al., 2020). Even the degraded polysaccharides of *G. lucidum* are a stronger hypolipidemic and antioxidant agent in hepatocytes (Xu et al., 2019). Furthermore, peptides from *G. lucidum* (180 mg/kg) were reported to possess strong antioxidant activities.

G. lucidum significantly protected against the decline of mitochondrial enzyme activities and thus maintained the electron transport chain complexes in carbon tetrachloride–treated liver in a dose-dependent manner (Sudheesh et al., 2012b). ROS has an important role in the fibrosis of the liver, and thus the antioxidants targeted to mitochondria can be a promising therapy in the prevention and management of liver fibrosis. The decreased oxidative stress can inhibit the hepatic stellate cell activation, expression of transforming growth factor beta 1 and activation of extracellular signal–regulated protein kinase (Rehman et al., 2016). *G. lucidum* aqueous extract can enhance the nuclear factor erythroid 2–related factor 2 and thus improve the antioxidant status in the cell (Aslaminabad et al., 2022). A study using the HepG2 cell line found that *Ganoderma* triterpenoids could increase the reduced glutathione and antioxidant enzyme, superoxide dismutase, and thus decrease the level of lipid peroxidation (Wu et al., 2016).

5.4 CLINICAL TRIALS OF HEPATOPROTECTION AND TOXICITY OF *G. LUCIDUM*

The capsules (225 mg twice daily for 6 months) of *G. lucidum* containing a few essential amino acids, 7% triterpenoid-ganoderic acids, 6% polysaccharide peptides and trace elements, when given to 42 healthy middle-aged volunteers, shown that mild liver dysfunction was found to be reversed to normal. The enhanced antioxidant enzymes and decreased oxidative stress markers were noticed as the mechanism (Chiu et al., 2017). Later, polysaccharide preparations of *G. lucidum* showed a curative effect in patients with chronic infection of hepatitis B. The serum ALT and AST activity in 33% of patients with chronic hepatitis B infection of 6years' duration was found to be normal when treated with Ganopoly (a capsule containing 600 mg extract of *G. lucidum*) orally three times day before food for 12 weeks. While serum hepatitis B surface antigen was cleared in 13%of patients compared to the control group (Gao et al., 2002). The study also revealed that Ganopoly is significantly active against hepatitis B virus (HBV). We have previously demonstrated that administration of *G. lucidum* (mycelia as well as fruiting bodies) up to 2500 mg/kg body wt. did not produce any lethal effect to animals, and thus LD_{50} could not be determined. No significant change in the activities of serum transaminases (glutamate oxaloacetate transaminase (GOT) and glutamate pyruvate transaminase (GPT)) and alkaline phosphatase (ALP) revealed that no liver damage was evident. Furthermore, no changes in total protein, albumin or globulin levels were found. The safety and tolerability of *G. lucidum* were tested in a randomized placebo-controlled clinical trial. Wicks et al. (2007) demonstrated that extract of *G. lucidum*, when administrated at dose of 2 g orally in 16 human volunteers twice daily for 10 consecutive days, produced no change in immunity. The reported findings also indicated that compared to the placebo group, no adverse effects were found after the administration of the mushroom extract. Wachtel-Galor et al. (2004), in a controlled human supplement study of four capsules of lingzhi powder (total 1.44 g of lingzhi extract per day for 1month), showed no hepatotoxic effect or drug accumulation. However, Wanmuang et al. reported a case of fatal toxic hepatitis associated with lingzhi powder in a patient who was taking 400 mg of lingzhi extract per day for 2months (Wanmuang et al., 2007). No specific mechanism of toxicity could be described.

5.5 CONCLUSION AND FUTURE PERSPECTIVES

Oxidative stress and free fatty acids have a central role in the pathophysiology of NASH. The spores, fruiting body and mycelia of *G. lucidum* contain approximately 400 various bioactive compounds such as triterpenoids, polysaccharides, fatty acids, nucleotides, protein/peptides, trace elements and sterols. Experimental studies have demonstrated that triterpenoids and polysaccharides in *G. lucidum* as two major categories of bioactive ingredients were effective in alleviating oxidative stress, inflammation, mitochondrial dysfunction, hypolipidemia and diabetes mellitus. Despite the favorable pharmacological effects in experimental studies, clinical trials are lacking to explore the beneficial application of *G. lucidum*.

REFERENCES

Anstee, Q.M., Targher, G., and Day, C.P., 2013. Progression of NAFLD to diabetes mellitus, cardiovascular disease or cirrhosis. *Nature Reviews Gastroenterology and Hepatology* 10:330–44.

Aslaminabad, R., Rahimianshahreza, N., Hosseini, S.A., et al., 2022. Regulation of Nrf2 and Nrf2-related proteins by *Ganoderma lucidum* in hepatocellular carcinoma. *Molecular Biology and Repair*49:9605–12.

Ballestri, S., Zona, S., Targher, G., et al., 2016. Nonalcoholic fatty liver disease is associated with an almost twofold increased risk of incident type 2 diabetes and metabolic syndrome: Evidence from a systematic review and meta-analysis. *Journal of Gastroenterology and Hepatology* 31:936–44.

Bazick, J., Donithan, M., Neuschwander-Tetri, B.A., et al., 2015. Clinical model for Nash and advanced fibrosis in adult patients with diabetes and NAFLD: Guidelines for referral in NAFLD. *Diabetes Care* 38:1347–55.

Boh, B., Berovic, M., Zhang, J., and Zhi-Bin, L., 2007.*Ganoderma lucidum* and its pharmaceutically active compounds. *Biotechnology Annual Revolution* 13:265–301.

Chen, J., He, X., Song, Y., Tu, Y., Chen, W., and Yang, G., 2022. Sporoderm-broken spores of *Ganoderma lucidum* alleviates liver injury induced by DBP and BaP co-exposure in rat. *Ecotoxicology and Environmental Safety* 241:113750.

Chiu, H.F., Fu, H.Y., Lu, Y.Y., et al., 2017. Triterpenoids and polysaccharide peptides—enriched *Ganoderma lucidum*: A randomized, double-blind placebo-controlled crossover study of its antioxidation and hepatoprotective efficacy in healthy volunteers. *Pharmceutical Biology* 55:1041–6.

Choudhury, J., and Sanyal, A.J., 2004. Insulin resistance and the pathogenesis of nonalcoholic fatty liver disease. *Clinical Liver Disease* 8:575–94.

Dudhgaonkar, S., Thyagarajan, A., and Sliva, D., 2009. Suppression of the inflammatory response by triterpenes isolated from the mushroom *Ganoderma lucidum. International Immuno Pharmacology* 9:1272–80.

Gao, Y., Zhou, S., Chen, G., Dai, X., Ye, J., and Gao, H., 2002. A phase I/II study of a *Ganoderma lucidum* (Curt.: Fr.) P. Karst. (Lingzhi, Reishi Mushroom) extract in patients with chronic hepatitis B. *International Journal of Medicinal Mushrooms* 4:7.

Green, C.J., and Hodson, L., 2014. The influence of dietary fat on liver fat accumulation. *Nutrients* 6:5018–33.

Green, C.J., Pramfalk, C., and Charlton, C.A., 2020. Hepatic de novo lipogenesis is suppressed and fat oxidation is increased by omega-3 fatty acids at the expense of glucose metabolism. *BMJ Open Diabetes Research Care* 8:e000871.

Guo, W.L., Guo, J.B., Liu, B.Y., et al., 2020. Ganoderic acid A from *Ganoderma lucidum* ameliorates lipid metabolism and alters gut microbiota composition in hyperlipidemic mice fed a high-fat diet. *Food Function* 11:6818–33.

Hu, Z., Du, R., Xiu, L., et al., 2020. Protective effect of triterpenes of *Ganoderma lucidum* on lipopolysaccharide-induced inflammatory responses and acute liver injury. *Cytokine* 127:154917.

Jong, S.C., and Birmingham, J.M., 1992. Medicinal benefits of the mushroom *Ganoderma. Advanced Applied Microbiology*37:101–34.

Karkucinska-Wieckowska, A., Simoes, I.C.M., Kalinowski, P., et al., 2022. Mitochondria, oxidative stress and nonalcoholic fatty liver disease: A complex relationship. *European Journal of Clinical Investigation* 52:e13622.

Kim, H.W., and Kim, B.K., 1999. Biomedical triterpenoids of *Ganoderma lucidum* (Curt.: Fr.) P. Karst. (Aphyllophoromycetideae). *International Journal of Medicinal Mushrooms*1:121–38.

Li, H.J., He, Y.L., Zhang, D.H., et al., 2016. Enhancement of ganoderic acid production by constitutively expressing *Vitreoscilla* hemoglobin gene in *Ganoderma lucidum*. *Journal of Biotechnology* 227:35–40.

Lindenmeyer, C.C. 2018. McCullough AJ, the natural history of nonalcoholic fatty liver disease: An evolving view. *Clinical Liver Disease* 22:11–21.

Malik, A.N., Simões, I.C.M., Rosa, H.S., Khan, S., Karkucinska-Wieckowska, A., and Wieckowski, M.R. 2019. A diet induced maladaptive increase in hepatic mitochondrial DNA precedes OXPHOS defects and may contribute to non-alcoholic fatty liver disease. *Cells*8:1222.

Portillo-Sanchez, P., Bril, F., Maximos, M., et al., 2015.High prevalence of nonalcoholic fatty liver disease in patients with type 2 diabetes mellitus and normal plasma aminotransferase levels. *Journal of Clinical Endocrinology and Metabolism* 100:2231–8.

Qiu, Z., Zhong, D., and Yang, B. 2019. Preventive and therapeutic effect of *Ganoderma* (Lingzhi) on liver injury. *Advanced Experimental Medicine and Biology* 1182:217–42.

Rehman, H., Liu, Q., Krishnasamy, Y., et al., 2016. The mitochondria—targeted antioxidant MitoQ attenuates liver fibrosis in mice. *International Journal of Physiology Pathophysiology and Pharmacology* 8:14–27.

Reyes-Gordillo, K., Segovia, J., Shibayama, M., et al., 2008. Curcumin prevents and reverses cirrhosis induced by bile duct obstruction or CCl4 in rats: Role of TGF-beta modulation and oxidative stress. *Fundamental Clinical Pharmacology*22:417–27.

Rinella, M.E., and Sanyal, A.J., 2016. Management of NAFLD: A stage-based approach. *Nature Review Gastroenterology and Hepatology* 13: 196–205.

Simões, I.C.M., Fontes, A., Pinton, P., Zischka, H., and Wieckowski, M.R., 2018. Mitochondria in non-alcoholic fatty liver disease. *International Journal of Biochemistry and Cell Biology* 95:93–9.

Singh, S., Allen, A.M., Wang, Z., Prokop, L.J., Murad, M.H., and Loomba, R. 2015. Fibrosis progression in nonalcoholic fatty liver vs nonalcoholic steatohepatitis: A systematic review and meta-analysis of paired-biopsy studies. *Clinical Gastroenterology and Hepatology* 13:643–54.

Sudheesh, N.P., Ajith, T.A., Mathew, J., Nima, N., and Janardhanan, K.K., 2012a. *Ganoderma lucidum* protects liver mitochondrial oxidative stress and improves the activity of electron transport chain in carbon tetrachloride intoxicated rats. *Hepatology Research* 42: 181–91.

Sudheesh, N.P., Ajith, T.A., Mathew, J., Nima, N., and Janardhanan, K.K., 2012b. *Ganoderma lucidum* protects liver mitochondrial oxidative stress and improves the activity of electron transport chain in carbon tetrachloride intoxicated rats. *Hepatology Research* 42:181–91.

Tariq, Z., Green, C.J., and Hodson, L., 2014. Are oxidative stress mechanisms the common denominator in the progression from hepatic steatosis towards non-alcoholic steatohepatitis (NASH)? *Liver International* 34:e180–90.

Wachtel-Galor, S., Tomlinson, B., and Benzie, I.F., 2004. *Ganoderma lucidum* ("Lingzhi"), a Chinese medicinal mushroom: Biomarker responses in a controlled human supplementation study. *British Journal of Nutrition*91: 263–9.

Wang, L., Li, J.Q., Zhang, J., Li, Z.M., Liu, H.G., and Wang, Y.Z., 2020. Traditional uses, chemical components and pharmacological activities of the genus *Ganoderma* P. Karst.: A review. *RSC Advances* 10:42084–97.

Wanmuang, H., Leopairut, J., Kositchaiwat, C., Wananukul, W., and Bunyaratvej, S., 2007. Fatal fulminant hepatitis associated with *Ganoderma lucidum* (Lingzhi) mushroom powder. *Journal of Medical Association Taiwan* 90:179–81.

Wicks, S.M., Tong, R., Wang, C.Z., et al., 2007. Safety and tolerability of *Ganoderma lucidum* in healthy subjects: A double-blind randomized placebo-controlled trail. *The American Journal of Chinese Medicine*35:407–41.

Wu, J.G., Kan, Y.J., Wu, Y.B., Yi, J., Chen, T.Q., and Wu, J.Z., 2016. Hepatoprotective effect of *Ganoderma* triterpenoids against oxidative damage induced by tert-butyl hydroperoxide in human hepatic HepG2 cells. *Pharmceutical Biology* 54:919–29.

Xu, J.B., Gao, G.C., Yuan, M.J., Huang, X., et al., 2020. Lignans from Schisandra chinensis ameliorate alcohol and CCl_4-induced long-term liver injury and reduce hepatocellular degeneration via blocking ETBR. *Journal of Ethnopharmacology* 258:112813.

Xu, Y., Zhang, X., Yan, X.H., et al., 2019. Characterization, hypolipidemic and antioxidant activities of degraded polysaccharides from *Ganoderma lucidum*. *International Journal of Biology and Macromolecule* 135:706–16.

Yanling, S., Jie, S., Hui, H., Hui, G., and Sheng, Z., 2008. Hepatoprotective effects of *Ganoderma lucidum* peptides against D-galactosamine-induced liver injury mice. *Journal of Ethnopharmacology*117: 415–19.

6 Antidiabetic Effects of *Ganoderma*

Prospects and Challenges

Chia Wei Phan[1], *Vikineswary Sabaratnam*[2], and *Umah Rani Kuppusamy*[3]

1 Department of Pharmaceutical Life Sciences, Faculty of Pharmacy, Mushroom Research Centre, Universiti Malaya, Kuala Lumpur, Malaysia

2 Institute of Biological Sciences, Faculty of Science, Mushroom Research Centre, Universiti Malaya, Kuala Lumpur, Malaysia

3 Department of Biomedical Science, Faculty of Medicine, Mushroom Research Centre, Universiti Malaya, Kuala Lumpur, Malaysia

6.1 INTRODUCTION

Type 2 diabetes, more commonly known as diabetes mellitus (DM), is characterized by hyperglycemia and glucose intolerance. It is one of the major killers in the world, with approximately 1.5 million deaths caused by it in 2019 according to the World Health Organization (WHO) (World Health Organization, 2022a). DM has gained attention as a global concern over the years with approximately 537 million adults between 20 and 79 years old having this illness in 2021, and it is projected to increase to 643 million in 2030 (International Diabetes Federation, 2021). DM is a complex metabolic disorder caused by the body's inability to secrete enough insulin, failure of insulin action, or both, resulting in dysregulation of carbohydrate, fat, and protein metabolism (Rehman and Akash, 2017). There are several types of DM which include type 1, type 2, gestational diabetes mellitus (GDM), and other types of diabetes caused by a variety of factors (American Diabetes Association, 2021). Type 1 DM is caused by autoimmune destruction of the β-cells in the pancreatic islets, which leads to total loss of insulin, and this accounts for 5–10% of all diabetes cases. Meanwhile, type 2 DM accounts for 90–95% of all diabetes cases and is primarily due to insulin resistance, attributed to an unhealthy lifestyle or a certain genetic predisposition that leads to overweight or obesity. DM is not transmissible between humans and is included among the 4 major non-communicable diseases by WHO (World Health Organization, 2022b), with the others being cardiovascular disease, cancers, and chronic respiratory illnesses. Figure 6.1 depicts the prevalence of DM in different regions of the world.

Diabetes is a disorder that affects blood glucose regulation. Blood glucose homeostasis is primarily mediated by 2 hormones, i.e. insulin and glucagon. Insulin is the hormone that acts when the blood glucose level is elevated. It is released by the β-cells in the pancreatic islets of Langerhans to lower the glucose level. Insulin is a dipeptide hormone comprising 51 amino acids with A and B subchains that are linked with disulfide bonds (Wilcox, 2005). Insulin works by binding to the insulin receptor, which is a tyrosine kinase family, causing skeletal muscle and adipose tissues to increase glucose uptake by translocating glucose transporter 4 (GLUT4) to the surface of the cells (Kaul et al., 2013). However, in the case of type 2 diabetes, this action is disrupted by an insensitivity to insulin, which results in the inability of the body to appropriately lower the blood glucose level.

Adding to its danger, DM is a condition that can lead to various life-threatening complications. According to Tripathi and Srivastava (2006), the complications of diabetes can be generally divided into acute and chronic complications, with the latter being the main killer in diabetes cases.

 DOI: 10.1201/9781003354789-6

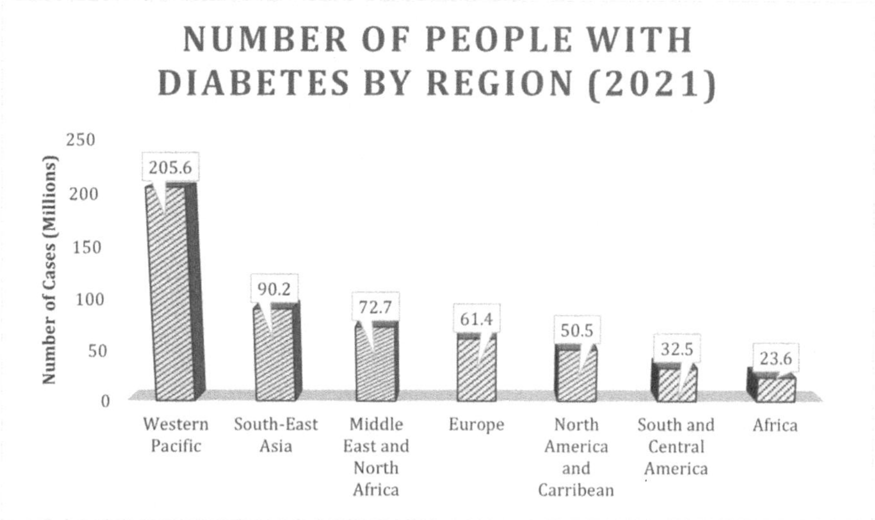

FIGURE 6.1 Prevalence of diabetes in the world by region in 2021 (International Diabetes Federation, 2021).

A common acute complication in elderly type 2 diabetic patients is hyperosmolar hyperglycemia (HHS), while the chronic complications include neuropathy (nerve damage), nephropathy (damaged kidney), diabetic retinopathy (retinal deterioration), vascular diseases, skin color changes, and even sexual dysfunction (Tripathi and Srivastava, 2006). HHS is a condition where the blood serum osmolality is too high, and this can lead to dehydration due to increased frequency of urination and glycosuria (presence of sugar in the urine) and also brain swelling (Adeyinka and Kondamudi, 2023). Moreover, patients with diabetes are at high risk of more severe and frequent infections due to the hyperglycemia-induced defects in cell-mediated immune responses and phagocyte functionality (Tripathi and Srivastava, 2006).

Currently, diabetes is treated with oral drugs and/or insulin therapies. Drugs that are prescribed for diabetic patients include the classes of sulfonylureas, glinides (meglitinides), biguanides, glitazones (thiazolidinediones), α-glucosidase inhibitors, and their combinations (Skyler, 2004). Meanwhile, insulin therapy is where synthetic insulin or an insulin analogue is injected into the patient's body regularly to mimic the activity of natural insulin to facilitate glycemic control. However, it should be noted that, as previously mentioned, type 2 DM mainly results from insulin resistance as a consequence of unhealthy lifestyles, such as obesity, alcohol usage (Tripathi and Srivastava, 2006), and a sedentary lifestyle (Skyler, 2004); hence, it is largely preventable or at least delayable.

Recent discoveries also demonstrate that there is a connection between DM and the diversity of the gut microbiota. Gurung et al. (2020) summarized the findings of multiple research that studied the effects of a certain genus of microbiota on type 2 diabetes. These studies suggest that some genera such as *Bacteroides, Akkermansia*, and *Bifidobacterium* have a beneficial effect against DM in humans. Meanwhile, the genera *Fusobacterium* and *Ruminococcus* display a positive association with DM (Gurung et al., 2020). The mechanism in which microbes play a role in DM varies greatly: some have been shown to promote inflammation, some do the opposite by stimulating anti-inflammatory agents, while others affect glucose metabolism by influencing gut hormones crucial to carbohydrate digestion (Gurung et al., 2020). However, this topic is still a fiercely debated one since multiple inconsistencies in the results were reported across different experiments. Although the notion of using gut microbiota in the treatment of diabetes is appealing and promising, more thorough research must be done to fully understand this subject.

6.2 ANTIDIABETIC ALTERNATIVES AND COMPLEMENTARY STRATEGIES

As implied earlier, leading a healthy lifestyle can be a key strategy in the prevention of diabetes. This includes a healthy diet, regular physical exercise, avoiding smoking, and moderate alcohol consumption. A study by Ley et al. (2016) also discussed that consuming fewer sugar-sweetened beverages (SSBs) and more intake of vegetables, fruits, legumes, and whole grains coupled with less red meat, as well as a moderate duration of sleep (between 5 and 9 hours a day) are associated with a lower risk of DM. Furthermore, some types of diets, such as vegetarian and Mediterranean diets, are also beneficial in glycemic control and overall diabetes management (Papamichou, Panagiotakos, and Itsiopoulos, 2019). Aside from prevention, lifestyle interventions are also able to help those who already have DM to manage their disease. Therapeutic fasting (intermittent fasting) is another proposed way of managing diabetes amidst the rising price of insulin. Fasting reduces the need for insulin, lowers the risk of further complications, and helps patients to lose weight, as obesity is one of the leading factors of DM (Furmli et al., 2018). This way, diabetic patients can manage their condition without excessive medications, thus decreasing the side effects of antidiabetic drugs at the same time.

In recent decades, people have started looking for natural substances and ingredients as alternatives to synthetic drugs for the treatment and prevention of diabetes, as they are cheaper and have fewer side effects than conventional antidiabetic drugs. For instance, some herbs have been shown to have the ability to treat diabetes. *Stevia rebaudiana* is one of the herbs that displayed its potential as a healthier substitute sweetener as well as for diabetic treatment (Ritu and Nandini, 2016). Other everyday plants such as garlic (*Allium sativum*), aloe vera (*Aloe barbadensis*), bitter gourd (*Momordica charantia*), ginger (*Zingiber officinale*), pomegranate (*Punica granatum*), and *Andrographis paniculata* have also been studied extensively for their active phytochemicals that have antidiabetic properties (Han et al., 2019).

Edible mushrooms are another group of natural sources that holds a promising prospect of being an alternative antidiabetic treatment. For millennia humans have cultivated mushrooms, and the demand for them is steadily increasing. Mushrooms are widely consumed as daily foods throughout the world, and they contain a multitude of vitamins such as B_2, niacin (B_3), and folates (B_9) as well as a good amount of minerals like K, P, Zn, and Cu (Mattila et al., 2001). The medicinal properties of mushrooms have also attracted a growing interest throughout the world in recent years, resulting in an increase in medicinal mushroom cultivation and consumption. Moreover, mushroom cultivation is relatively easy and has a low negative impact on the environment; hence, it is abundantly available. Research showed that some mushrooms contain compounds beneficial as antidiabetic agents. Dubey et al. (2019) summarized previous research on the antidiabetic properties of various dietary mushroom species and concluded that comatin, β-glucan, tremellastin, and lentinan KS-2 are examples of active compounds found in mushrooms that benefit diabetes management and obesity. Those species, among others, are *Agaricus campestris* (field mushroom), *Grifola frondosa* (maitake), *Cordyceps sinensis* (caterpillar fungus), *G. lucidum*, etc. (Dubey et al., 2019). The genus *Ganoderma*, which is also mentioned in Dubey et al. (2019), will be discussed in detail in the following sections of this chapter.

6.3 ANTIDIABETIC EFFECTS OF *GANODERMA* SPP.

Ganoderma is a genus of basidiomycete that has been used for its medicinal properties for millennia by mainly Asian civilizations. It is known by many names across different cultures, lingzhi in Chinese and munertake, reishi, and sachitake in Japanese culture, while Koreans call it youngzhi (Paterson, 2006).

Multiple research as summarized by Paterson (2006) demonstrated that the main bioactive compounds in various *Ganoderma* species are polysaccharides and triterpenoids, as well as some steroids. Most of these compounds are found highly concentrated in the basidiocarp (the cap) of the mushroom, but other parts of the fruiting body also contain it in relatively lower amounts. For example, in *G. lucidum*, the species believed to be referred to as lingzhi, along with *G. sinense*, is found to contain ganosporeric acid A, lucidenic acid A, ganoderic acids B and C, and ganoderic alcohols (Paterson, 2006).

Meanwhile, still referring to the same source (Paterson, 2006), researchers managed to isolate ganoderic acids E, C5, C6, G, and D from *G. tsugae*. Ganoderic acids are triterpenoids proposed as anticancer agents, as they were demonstrated to be able to induce apoptosis in cancer cells by inhibiting topoisomerase, which is crucial in DNA replication (Li et al., 2005). As for the polysaccharide content, Huie and Di (2004), determined from water extracts of 4 different *Ganoderma* species using high-performance liquid chromatography (HPLC) that the polysaccharides contained glucose, galactose, mannose, galacturonic acid, glucuronic acid, arabinose, xylose, fucose, and rhamnose. However, the most prominent one is glucose.

In traditional medicine, *Ganoderma* is claimed to be able to remedy all kinds of diseases. These diseases range from cancer to diabetes and cardiovascular disease, and even viral diseases such as HIV, although it should be noted that each claim must be studied clinically to be proven. In ancient China, the usage of lingzhi was recorded in *Shen Nong Ben Cao Jing*, or Shen Nong's Herbal Classics, a text compiling traditional medicines believed to be written during the Han Dynasty (between 206 BC and 220 AD), in which lingzhi is considered a precious medicine with the capability to prolong life, treat amnesia, and strengthen the stomach and spleen (Wang et al., 2020). In the past few decades, the myriad therapeutic prospects of *Ganoderma* were demonstrated scientifically, including its antidiabetic properties, which are the main focus of this chapter. Research and studies showing the potential of several *Ganoderma* species for the treatment of type 2 DM are summarized in Table 6.1.

6.3.1 Hypoglycemic Effect of *Ganoderma* spp.

Numerous studies demonstrated that the extracts and bioactive compounds from *Ganoderma* spp. are promising alternatives to be used as antidiabetic agents, and they came to a similar conclusion about the hypoglycemic effect of *Ganoderma* spp., especially *G. lucidum*. Shao et al. (2022) demonstrated that polysaccharide F31, extracted from *G. lucidum* which was determined to be nontoxic, displayed a hypoglycemic effect on streptozotocin-induced mice by observing the decrease in fasting blood glucose level. This result is in accordance with the findings of the experiments using different compounds from the respective mushroom such as the GLP, PsP, and GI powder.

The hypoglycemic effect is also shown in case of aqueous and methanolic extracts of *G. applanatum*, another species of *Ganoderma* spp., as reported by Hossain et al. (2021). Hyperglycemia in type 2 DM patients is mainly caused by insulin resistance in which the tissues become insensitive to insulin, causing less glucose uptake and utilization. *G. lucidum* in particular is found to have the ability to improve the insulin resistance condition in diabetic rats, which is indicated by the decrease in the Homeostatic Model Assessment for Insulin Resistance (HOMA-IR) and increase in the insulin sensitivity index (ISI) score (Heriansyah et al., 2019; Shao et al., 2022), as well as elevating insulin secretion by β-cells (Hossain et al., 2021; Shao et al., 2022). Liang et al. (2020) noted that the elevation of insulin secretion in INS-1 cells in vitro after the treatment using Fudan-Yueyang *G. lucidum* (FYGL), a proteoglycan, is due to the stimulation of PDX-1 protein expression.

Ma et al. (2015), summarized the antidiabetic mechanisms of polysaccharides from *G. lucidum*. They noted that the bioactive polysaccharide of *G. lucidum* stimulates glucokinase, phosphofructokinase, and glucose-6-phosphate dehydrogenase, which inhibits glucose production from the liver (Ma, Hsieh, and Chen, 2015). Overall, *G. lucidum* inhibits the process of glucose release from the liver; hence, the hypoglycemic effect. Moreover, research by Liu et al. (2019) demonstrated that the combination of polysaccharides from *G. lucidum* coupled with inulin, another proposed antidiabetic agent, is more efficient in upregulating the PI3K/Akt pathway than treatment only with inulin and has the ability to improve glycogen synthesis, as shown by Shao et al. (2022). PI3K/Akt is a signaling pathway that regulates many metabolic processes, including glucose homeostasis. The expression, or phosphorylation, of Akt increases glucose uptake by inducing GLUT4 translocation; in this case, polysaccharides of *G. lucidum* were shown to be able to stimulate Akt phosphorylation, hence enhancing glucose utilization and reducing insulin resistance (Liu et al., 2019).

TABLE 6.1
Summary of Previous Research on the Antidiabetic Effect of *Ganoderma* spp.

Species	Tested Compound	Experimental Model	Outcome Measurements	Results	Postulated Activity	Reference
G. lucidum	Nontoxic heteropolysaccharide F31	Male kunming mice induced with high-fat diet and streptozotocin (in vivo)	Fasting blood glucose Body and organ weight Serum insulin and biochemical parameters Hepatic glycogen and activities of antioxidant enzymes in liver Histopathological and immunohistochemical staining Gut microbiota analysis	F31 decreased fasting blood glucose F31 increased liver and kidney weight F31 increased HDL-C, lowered serum insulin, HOMA-IR, LDL-C, TC, TG, ALT, and AST Increased glycogen and GSH-Px and SOD activity F31 repaired liver cells and islet cells and improved insulin secretion F31 normalized activity of gut microbiota	Hypoglycemic, hepatoprotective, pancreatic protective and reparative, antioxidant effects	Shao et al., 2022
G. lucidum	*Ganoderma lucidum* polysaccharides (GLP)	8-week-old male SPF SD rats induced with streptozotocin (in vivo)	Blood test Urine protein, serum creatinine, and blood urea nitrogen Renal specimen staining (hematoxylin-eosin, Masson, periodic acid-Schiff) Western blotting (WB) and immunohistochemical staining (IHC)	GLP decreased fasting blood glucose and glycated hemoglobin GLP lowered 24 h urine protein, serum creatine, and blood urea nitrogen GLP reduced renal tissue injuries GLP decreased α-SMA, BNP, p-PI3K, p-Akt, p-mTOR, caspase-3, caspase-9, IL-6, IL-1β, TNF-α, and P62 expression in the kidney; meanwhile, beclin-1 and LC3 increased	Hypoglycemic effect, renal tissue reparative activity, anti-apoptosis, anti-inflammatory	Hu et al., 2022
G. lucidum	*Ganoderma lucidum* (Gl) powder suspended in water	Pregnant Wistar rats (*Rattus norvegicus*) induced with streptozotocin (in vivo)	Blood test Reproductive performance Maternal biochemical parameters Redox status Embryo-fetal development	Gl lowers glycemic level minutes after administration Rats treated with Gl have a heavier weight of the uterus, higher offspring vitality, and less post-implantation percentage of loss Gl lowered the level of AST and ALT Gl raised catalase activity, lowered reduced glutathione and TBARS GI protected the embryos from abnormal fetal measurements and hippocampal due to the induced diabetes	Inhibition of lipid peroxidase, antioxidant, hepatoprotective effect, apoptosis prevention by β-glucan, hypoglycemic activity	Viroel et al., 2022
G. lucidum	Polysaccharide peptide (PsP) of *G. lucidum*	8-week-old Wistar rats weighing 150–200 g induced with high-cholesterol diet and streptozotocin (in vivo)	H_2O_2 measurement Insulin resistance Lipid profile level EPC and CEC	PsP lowered H_2O_2 level PsP reduced ISI score in the DM rats PsP lowered TC and TG PsP lowered CEC and improved EPC, increased the ratio of EPC:CEC	Antioxidant, anti-inflammatory, hypolipidemic activity	Heriansyah et al., 2019

G. lucidum	Polysaccharide peptide (PsP) of *G. lucidum*	6- to 8-week-old Wistar rats weighing 100–200 g induced with a high-fat diet and streptozotocin (in vivo)	Lipid profile level H_2O_2 measurement Hematoxylin-eosin (HE) staining of vasa vasorum Lp-PLA2 analysis	PsP decreased TC and TG level PsP decreased H_2O_2 PsP reduced the formation of vasa vasorum in the rat's aorta PsP did not display a considerable reducing effect on Lp-PLA2 level	Antioxidant, antiangiogenic (antiatherogenic), hypolipidemic activity	Wihastuti et al., 2020
G. lucidum	Fudan-Yueyang *G. lucidum* (FYGL) proteoglycan	INS-1 rat insulinoma beta cells induced with streptozotocin (in vitro)	Cell viability Cell apoptosis assay Intracellular ROS assay Nitric oxide level ELISA for insulin measurement Western blot analysis	FYGL improved cell viability and growth FYGL protected INS-1 cells from apoptosis induced by STZ FYGL decreased the formation of ROS FYGL lowered the formation of NO in a dose-dependent manner FYGL improved insulin secretion FYGL inhibited the phosphorylation of NF-κB, inhibited the expression of cleaved caspase-3, Bax, and stimulated Bcl-2, inhibited the activation of p38 MAPK and JNK pathway, raised PDX-1 expression	Antioxidant, antiapoptotic effect	Liang et al., 2020
G. applanatum	Methanolic and aqueous extract of *G. applanatum*	Albino Wistar rats from both sexes weighing 180–200 g were injected with alloxan (in vivo)	Acute toxicity test Blood glucose level Body weight Serum lipid profile Serum liver function parameters Serum kidney function parameters Molecular docking	No rats died due to the extract and no behavioral, neurological, or autonomic changes were observed Both *G. applanatum* extracts lowered the blood glucose level The extracts elevated the body weight of the rats Both extracts reduced the level of TC, TG, and LDL *G. applanatum* extracts lowered the level of AST, ALT, and ALP No significant effect is observed in the kidney function parameters The bioactive compounds in *G. applanatum* have a good docking affinity toward dual PPARα/γ agonist protein, farnesoid X receptor (FXR) agonist protein, hepatitis C virus NS3/4A protease inhibitors, and human IgG Fc domain	Hypolipidemic, hypoglycemic, hepatoprotective activity	Hossain et al., 2021
G. australe & *G. neo–japonicum*	Ethanolic extract of wheat inoculated with mushroom mycelia	3T3–L1 mouse preadipocyte	MTT assay Oil Red O quantification assay Glycerol quantification assay RT-PCR	The extracts of fermented wheat stimulated cell proliferation The fermented grain extracts stimulated adipogenesis Inhibited epinephrine-induced lipolysis, displayed a slight lipolytic activity The ethanolic extract of wheat grains fermented with *G. neo-japonicum* raised the expression of PPARγ, GLUT4, and various adiponectin genes	Adipogenic, insulin-like properties	Subramaniam, Sabaratnam, and Kuppusamy 2015

Note: α-SMA: alpha-smooth muscle actin; ALP: alkaline phosphatase; ALT: alanine transaminase; AST: aspartate aminotransferase; BNP: brain natriuretic peptide; CEC: circulating endothelial cells; EPC: endothelial progenitor cell; GLUT4: glucose transporter type 4; GSH-px: glutathione peroxidase; HDL-C: high-density lipoprotein-cholesterol; HOMA-IR: Homeostatic Model Assessment for Insulin Resistance; IL: interleukin; ISI: insulin sensitivity index; LC3: microtubule-associated protein light chain 3; LDL-C: low-density lipoprotein–cholesterol; Lp-PLA2: lipoprotein-associated phospholipase A2; p-Akt: phosphorylated Akt (protein kinase B); PDX-1: pancreatic and duodenal homeobox 1; p-mTOR: phosphorylated mammalian target of rapamycin; PPARγ: peroxisome proliferator-activated receptors γ; p-PI3K: phosphorylated phosphoinositide 3-kinase; RT-PCR: real-time polymerase chain reaction; SOD: superoxide dismutase; TBARS: thiobarbituric acid reactive substances; TC: total cholesterol; TG: triglyceride; TNF: tumor necrosis factor

6.3.2 Protective and Reparative Effects on Tissues in Diabetic Conditions by *Ganoderma* spp.

Most of the studies mentioned in Table 6.1 also indicate that *Ganoderma* spp. has both protective and reparative effects on tissues in diabetic conditions. For instance, an elevation in aspartate alanine (ALT) and aspartate aminotransferase (AST) levels is expected in an injured liver. Treatment with *Ganoderma* lowered the ALT and AST levels in diabetic rats, indicating hepatoprotective activity (Hossain et al., 2021; Shao et al., 2022; Viroel et al., 2022). Research by Hu et al. (2022) on the effect of *G. lucidum* polysaccharides (GLP) in the treatment of diabetic nephropathy found that *G. lucidum* has renoprotective and reparative properties, evident by the improvement of the renal tissue, reduction of abnormal urine contents, and the adjustments of some biomarkers of renal tissue injury. The mechanisms are that the meroterpenoids in GLP reduced the expression of BNP and α-SMA which indicates the renoprotective activity,

GLP also inhibited the PI3k/Akt/mTOR pathway, hence promoting kidney autophagy (Hu et al., 2022). GLP also promotes renal tissue autophagy by stimulating the expression of beclin-1 and LC3, while inhibiting P62. Autophagy, or the cell's "cleaning out" process, is essential to the wellness of tissue by regulating damaged cell degradation. Next, GLP also inhibits proinflammatory agents such as IL-6, IL-1β, and TNF-α, showing that it has an anti-inflammatory response. This anti-inflammatory activity is also reported in other experiments, including by Heriansyah et al. (2019), who studied the antidiabetic effect of polysaccharides peptide (PsP) extracted from the same mushroom.

6.3.3 Anti–Apoptotic Effects by *Ganoderma* spp.

Apoptosis, or programmed cell death, is essentially a good mechanism to get rid of abnormal cells. However, hyperglycemia-induced apoptosis is a disadvantageous occurrence since it destroys cells impaired by the high glucose in the blood, such as the β-cells of the pancreas that leads to pancreatic dysfunction. Anti-apoptotic activities of the bioactive compounds of *G. lucidum* can be achieved through various mechanisms. According to Liang et al. (2020), FYGL prevents streptozotocin-induced cell apoptosis of INS-1 cells by inhibiting the NF-κB phosphorylation. Compounds in *G. lucidum* also display anti-apoptotic effects by inhibiting the caspase-3, caspase-9, and Bax expression, which is related to oxidative stress–induced apoptosis, as well as stimulating Bcl-2 protein, which inhibits apoptosis (Hu et al., 2022; Liang et al., 2020). ROS-induced apoptosis is also reduced by FYGL, by the inhibition of JNK and p38 MAPK pathway (Liang et al., 2020). Viroel et al. (2022), who studied the effect of *G. lucidum* powder in diabetic pregnant rats, stated that the β-glucan from *G. lucidum* exhibited an anti-apoptotic effect.

6.3.4 Beneficial Effects of *Ganoderma* spp. on Gestational Diabetes

Gestational diabetes is a condition where diabetes develops during pregnancy. This combination of diabetes and pregnancy is a dangerous one, as it may impose serious complications on the fetus and the mother as well. According to the Centers for Disease Control and Prevention (2022), gestational diabetes may cause birth defects, stillborn, premature birth, macrosomic baby, and a heightened risk of obesity and developing type 2 DM later in life. Viroel et al. (2022) reported that *G. lucidum* (GI) powder exhibited beneficial effects on gestational diabetic rats such as increasing offspring vitality and preventing abnormalities in the fetus's body measurements and brain. GI powder also increased the maternal reproductive capability, as shown by the increase in the ovary and uterus weight and the reduced percentage of post-implantation loss (Viroel et al., 2022). Furthermore, GI powder improved the mother's diabetic conditions by inducing hypoglycemic; hepatoprotective; and reparative, antioxidant, and hypolipidemic activity (Viroel et al., 2022).

6.3.5 Inhibitory Effects of Oxidative Stress by *Ganoderma* spp.

Many complications of diabetes are caused by oxidative stress and elevation of lipid profiles such as cholesterol and triglycerides. Bioactive compounds extracted from *Ganoderma* spp. are shown to combat these occurrences (Table 6.1). Oxidative stress is caused by the hyperproduction of superoxide by defective mitochondria due to hyperglycemia; this will lead to the overproduction of other reactive oxygen species (ROS), which will activate 5 pathways that are responsible for various diabetic complications. The said pathways are polyol pathway, advanced glycation end products (AGEs) formation, stimulation of AGE receptor (RAGE) ligand binding, increased activity of protein kinase C, and hexosamine pathway (Giacco and Brownlee, 2010). A major complication caused by ROS is cardiovascular diseases in which the macrovascular and microvascular systems are damaged. ROS is able to induce the formation of atherosclerosis plaque, which leads to endothelial dysfunction and heart disorders. Atherosclerosis is a plaque on the blood vessel wall formed by buildups of fats and cholesterols. Hence, the antioxidant and antilipidemic effects of *Ganoderma* are extensively studied.

Shao et al. (2022) demonstrated that F31 from *G. lucidum* successfully combats oxidative stress by stimulating the activity of glutathione peroxidase (GSH-Px) and superoxide dismutase (SOD), which are antioxidant enzymes. This effect is also shown by the administration of Gl powder in diabetic pregnant rats. Gl managed to inhibit lipid peroxidation, or the oxidation damage to lipids, by observing the decrease of thiobarbituric acid reactive substances (TBARSs), as well as increasing catalase (CAT) and GSH-Px activity (Viroel et al., 2022). Both PsP (Heriansyah et al., 2019; Wihastuti et al., 2020) and FYGL (Liang et al., 2020) from *G. lucidum* also displayed this antioxidant activity; hence it can be concluded that *G. lucidum* is a good alternative to combat ROS-induced diabetic complications.

6.3.6 Hypolipidemic Effects by *Ganoderma* spp.

Dyslipidemia, which is a condition of lipid imbalance, is quite common in diabetic patients, especially in DM. It is believed that the main cause of diabetic dyslipidemia is insulin resistance, which disrupts normal lipid metabolism (Wu and Parhofer, 2014). This may lead to complications such as atherosclerosis and elevated oxidative stress due to the increase of lipid oxidation. Extracts from *Ganoderma* spp., aside from improving insulin resistance, are proven to exhibit hypolipidemic properties by various studies. Both PsP from *G. lucidum* (Heriansyah et al., 2019; Wihastuti et al., 2020) and the extract of *G. applanatum* (Hossain et al., 2021) demonstrated that they are able to lower the total cholesterol (TC) and total triglyceride (TG) levels in diabetic mice. Huang et al. (2021) suggested that the hypolipidemic activities of *G. lucidum* are inhibiting acetyl-CoA carboxylase (ACC) and fatty acid synthase (FAS) in the liver, stimulating hormone-sensitive lipase (HSL) in the adipocytes, suppressing lipoprotein lipase (LPL) activity, and increasing triglyceride excretion through the feces. In short, *Ganoderma* works by slowing down lipid synthesis while speeding up lipid metabolism and excretion.

Wihastuti et al. (2020) reported that treatment with PsP managed to decrease vasa vasorum angiogenesis in the rat's aortic layer, indicating a lowered lipid profile. Vasa vasorum angiogenesis is caused by cell hypoxia, i.e., lack of oxygen supply, induced by the atherosclerotic plaque (Wihastuti et al., 2020). In turn, the body forms vasa vasorum, which are small blood vessels to compensate for the insufficiency of oxygen. Thus, PsP can be proposed as a beneficial compound for its antiatherosclerotic activity. This conclusion is further supported by the result of the study by Heriansyah et al. (2019). They showed that administration of PsP reduced the number of circulating endothelial cells (CECs) while increasing the endothelial progenitor cells (EPCs), effectively increasing the ratio of EPC:CEC. This ratio increase indicates that there is an improvement in the vascular repair process. A decrease in EPC can be associated with diabetic-induced insulin resistance complications such as increased ROS, lowered NO bioavailability, and defects in cellular signaling pathways (Desouza et al., 2011).

6.3.7 Regulation of Gut Microbiota by *Ganoderma* spp.

As discussed earlier, there is growing evidence of connections between gut microbiota with diabetes, especially DM. Shao et al. (2022) found that treatment with P31 of *G. lucidum* can reverse the changes induced by the high-fat diet and streptozotocin in the gut microbiota diversity of the mice and affects it to become similar to the normal gut microbiota composition. Bacteria have different metabolic activities; some can promote conditions associated with diabetes such as obesity and hyperlipidemia, and others can exhibit anti-inflammation and insulin resistance (Shao et al., 2022). In short, the treatment with F31 suppressed the growth of microbes that elevate diabetes and promote those that have an improving effect on diabetic symptoms. For instance, in the phylum level, F31 stimulates the abundance of *Bacteroides* and decreases *Firmicutes*, resulting in the elevated B/F ratio, which is associated with a lower risk of obesity (Shao et al., 2022). This highlights that other bioactive compounds in *Ganoderma* should be further researched for their effect on the gut microbiota composition to further understand this intriguing correlation.

6.3.8 Tapping Into *Ganoderma* Species Other than *G. lucidum*

The summary of the diabetes-combating activity of *Ganoderma* spp. is depicted in Figure 6.2. *Ganoderma applanatum*, another species of *Ganoderma*, is not as well studied as *G. lucidum*.

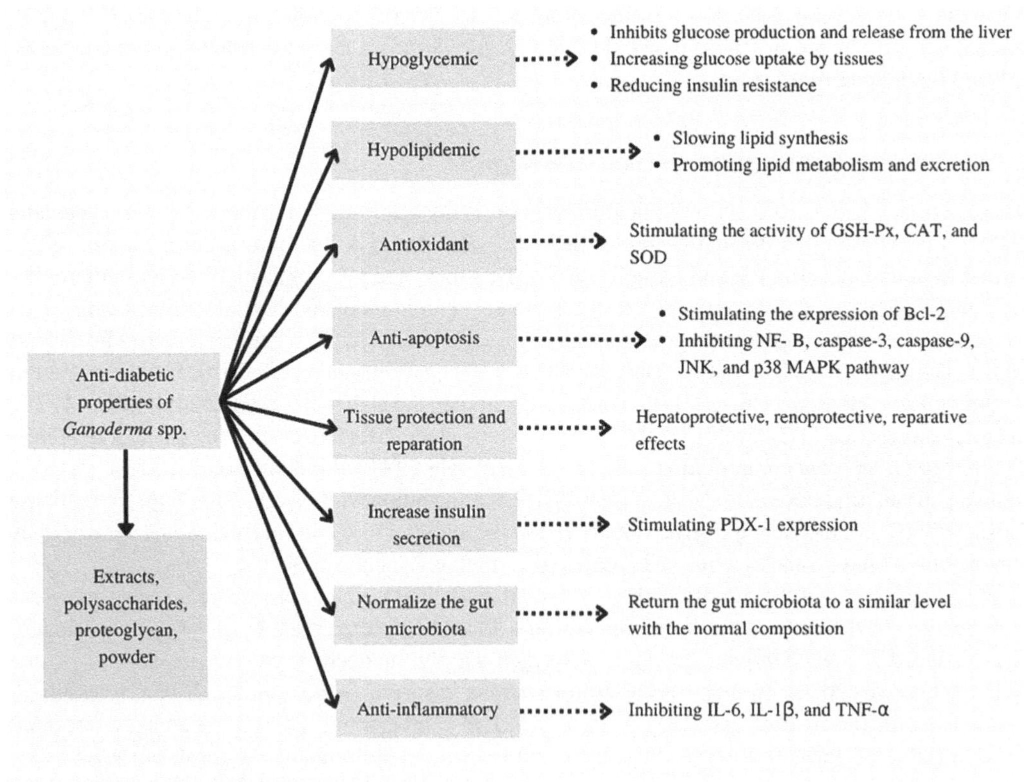

FIGURE 6.2 Summary of some proposed antidiabetic activities and mechanisms of bioactive compounds from *Ganoderma* spp. CAT, catalase; GSH-Px, glutathione peroxidase; IL-1β, interleukin 1 beta; IL-6, interleukin 6; JNK, c-Jun N-terminal kinase; MAPK, mitogen-activated protein kinase; NF-κB, nuclear factor kappa-light-chain-enhancer of activated B cells; PDX-1, pancreatic and duodenal homeobox 1; SOD, superoxide dismutase; TNF-α, tumor necrosis factor-alpha.

Nonetheless, it also exhibits some beneficial effects on health, including antidiabetic. Hossain et al. (2021) reported that both methanolic and aqueous extracts of *G. applanatum* induce hepatoprotective, hypolipidemic, and hypoglycemic activities in alloxan-induced diabetic mice. Thus, more research on the antidiabetic properties of *G. applanatum* is expected to be done in the future.

Hossain et al. (2021) further expanded their research by performing a molecular analysis to determine the target proteins for *G. applanatum*'s antidiabetic activity. They found that the bioactive compounds in *G. applanatum*, especially Myrocin C, have a good affinity to the 4 tested protein targets, namely dual PPARα/γ agonist protein, farnesoid X receptor (FXR) agonist protein, hepatitis C virus NS3/4A protease inhibitors, and human IgG Fc domain (Hossain et al., 2021). The dual PPARα/γ agonist has a key role in elevating insulin sensitivity and preventing the occurrence of diabetic-induced cardiovascular diseases (Balakumar et al., 2007). Next, FXRs are receptors that play a crucial part in lipid metabolism; its activation showed improving effects for hyperlipidemia and insulin resistance (Ali et al., 2015). Meanwhile, the hepatitis C virus NS3/4A protease inhibitors and human IgG Fc domain were both tested for their hepatoprotective properties (Hossain et al., 2021). These findings imply that phytochemicals found in *Ganoderma* are suitable to be used as agonists for antidiabetic agents, and thus their usage as alternative treatments may help reduce the side effects caused by regular drugs.

Other members of the *Ganoderma* family that are less known are also highly promising to be used for potential treatment. A study by Subramaniam et al. (2015) showed that *G. australe* and *G. neo-japonicum*, which are rare mushrooms found in Malaysia, are beneficial in terms of diabetic mitigation, especially *G. neo-japonicum*. As mentioned in Table 6.1, the ethanolic extract of wheat that had been inoculated with mushroom mycelia showed insulin-like properties, which are useful for the treatment of adipocyte dysfunction and, ultimately, DM. It is postulated that phenolic compounds are the bioactive compounds responsible for adipogenic activity since wheat fermented with *G. neo-japonicum* mycelia showed a higher level of phenolics (Subramaniam, Sabaratnam, and Kuppusamy, 2015). Even though the wheat grains fermented with *G. neo-japonicum* demonstrated that it could inhibit epinephrine-induced lipolysis, just like insulin, it is also by itself moderately lipolytic. A strong epinephrine-induced lipolysis-inhibiting agent is also not preferred, since it can lead to obesity and insulin resistance due to the tight prevention of lipid mobilization (Subramaniam, Sabaratnam, and Kuppusamy, 2015). This peculiar characteristic shown by the ethanolic extract of wheat grain fermented with *G. neo-japonicum* is favorable for use as a therapeutic agent, since it can potentially both stimulate lipolysis in mature adipocytes and also promote lipogenesis in differentiating adipocytes (Subramaniam, Sabaratnam, and Kuppusamy, 2015). Overall, it can be concluded that *G. neo-japonicum* is a better candidate for the treatment of DM as compared to *G. australe*. That being said, more extensive studies need to be carried out to deduce the exact molecular mechanism of the mentioned beneficial activities.

6.4 CHALLENGES

There are numerous issues and challenges common to most bioactives, including those from *Ganoderma* spp., such as the bioavailability, secondary metabolites, toxicity, and adverse effects that exist when considering the efficacies in humans. Additional challenges include mushroom authentication, acquisition/source, cultivation, extract preparation, and other related issues.

6.4.1 Issues Pertaining to the Mushroom

Due to the high morphological variability of *Ganoderma* caused by geographic and climatic influences, there is controversy over its taxonomical status resulting in a confusing nomenclature within the genus (Richter et al., 2015). Furthermore, the source of the mushroom is an issue, such as the environment where it is foraged or cultivated and the maturity of the fruiting body, which can determine the level of the bioactive compounds. For example, Tan et al. (2015) showed that the aqueous and ethanol extracts of the wild *G. neo-japonicum* mushroom showed stronger antioxidant properties than the cultivated mushroom. Generally, the polysaccharides of *Ganoderma* spp. have shown promising effects against hyperglycemia and associated complications such as dyslipidemia, inflammation, and oxidative stress (Hossain et al., 2021; Shao et al., 2022; Viroel et al., 2022). Polysaccharides such as the β-glucans have been shown to possess different degrees of branching, monosaccharide composition, linkage ratio, and type attributed to the different molecular weights and concentrations of the active component, depending on the extraction methods (Murphy et al., 2021). The feature and the composition of the polysaccharides extracted can vary due to the difference in the extraction method despite being from the same source of mushroom. It has been reported that the yield of β-glucan obtained by hot water extraction from *G. lucidum* is low but increases when subjected to a higher temperature, supercritical fluids, and high-pressure steaming (Hwang et al., 2018). This may have a very significant effect on the intestinal gut microbiota, bioavailability, and potency. Aqueous or polysaccharide extracts of *Ganoderma* spp. rich in β-1,3/1,6 glucan as well as proteoglycans and heteroglycans are generally considered to be indigestible and are known to act as a probiotic (Carrieri et al., 2017). However, it is also known that since β-glucans are pathogen–associated molecular pattern (PAMP) molecules, a small fraction of soluble β-glucans can be translocated into the blood circulation via antigen-presenting cells (APCs) and macrophages present in Peyer's patches (PPs) of the intestinal mucosa (Rossi et al., 2018; Sandvik et al., 2007) One important consideration to ensure a consistent composition and concentration of the active components is standardized cultivation, the use of mycelia grown under specific standardized conditions, and a standardized extraction procedure. A quality control check on the concentration of the active component would prevent the potential excessive variability in the efficacy observed. Other issues pertaining to storage, shelf life, and contamination with toxic/hazardous compounds leached from the cultivation bags have to be avoided and monitored to ensure safety. Standardization of mushroom supplement production throughout the supply chain, from cultivation to extraction and commercial formulation preparation, as well as precise monitoring and regulation, are also required to ensure high-quality levels and consistent results.

6.4.2 Lack of Clinical Trials and Challenges Involved

Despite the identification of numerous bioactive compounds in extracts from *Ganoderma* spp., with antidiabetic potencies determined in laboratory models, clinical research on the beneficial effects of *Ganoderma* spp. is limited, especially for its antidiabetic effects.

Most of the clinical trials of *Ganoderma* are pertaining to antiaging, immunomodulatory, anticancer, or antidyslipidemia effects (Henao et al., 2018; Zeng et al., 2018). Ideally, placebo-controlled, blind, multicenter clinical studies with a high number of participants must be conducted for better and wider acceptance of this mushroom for the management of diabetes. The appropriate effective dose is not clear, as the level of tolerance of each individual can vary (Parveen et al., 2015). The adverse events following treatment with *Ganoderma* have not been well documented. The right doses and toxicity profiles of various species of *Ganoderma* have not been ideally worked out, and this gap needs to be filled. A pilot study on healthy subjects would be necessary to ensure safety and tolerance. The results obtained can determine the appropriate dose for interventional clinical trials. Currently, regulatory bodies require that clinical trials include double-blinded, randomized, placebo-controlled, long-term trials with large numbers of patients, taking into consideration the appropriate inclusion and exclusion criteria (Brown et al., 2018). Such trials are not only costly but the chances for successful completion are challenging due to poor compliance, high drop-out rates, logistic issues, and potential issues in maintaining the stability of the product (Brown et al., 2018; Parveen et al., 2015)

6.5 CONCLUSION

The polysaccharides of *Ganoderma* spp. are generally useful in the prevention and/or management of DM. *Ganoderma lucidum* is the most widely studied and established as a health supplement, mainly as an immunomodulator. However, its use as an antidiabetic agent will gain better credibility only with a well-controlled clinical trial with a large number of participants. Other *Ganoderma* spp. can also be explored and cultivated to upscale their production. To date the efficacy of the secondary metabolites from *Ganoderma* spp. on the antidiabetic property is underexplored, and this deserves attention. It is also imperative that while advocating for the use of *Ganoderma* as a functional food or nutraceutical in the management of diabetes (as an alternative or adjuvant therapy), the public must be made aware of any potential safety issues and adverse effects. In conclusion, to successfully manage diabetes and delay the onset of complications, it is crucial to educate the public that other health practices, such as exercise, calorie restriction, appropriate stress management, and adequate sleep, are just as important.

REFERENCES

Adeyinka, A., and N. P. Kondamudi. 2023. Hyperosmolar Hyperglycemic Syndrome. *StatPearls*. www.ncbi.nlm.nih.gov/pubmed/25342831.

Ali, A. H., E. J. Carey, and K. D. Lindor. 2015. Recent Advances in the Development of Farnesoid X Receptor Agonists. *Ann Transl Med* 3 (1):5. doi:10.3978/j.issn.2305–5839.2014.12.06.

American Diabetes Association. 2021. Classification and Diagnosis of Diabetes: Standards of Medical Care in Diabetes. *Diabetes Care* 44 (Suppl.):S15–S33.

Balakumar, P., M. Rose, S. Ganti, P. Krishan, and M. Singh. 2007. PPAR Dual Agonists: Are They Opening Pandora's Box? *Pharmacol Res* 56 (2):91–98. doi:10.1016/j.phrs.2007.03.002.

Brown, L., S. P. B. Caligiuri, D. Brown, and G. N. Pierce. 2018. Clinical Trials Using Functional Foods Provide Unique Challenges. *J Funct Foods* 45 (June):233–238. doi:10.1016/j.jff.2018.01.024.

Carrieri, R., R. Manco, D. Sapio, M. Iannaccone, A. Fulgione, M. Papaianni, B. de Falco, L. Grauso, P. Tarantino, F. Ianniello, et al. 2017. Structural Data and Immunomodulatory Properties of a Water-Soluble Heteroglycan Extracted from the Mycelium of an Italian Isolate of *Ganoderma lucidum*. *Nat Prod Res* 31 (18):2119–2125. doi:10.1080/14786419.2017.1278593.

Centers for Disease Control and Prevention. 2022. Diabetes During Pregnancy. *CDC*. www.cdc.gov/reproductivehealth/maternalinfanthealth/diabetes-during-pregnancy.htm#:~:text=Diabetes during pregnancy-including type,%2C stillbirth%2C and preterm birth (accessed September 1, 2022).

Desouza, C. V., F. G. Hamel, K. Bidasee, and K. O'Connell. 2011. Role of Inflammation and Insulin Resistance in Endothelial Progenitor Cell Dysfunction. *Diabetes* 60 (4):1286–1294. doi:10.2337/db10–0875.

Dubey, S. K., V. K. Chaturvedi, D. Mishra, A. Bajpeyee, A. Tiwari, and M. P. Singh. 2019. Role of Edible Mushroom as a Potent Therapeutics for the Diabetes and Obesity. *3 Biotech* 9 (12):450. doi:10.1007/s13205-019-1982-3.

Furmli, S., R. Elmasry, M. Ramos, and J. Fung. 2018. Therapeutic Use of Intermittent Fasting for People with Type 2 Diabetes as an Alternative to Insulin. *BMJ Case Rep* (October), bcr–2017–221854. doi:10.1136/bcr–2017–221854.

Giacco, F., and M. Brownlee. 2010. Oxidative Stress and Diabetic Complications. Edited by Ann Marie Schmidt. *Circulation Res* 107 (9):1058–1070. doi:10.1161/CIRCRESAHA.110.223545.

Gurung, M., Z. Li, H. You, R. Rodrigues, D. B. Jump, A. Morgun, and N. Shulzhenko. 2020. Role of Gut Microbiota in Type 2 Diabetes Pathophysiology. *EBioMedicine* 51 (January):102590. doi:10.1016/j.ebiom.2019.11.051.

Han, D.-G., S. -S. Cho, J.-H. Kwak, and I.-S. Yoon. 2019. Medicinal Plants and Phytochemicals for Diabetes Mellitus: Pharmacokinetic Characteristics and Herb-Drug Interactions. *J Pharm Investig* 49 (6):603–612. doi:10.1007/s40005-019-00440-4.

Henao, S. L. D., S. A. Urrego, A. M. Cano, and E. A. Higuita. 2018. Randomized Clinical Trial for the Evaluation of Immune Modulation by Yogurt Enriched with β-Glucans from Lingzhi or Reishi Medicinal Mushroom, *Ganoderma lucidum* (Agaricomycetes), in Children from Medellin, Colombia. *Int J Med Mushrooms* 20 (8):705–716. doi:10.1615/IntJMedMushrooms.2018026986.

Heriansyah, T., W. Nurwidyaningtyas, D. Sargowo, C. T. Tjahjono, and T. A. Wihastuti. 2019. Polysaccharide Peptide (PsP) *Ganoderma lucidum*: A Potential Inducer for Vascular Repair in Type 2 Diabetes Mellitus Model. *Vascular Health Risk Manag* 15 (October):419–427. doi:10.2147/VHRM.S205996.

Hossain, M. S., A. Barua, M. A. H. Tanim, M. S. Hasan, M. J. Islam, M. R. Hossain, N. U. Emon, and S. M. M. Hossen. 2021. *Ganoderma applanatum* Mushroom Provides New Insights into the Management of Diabetes Mellitus, Hyperlipidemia, and Hepatic Degeneration: A Comprehensive Analysis. *Food Sci Nutr* 9 (8):4364–4374. doi:10.1002/fsn3.2407.

Hu, Y., S.-X. Wang, F.-Y. Wu, K.-J. Wu, R.-P. Shi, L.-H. Qin, C.-F. Lu, S.-Q. Wang, F.-F. Wang, and S. Zhou. 2022. Effects and Mechanism of *Ganoderma lucidum* Polysaccharides in the Treatment of Diabetic Nephropathy in Streptozotocin-Induced Diabetic Rats. Edited by Abdul Rehman Phull. *BioMed Res Int* 2022 (March):1–13. doi:10.1155/2022/4314415.

Huang, C.-H., W.-K. Lin, S.-H. Chang, and G.-J. Tsai. 2021. *Ganoderma lucidum* Culture Supplement Ameliorates Dyslipidemia and Reduces Visceral Fat Accumulation in Type 2 Diabetic Rats. *Mycology* 12 (2):94–104. doi:10.1080/21501203.2020.1740409.

Huie, C. W., and X. Di. 2004. Chromatographic and Electrophoretic Methods for Lingzhi Pharmacologically Active Components. *J Chromatogr B* 812 (1–2):241–257. doi:10.1016/j.jchromb.2004.08.038.

Hwang, I.-W., B.-M. Kim, Y.-C. Kim, S.-H. Lee, and S.-K. Chung. 2018. Improvement in β-Glucan Extraction from *Ganoderma lucidum* with High-Pressure Steaming and Enzymatic Pre-Treatment. *Appl Biol Chem* 61 (2):235–242. doi:10.1007/s13765-018-0350-z.

International Diabetes Federation. 2021. *IDF Diabetes Atlas*. Edited by IDF. 10th ed. www.diabetesatlas.org/ (accessed September 1, 2022).

Kaul, K., J. M. Tarr, S. I. Ahmad, E. M. Kohner, and R. Chibber. 2013. Introduction to Diabetes Mellitus. In *Diabetes. Advances in Experimental Medicine and Biology*, 1–11. Springer. doi:10.1007/978-1-4614-5441-0_1.

Ley, S. H., A. V. A. Korat, Q. Sun, D. K. Tobias, C. Zhang, L. Qi, W. C. Willett, J. E. Manson, and F. B. Hu. 2016. Contribution of the Nurses' Health Studies to Uncovering Risk Factors for Type 2 Diabetes: Diet, Lifestyle, Biomarkers, and Genetics. *Am J Public Health* 106 (9):1624–1630. doi:10.2105/AJPH.2016.303314.

Li, C.-H., P.-Y. Chen, U.-M. Chang, L.-S. Kan, W.-H. Fang, K.-S. Tsai, and S.-B. Lin. 2005. Ganoderic Acid X, a Lanostanoid Triterpene, Inhibits Topoisomerases and Induces Apoptosis of Cancer Cells. *Life Sci* 77 (3):252–265. doi:10.1016/j.lfs.2004.09.045.

Liang, H., Y. Pan, Y. Teng, S. Yuan, X. Wu, H. Yang, and P. Zhou. 2020. A Proteoglycan Extract from *Ganoderma lucidum* Protects Pancreatic Beta-Cells against STZ-Induced Apoptosis. *Biosci Biotechnol Biochem* 84 (12):2491–2498. doi:10.1080/09168451.2020.1805718.

Liu, Y., Y. Li, W. Zhang, M. Sun, and Z. Zhang. 2019. Hypoglycemic Effect of Inulin Combined with *Ganoderma lucidum* Polysaccharides in T2DM Rats. *J Funct Foods* 55 (April):381–390. doi:10.1016/j.jff.2019.02.036.

Ma, H.-T., J.-F. Hsieh, and S.-T. Chen. 2015. Anti-Diabetic Effects of *Ganoderma Lucidum*. *Phytochem* 114 (June):109–113. doi:10.1016/j.phytochem.2015.02.017.

Mattila, P., K. Könkö, M. Eurola, J.-M. Pihlava, J. Astola, L. Vahteristo, V. Hietaniemi, J. Kumpulainen, M. Valtonen, and V. Piironen. 2001. Contents of Vitamins, Mineral Elements, and Some Phenolic Compounds in Cultivated Mushrooms. *J Agric Food Chem* 49 (5):2343–2348. doi:10.1021/jf001525d.

Murphy, E. J., E. Rezoagli, I. Major, N. Rowan, and J. G. Laffey. 2021. β-Glucans. *Encyclopedia* 1 (3):831–847. doi:10.3390/encyclopedia1030064.

Papamichou, D., D. B. Panagiotakos, and C. Itsiopoulos. 2019. Dietary Patterns and Management of Type 2 Diabetes: A Systematic Review of Randomised Clinical Trials. *Nutr Metab Cardiovasc Dis* 29 (6):531–543. doi:10.1016/j.numecd.2019.02.004.

Parveen, A., B. Parveen, R. Parveen, and S. Ahmad. 2015. Challenges and Guidelines for Clinical Trial of Herbal Drugs. *J Pharm Bioallied Sci* 7 (4):329. doi:10.4103/0975–7406.168035.

Paterson, R. R. M. 2006. *Ganoderma*—A Therapeutic Fungal Biofactory. *Phytochem* 67 (18):1985–2001. doi:10.1016/j.phytochem.2006.07.004.

Rehman, K., and M. S. H. Akash. 2017. Mechanism of Generation of Oxidative Stress and Pathophysiology of Type 2 Diabetes Mellitus: How Are They Interlinked? *J Cell Biochem* 118 (11):3577–3585. doi:10.1002/jcb.26097.

Richter, C., K. Wittstein, P. M. Kirk, and M. Stadler. 2015. An Assessment of the Taxonomy and Chemotaxonomy of *Ganoderma*. *Fungal Divers* 71 (1):1–15. doi:10.1007/s13225-014-0313-6.

Ritu, M., and J. Nandini. 2016. Nutritional Composition of *Stevia rebaudiana*, a Sweet Herb, and Its Hypoglycaemic and Hypolipidaemic Effect on Patients with Non-Insulin Dependent Diabetes Mellitus. *J Sci Food Agric* 96 (12):4231–4234. doi:10.1002/jsfa.7627.

Rossi, P., R. Difrancia, V. Quagliariello, E. Savino, P. Tralongo, C. L. Randazzo, and M. Berretta. 2018. B-Glucans from *Grifola frondosa* and *Ganoderma lucidum* in Breast Cancer: An Example of Complementary and Integrative Medicine. *Oncotarget* 9 (37):24837–24856. doi:10.18632/oncotarget.24984.

Sandvik, A., Y. Y. Wang, H. C. Morton, A. O. Aasen, J. E. Wang, and F.-E. Johansen. 2007. Oral and Systemic Administration of β-Glucan Protects Against Lipopolysaccharide-Induced Shock and Organ Injury in Rats. *Clin Exp Immunol* 148 (1):168–177. doi:10.1111/j.1365–2249.2006.03320.x.

Shao, W., C. Xiao, T. Yong, Y. Zhang, H. Hu, T. Xie, R. Liu, L. Huang, X. Li, Y. Xie, et al. 2022. A Polysaccharide Isolated from *Ganoderma Lucidum* Ameliorates Hyperglycemia Through Modulating Gut Microbiota in Type 2 Diabetic Mice. *Int J Biol Macromolecules* 197 (February):23–38. doi:10.1016/j.ijbiomac.2021.12.034.

Skyler, J. S. 2004. Diabetes Mellitus: Pathogenesis and Treatment Strategies. *J Med Chem* 47 (17):4113–4117. doi:10.1021/jm0306273.

Subramaniam, S., V. Sabaratnam, and U. R. Kuppusamy. 2015. Solid-Substrate Fermentation of Wheat Grains by Mycelia of Indigenous *Ganoderma* Spp. Enhanced Adipogenesis and Modulated PPARγ Expression in 3T3-L1 Cells. *Chiang Mai J Sci* 42 (2):269–281.

Tan, W.-C., U. R. Kuppusamy, C.-W. Phan, Y.-S. Tan, J. Raman, A. M. Anuar, and V. Sabaratnam. 2015. *Ganoderma Neo-Japonicum* Imazeki Revisited: Domestication Study and Antioxidant Properties of Its Basidiocarps and Mycelia. *Sci Rep* 5. doi:10.1038/srep12515.

Tripathi, B. K., and A. K. Srivastava. 2006. Diabetes Mellitus: Complications and Therapeutics. *Med Sci Monit* 12 (7):RA130–47. www.ncbi.nlm.nih.gov/pubmed/16810145.

Viroel, F. J. M., L. F. Laurino, É. L. A. Caetano, A. F. Jozala, S. R.V. Spim, T. B. Pickler, M. K. Sercundes, M. C. Gomes, A. Hataka, D. Grotto, et al. 2022. *Ganoderma Lucidum* Modulates Glucose, Lipid Peroxidation and Hepatic Metabolism in Streptozotocin-Induced Diabetic Pregnant Rats. *Antioxidants* 11 (6):1035. doi:10.3390/antiox11061035.

Wang, L., J.-Q. Li, J. Zhang, Z.-M. Li, H.-G. Liu, and Y.-Z. Wang. 2020. Traditional Uses, Chemical Components and Pharmacological Activities of the Genus *Ganoderma* P. Karst.: A Review. *RSC Advances* 10 (69):42084–42097. doi:10.1039/D0RA07219B.

Wihastuti, T. A., R. Amiruddin, F. Y. Cesa, A. I. Alkaf, M. Setiawan, and T. Heriansyah. 2020. Decreasing Angiogenesis Vasa Vasorum through Lp-PLA2 and H2O2 Inhibition by PSP from *Ganoderma Lucidum* in Atherosclerosis: *In vivo* Diabetes Mellitus Type 2. *J Basic Clin Physiol Pharmacol* 30 (6). doi:10.1515/jbcpp–2019–0349.

Wilcox, G. 2005. Insulin and Insulin Resistance. *Clin Biochem Rev* 26 (2):19–39. doi:10.1007/BF01945767.

World Health Organization. 2022a. *Diabetes*. www.who.int/news-room/fact-sheets/detail/diabetes (accessed September 1, 2022).

World Health Organization. 2022b. *Non Communicable Diseases*. www.who.int/news-room/fact-sheets/detail/noncommunicable-diseases#:~:text=The main types of NCD, disease and asthma (accessed September 1, 2022).

Wu, L., and K. G. Parhofer. 2014. Diabetic Dyslipidemia. *Metabolism* 63 (12):1469–1479. doi:10.1016/j.metabol.2014.08.010.

Zeng, P., Z. Guo, X. Zeng, C. Hao, Y. Zhang, M. Zhang, Y. Liu, H. Li, J. Li, and L. Zhang. 2018. Chemical, Biochemical, Preclinical and Clinical Studies of *Ganoderma Lucidum* Polysaccharide as an Approved Drug for Treating Myopathy and Other Diseases in China. *J Cell Mol Med* 22 (7):3278–3297. doi:10.1111/jcmm.13613.

7 Ganoderma
A Pharmacological Mushroom with Remarkable Potency in Human Gut Microflora Dysbiosis

Supratim Mandal[1] *and Adhiraj Roy*[2]
1 University of Kalyani, Kalyani, India
2 Amity University Noida, Noida, India

7.1 INTRODUCTION

The successful establishment of a human-microbial relationship benefits the maintenance of a healthy life for decades. Since ancient times, the investigation of such companionship has been deeply studied, and the success finally came out by developing the Human Microbiome Project (Human Microbiome Project Consortium, 2012). The detailed analysis reveals that the microbial consortium in the human intestinal region significantly contributes to human health and different disease circumstances. The human gut microenvironment accounts for around 100 trillion microorganisms that outnumber the human body by 10 factors (Li et al., 2016). Genetic factors, dietary patterns, age, and even geographical locations modulate the establishment of varied gut microflora. *Bacteroidetes*, *Firmicutes*, *Acinetobacteria*, *Proteobacteria*, and *Verrumicrobia* are the predominant phyla within the human gut regions. Among these, the *Firmicutes/Bacteroidetes* ratio acts as a key marker of a healthy individual in community settings (Patterson et al., 2016; Gomaa, 2020). Disruption of gut microenvironments restricts the secretion of important metabolites and the digestion process, leading to the development of lethal maladies. Mounting evidence has reported that a decreased *Firmicutes/Bacteroidetes* ratio is associated with several disease ailments such as obesity, diabetes, inflammation, and cancer. On top of that, the indiscriminate use of antibiotics has modified the gut microflora community and also their phenotypic characteristics, rendering them resistant to multiple antibiotics (multidrug resistant) (Sittipo et al., 2018; Hollister and Gao, 2012). Such complex scenarios have attracted scientists to explore novel natural alternatives in the management of stable human intestinal microbial habitats.

Ganoderma, a medicinal mushroom, has been traditionally used for centuries in Asia to combat different public health hazards with improved pharmacokinetic efficacies. *Ganoderma* is considered a woody Basidiomycetes mushroom under the Ganodermaceae family of Aphyllophoral (Hsu and Cheng, 2018). The extraction of bioactive polysaccharides, triterpenoids from fruit bodies, spores, and mycelium of *Ganoderma lucidum* and *Ganoderma sinense* has been proven for its significant activity against metabolic alterations, diabetes, obesity, tumor cell proliferation, invasion, metastasis, etc. Accumulating reports have evinced that bioactive chemical compounds from *Ganoderma* could modulate the human gut microflora and thereby limit the progression of different lethal maladies, including cancer (Kladar et al., 2016; Hsu and Cheng, 2018). For instance, a triterpenoid ganoderic acid (GA) derived from *Ganoderma* could control the abnormal lipid metabolism and oxidative stress in the liver of mice exposed to high alcohol consumption. GA could attenuate the accumulation of fatty liver and alcohol-induced liver injury by decreasing the amount of lactate dehydrogenase and maleic dialdehyde while stimulating the proportions of catalase, glutathione, superoxide dismutase, and alcohol dehydrogenase (Cao et al., 2022). Accordingly, the GA supplementation stimulated the

DOI: 10.1201/9781003354789-7

abundance of *Fecalibaculum*, *Bifidobacterium*, *Lactobacillus*, and *Romboutsia* while reducing the presence of *Heliobacter*, which was completely reversed during excessive alcohol intake (Cao et al., 2022). In accordance with other reports, it can be inferred that *Ganoderma* serves as a potent alternative in the prevention and treatment of different types of pathogenesis ranging from metabolic disorders to carcinoma. Here, we have discussed the overall chemical constituents of *Ganoderma*, including polysaccharides, terpenoids, and proteins. The overall idea about gut flora and its importance in the maintenance of human health has also been discussed. More importantly, we have summarized the role of *Ganoderma* and its chemical compounds in the modulation of human gut microbiota and their efficacy in restricting the progression of lethal maladies.

7.2 CHEMICAL CONSTITUENTS OF *GANODERMA*

While 90% of the mushroom's mass is derived from watery contents, the remaining 10% constitutes carbohydrate (3–28%), proteins (10–40%), fiber (3–32%), fat (28%), ash (8–10%), minerals, and vitamins. Mounting evidence has revealed that *Ganoderma*, an important pharmacological mushroom, possesses an array of bioactive molecules such as polysaccharides, terpenoids, steroids, nucleotides, phenolic compounds, etc. Moreover, the higher percentage of polyunsaturated fatty acids in *Ganoderma* provides significant health benefits to consumers (Cör et al., 2018; Kladar et al., 2016). This section discusses the major chemical compounds of *Ganoderma* bearing potential nutraceutical values.

7.2.1 Polysaccharides

Polysaccharides are integral constituents of *Ganoderma* with potential bioactivities such as antimicrobial, antitumor, antidiabetic, and immunomodulation. Most of the polysaccharides are derived from fruiting bodies, spores, and mycelia of *Ganoderma*, termed intracellular polysaccharides (from fruit bodies and spores), and exopolysaccharides (from mycelia) (Hsu and Cheng, 2018). Hot water extraction is one of the most prominent methods for polysaccharide isolation from *Ganoderma*, which further follows methanol or ethanol precipitation as well as alkali and water extraction. However, other practices, including microwave, ultrasonic, and enzymatic approaches, are increasingly used that could permeate the cell wall of *Ganoderma* and enhance the yield of polysaccharides (Ferreira et al., 2015). Heteropolysaccharides belong to one of the largest classes of *Ganoderma* polysaccharides wherein glucose, mannose, galactose, and fucose serve as backbones. Most of them possess 1-3, 1-6, and 1-4 α or β linkages between monomers, developing glucans, glycoproteins, and glycopeptides. Among them β 1-3 glucans are receiving a lot of attention as they can modulate immune functions by activating macrophages and natural killer cells and acting as a pathogen-associated molecular pattern (Ferreira et al., 2015; Hsu and Cheng, 2018). Several epidemiological studies have reported that *Ganoderma* polysaccharides could exert potential anticancer activity by being administered to patients on chemotherapy with immune-suppressed circumstances. Zhong *et al.*, found that sporoderm-removed spores of *Ganoderma lucidum* attenuated gastric cancer cell viability by impairing the expression of pro-apoptotic factors (Zhong et al., 2021). In another study, polysaccharides isolated from *Ganoderma* have shown remarkable antidiabetic activity by modulating the gut microbiome in type 2 diabetic mice (Shao et al., 2022). However, further research is highly warranted to explore more bioactive polysaccharides and their detailed characterization with improved pharmaceutical efficacies on human health.

7.2.2 Triterpenes

Triterpenes are another subclass of *Ganoderma* chemical compounds with potential health benefits. To date, more than 130 triterpenes have been isolated and extracted with methanol, chloroform, and ethanol from mycelia, spores, and fruit bodies of *Ganoderma*. Six isoprenes interact with each other to develop five- or six-membered ring conformations of triterpene compounds and are categorized into two groups: ganoderic acids and lucidenic acids (Kladar et al. 2016; Hsu and Cheng, 2018). According to several published research studies, ganoderic acids could impart significant immunomodulatory and anticancer activities such as downregulating prostate cancer cell viability, impairing tumor invasion, and stimulating the apoptosis machinery in cervical cancer (Hsu and Cheng, 2018). Therefore, ganoderic acids could be used as a quality control marker of different preparations from *Ganoderma*. Around 0.6 to 11 mg/g of dry powder of *G. lucidum* comprises the total triterpenoid content, which is significantly 10 times higher compared to *Ganoderma sinense*. Mechanistically, a carbonyl group at the C-3 position potentiates the cytotoxic activity of these terpenoid acids (Hsu and Cheng, 2018; Kladar et al., 2016). Chen et al. (2010) identified that ganoderic acid T could impair the invasion and metastasis of human colon cancer cell lines both in vitro and in vivo, hinting toward its use in anticancer therapies. Lucidenic acids are another important triterpenoid found in *G. lucidum* with potential immune stimulatory activity (Cör et al., 2018). Lucidenic acid was found to hamper the growth of human leukemia cells Hl 60 by triggering apoptosis (Hsu et al., 2008). Therefore, triterpenoids could be used as a promising pharmaceutical compound with increased bioactivity. More findings are needed to explore newer triterpenoids with significantly higher efficacy in biological systems.

7.2.3 Proteins

Ling Zhi-8 (M.W 13KDa) was the first fungal immune modulatory protein isolated from *Ganoderma*, protecting systemic anaphylaxis through decreasing antibody production. LZ-8 could be categorized as one of the immunoglobulin superfamilies, as it shares more than 50% homology with the immunoglobulin heavy chains including AVMS67, G1HUHL, G5A, and A1MS47 (Li et al., 2019; Kino et al., 1989). Several epidemiological data reveal that LZ-8 could significantly stimulate the activation of T lymphocytes and macrophages, and it can serve as a potent adjuvant for the activation of a DNA vaccine through dendritic cell stimulation (Hsu and Cheng, 2018). *Ganoderma* comprises a significant proportion of glycoproteins and glycopeptides. The covalent interaction between the sugar and peptide chain leads to the generation of the complex organization of glycoproteins, while glycopeptide is considered a type of glycoprotein with a higher amount of polysaccharides with improved immune-stimulatory functions (Li et al., 2019). Protease digestion of *G. lucidum* fruit bodies generates *G. lucidum* polysaccharide peptides, one of the major components of *Ganoderma*. Lectins are classified as a group of non-immuno-derived proteins with potent bioactivity. Most of them show efficient antibacterial, antioxidant, antiviral, and immune-modulatory activities (Li et al., 2019). Therefore, proteins from *Ganoderma* impart remarkable health benefits that further warrant more findings to explore their mechanistic insights in detail.

7.3 OVERVIEW OF THE HUMAN GUT MICROFLORA

The human gut microenvironment facilitates the colonization of most diversified microorganisms, including bacteria, fungi, and even some eukaryotes. Diet, age, antibiotic usage, type of delivery at birth, and lifestyle patterns are the leading factors in the establishment of the gut microenvironment (Patterson et al., 2016; Gomaa, 2020). The gut microbial composition differs greatly along the gastrointestinal tract. The colon comprises more than 10^{12} bacterial species, whereas the stomach and small intestine harbor comparatively fewer bacterial species. *Bacteroidetes* and *Firmicutes* constitute more than 90% of the gut microbial population followed by other phyla such as Fusobacteria, Acinetobacteria, Verrucomicrobia, Proteobacteria, and Cyanobacteria. *Bacteroides* and *Prevotella*, under the phylum Bacteroidetes, as well as *Clostridium*, *Ruminococcus*, and *Eubacterium*, under Firmicutes phylum, predominantly inhabit the human gut (Adak and Khan, 2019). Diet pattern is one of the important determinants of a diversified bacterial presence within the digestive tract. Studies have found that the intake of a protein-enriched diet enhances the colonization of *Bacteroides*, whereas a carbohydrate-rich diet potentiates the establishment of *Prevotella* (Adak and Khan, 2019). The healthy or diseased human health condition can be determined by the *Firmicutes/Bacteroides* ratio, acting as a remarkable marker in the human gut (Mariat et al., 2009). The mucosa of the healthy human gut comprises the following bacterial genera: *Enterococcus*, *Lactobacillus*, *Clostridium*, and *Akkermansia*. In contrast, the lumen includes *Bacteroides*, *Streptococcus*, *Bifidobacterium*, *Clostridium*, *Enterococcus*, *Ruminococcus*, and *Lactobacillus*. However, a few pathogens such as *Salmonella enterica*, *Campylobacter jejuni*, *Bacteroides fragilis*, and *Escherichia coli* are also found within the gut microenvironment (Hollister et al., 2012; Adak and Khan, 2019).

Dysbiosis is denoted as a disruption of the microbial community from its native habitat, leading to the development of different disease conditions. Studies have found that the hampering of a beneficial relationship between microorganisms and the host environment is associated with the emergence of several pathogenic conditions, even carcinoma. For instance, a high *Firmicutes/Bacteroides* ratio with reduced microbial diversity and enrichment is one of the leading causes of hypertension. Mechanistically, the dysbiosis of a hostile microenvironment can hamper the conventional synthesis of different metabolites and the digestion process within the gastrointestinal tract. For example, the native bacterial species within the human intestine, including *Eubacterium rectale*, *Clostridium* groups IV and XIVa, and *Roseburia* spp. are well known for the synthesis of short-chain fatty acids (SCFs) (formate, butyrate, propionate, acetate, etc.). SCF production could increase energy metabolism and reduce the chronic inflammation of different host tissue (Adak and Khan, 2019). Therefore, the disruption of such a microbial community is significantly associated with several metabolic disorders, including obesity, hyperlipidemia, diabetes, and even colon carcinoma (Aron-Wisnewsky et al., 2021). The excessive intake of a high-fat diet reduces the abundance of *Bacteroides* while increasing the presence of *Firmicutes* in the human colonic region. More specifically, a high-fat diet enhances the *Firmicutes/Bacteroides* ratio in the human intestine, which is correlated with the onset of obesity and type 2 diabetes mellitus (Sittipo et al., 2018). In support of this observation, Ridaura *et al.* found that the transmission of *Bacteroides* from normal mice to obese mice (twin mice) could decrease the high weight gain and obesity in diseased mice (Ridaura et al., 2013). Therefore, the stable maintenance of human intestinal flora is of the utmost importance for healthy individuals, and the dysbiosis of such a native companionship could lead to several health hazards, even those associated with high mortality.

7.4 THE REGULATION OF GUT MICROFLORA BY *GANODERMA* AND ITS CHEMICAL COMPOSITIONS

Different chemical extracts derived from *Ganoderma* have shown significant efficacies against different lethal human health conditions. Here, we have summarized the role of *Ganoderma* supplementation in the management of such lethal stages ranging from metabolic disorders to colon cancer. Table 7.1 depicts the overall role of *Ganoderma* chemical compounds in several maladies.

7.4.1 Obesity

Over the past few decades, obesity has become one of the major public health threats, accounting for approximately 1.4 billion overweight persons and 500 million obese people across the globe (El-Sayed Moustafa and Froguel, 2013). Excessive fat accumulation and unregulated weight gain are the specific attributes of obese individuals, which could lead to different disease conditions including chronic and low-grade inflammation, type 2 diabetes, insulin resistance, hyperlipidemia, cardiovascular disorders, and carcinoma. An unhealthy lifestyle and inappropriate diet intake are the major contributing factors to obesity development. Several epidemiological data reveal that human intestinal gut microflora could restrict the risk of obesity development by regulating energy metabolism and host inflammatory responses (El-Sayed Moustafa and Froguel, 2013; Chang et al., 2015). A significant difference can be observed in the gut microbial consortium between obese and nonobese individuals, particularly in the increased ratio of *Firmicutes/Bacteroidetes* bacterial phyla. The dysbiosis in gut microflora could accumulate lipopolysaccharides (LPSs), secreted by gram-negative bacterial species. Such LPS production could gain access into the mesenteric vein, followed by the internal circulation system and thereby generating endotoxemia, inflammation of the targeted organ and tissues within the host (Chang et al., 2015). The chemical extracts derived from the medicinal mushroom *Ganoderma* could

TABLE 7.1
The Role of *Ganoderma* in Modulation of Human Gut Microflora in Different Disease Conditions

Disease	*Ganoderma* Extract	Role in Gut Flora Microenvironment	Reference
Obesity	Water extract of mycelium	• Increased the abundance of specific bacterial species such as *Anaerotruncus colihominis*, *Parabacteroides goldsteinii*, *Roseburia hominis*, and *Eubacterium coprostanoligenes*; associated with reduced the risk of colon cancer. • Enhanced the accumulation of SCFs as fermented by gut microbiota, specifically by *Bacteroides*, reducing inflammation and obesity.	(Chang et al., 2015)
	Sporoderm broken spores of *Ganoderma lucidum* polysaccharide (GLP)	Abundance of a few bacterial species such as *Rikenella*, *Lachnospiraceae*, *Ruminiclostridium*, and *Blautia* and a notable decrease in other bacterial species such as *Allobaculum*, *Ruminococcaceae*, *Bifidobacterium*, *Turicibacter*, etc., in HFD-induced mice. Increase in SCF production with stimulation of immune responses in adipose tissues by activating GPCR 43 cascade. Downregulation of the chronic inflammatory responses by reducing LPS production level and impairing TLR4/NF-κβ signaling cascades.	(Sang et al., 2021)
Diabetes mellitus	F31 polysaccharide	• Increased *Bacteroidetes/Firmicutes* ratio. • Enriched the presence of *Lactobacillus*, *Ruminococcaceae*, and *Bacteroides*, restricting the uncontrolled entry of endotoxin to the circulation system, fermentation of carbohydrates, and stimulation of the intestine-brain axis.	(Shao et al., 2022)
	Polysaccharide extract	• Reduction in the colonization of pathogenic bacteria such as *Aerococcus*, *Corynebacterium*, *Ruminococcus*, and *Proteus* in the human intestine. • Increased the abundance of *Bacteroides*, *Balutia*, *Parabacteroides*, and *Dehalobacter*. • Stimulation of the metabolism of carbohydrates, amino acids, nucleic acids, and other inflammatory components thereby restoring the blood glucose level to a normal level.	(M. Chen et al., 2020)
	Spore powder from *Ganoderma* (SPG)	• Administration of SPG encapsulated with RS regulated the hypolipidemic and hypoglycemic efficacies in the diabetic rat's liver by upregulating the gene expression of glycogen generation (*GS2* and *GYG1*), lipid oxidation (*Acox1*), glucose and cholesterol homeostasis (*Insig1* and *Insig2*), and lipogenesis suppression (*ACC*, *Fads1*). • Reduced the occurrence of *Proteobacteria* in the gut, hampering inflammation and limiting the recurrence of diabetic conditions.	(Jiang et al., 2021)

Lipid metabolic disorder	GL95 polysaccharide	• Downregulated triglyceride and total cholesterol amount in hyperlipidemic rats. • Elevated the abundance of *Prevotella*, *Alloprevotella*, *Desulfovibrio*, *Alistipes*, *Butyricimonas*, *Sutterella*, and *Holdemania* and *Bacteroidetes*/*Firmicutes* ratio. • *Alistipes* abundance reduced the triglyceride and total cholesterol levels in serum and liver. • *Desulfovibrio*, *Alloprevotella* prevalence enhanced the production of SCFs, sensitizing the peripheral tissues toward insulin via GPCR 41 and GPCR 43 cascades after intestinal metabolization.	(W. L. Guo et al., 2018)
	Ganoderic acid A (GA), a triterpenoid from the fruit body of *Ganoderma*	• GA supplementation at 75 mg/kg enhanced the abundance of *Alistipes*, which increased the energy metabolism within the host and stimulated the generation of antimicrobial peptides. • Reduced the colonization of pathogenic organisms and downregulated the lipid biochemical parameters such as triglyceride and total cholesterol levels in serum and liver. • *Bacteroides* abundance by GA supplementation upregulated SCF formation and reduced the pH of the lumen, leading to the disruption of pathogenic bacterial colonization.	(W. L. Guo et al., 2020)
	Ganoderma polysaccharide	• *Ganoderma* polysaccharide and chitosan (PC) synergistically increased the abundance of *Bifidobacterium*, *Prevotella*, *Alistipes*, and *Alloprevotella* that could reduce the total cholesterol and triglyceride levels in serum and liver of HFD-stimulated golden hamster model.	(Tong et al., 2020)
Colon carcinoma	Water extract of either the *Ganoderma lingzhi* or the autodigested *Ganoderma lingzhi*	• Reduced the synthesis of secondary bile acids such as lithocholic and deoxycholic acids in the colon, lowering the risk of colon malignancy. • Decreased the abundance of *Clostridium leptum* and *Clostridium coccoides* per gram of cecal digesta-ameliorated the inflammation and malignant transformation of colon cells.	(Yang et al., 2017)
	Polysaccharide extract from sporoderm broken spores of *Ganoderma lucidum*	• Obstructing colon carcinoma by hampering gut flora dysbiosis through increasing SCF production and TLR4/MyD88, NF-κβ signaling machinery.	(C. Guo et al., 2021)
	The polysaccharide extract from *Ganoderma lucidum* and *Ganoderma sinense*	• Elevated the abundance of *Bacteroides* and *Alistipes*, reducing the progression of colon cancer. • Increased the abundance of *Bacteroides*/*Firmicutes*. • Inhibited cancer progression by enhancing SCF production and reducing the inflammatory response.	(L. F. Li et al., 2018)

limit the development of obesity conditions by modulating the intestinal gut microflora dysbiosis. In an innovative study by Chang *et al.*, the water extract from *Ganoderma* mycelium could standardize the gut microbial composition in HFD-stimulated mice, thereby reducing obesity and other inflammatory disorders (Chang et al., 2015). Eight percent supplementation of *Ganoderma* mycelium extract (GME) induced the abundance of specific bacterial species such as *Anaerotruncus colihominis*, *Parabacteroides goldsteinii*, *Roseburia hominis*, and *Eubacterium coprostanoligenes* that are associated with reducing the risk of obesity development. Mechanistically, GME (8%) administration enhanced the accumulation of SCFs such as acetate and butyrate that are synthesized from indigestible polysaccharides and dietary fibers via fermentation by gut microbiota, specifically by *Bacteroidetes*. Hence, it could be hypothesized that GME could reduce inflammatory responses and obesity by reversing gut flora dysbiosis in HFD-induced mice (Chang et al., 2015). In another current report, Sang *et al.* found that *Ganoderma lucidum* polysaccharide (GLP) isolated from sporoderm broken spores of *Ganoderma* could significantly reduce the obesity, hyperlipidemia, and inflammation in high-fat diet (HFD)–induced mice, mainly by controlling the dysbiosis of the gut microbiome (Sang et al., 2021). The 16S ribosomal RNA (rRNA) sequencing between HFD- and non-HFD–induced mice revealed a significant abundance of a few bacterial species such as *Rikenella*, *Lachnospiraceae*, *Ruminiclostridium*, and *Blautia* and a notable decrease in other bacterial species such as *Allobaculum*, *Ruminococcaceae*, *Bifidobacterium*, *Turicibacter*, etc., in HFD-induced mice. The administration of GLP at 300 mg/kg could significantly alter the microbial consortium in HFD-induced mice, associated with decreased obesity (Sang et al., 2021). In corroboration with the previous study (Chang et al., 2015), GLP supplementation was also found to increase the buildup of SCF that potentiated the immune responses in adipose tissues by stimulating the activation of G-protein–coupled receptor (GPCR) 43. Moreover, the GLP also downregulated the chronic inflammatory responses by reducing LPS production level and impairing TLR4/NF-κβ signaling cascades (Sang et al., 2021). Hence, *Ganoderma* could serve as a potential antiobese compound with improved pharmaceutical attributes. More research is required to deeply elucidate the mechanisms of more chemical constituents from *Ganoderma* for the betterment of obese conditions across the globe.

7.4.2 Diabetes Mellitus

Diabetes mellitus is another public health hazard that is characterized by high blood glucose, low-grade chronic inflammation, insulin resistance, and a very minute amount of insulin secretion. Different environmental and genetic factors, obesity, hypertension, and an unhealthy lifestyle are the primary reasons for diabetic conditions. Currently, recombinant insulin and its associated pharmacotherapy are the major means of diabetes therapy throughout the world. However, these strategies have unfortunately led to excessive weight gain and hypoglycemia with increasing drug resistance (Ma et al., 2019; Harrigan, 2007). Such drawbacks have attracted scientists to explore different phytochemicals against which insulin resistance and other inflammatory responses have not been reported so far. Studies reveal that the intake of polysaccharides from medicinal mushroom such as *Ganoderma* could ameliorate the risk of diabetes. Shao *et al.* found that F31 polysaccharide isolated from *Ganoderma* could attenuate type 2 diabetes by increasing insulin secretion by repairing the islet cells and sensitizing insulin through the modulation of gut microflora (Shao et al., 2022). Mechanistically, F31 was found to obstruct the gut flora dysbiosis by increasing the *Bacteroidetes/Firmicutes* ratio. More specifically, this isolated polysaccharide could enrich the presence of *Lactobacillus*, *Ruminococcaceae*, and *Bacteroides* that restricted the uncontrolled entry of endotoxin to the circulation system, fermentation of carbohydrates, and stimulation of the intestine-brain axis. Such significant functions by stable gut flora organization could reduce the low-grade inflammation and insulin resistance that lead to recovery from type 2 diabetes (Shao et al., 2022). In line with this evidence, Chen et al. (2020) determined that GLP could restore gut flora dysbiosis and reduce the excessive high glucose level in the blood and chronic inflammation. The 16S rRNA sequencing reveals that GLP reduced the occurrence of harmful bacteria such as *Aerococcus*, *Corynebacterium*, *Ruminococcus*, and *Proteus* within the human intestine. Similarly, they followed a significant upregulation of other advantageous bacteria such as

Bacteroides, *Balutia*, *Parabacteroides*, and *Dehalobacterium*. Such normal gut flora composition could enhance the metabolism of carbohydrates, amino acids, nucleic acids, and other inflammatory components, thereby restoring the blood glucose level to a normal level and downregulating type 2 diabetes in a mice model (Chen et al., 2020). In an earlier report, it has been observed that spore powder from *Ganoderma* (SPG) could significantly decrease the blood glucose level by around 21% in diabetic mice induced by streptozocin injection. Moreover, SPG could downregulate the total cholesterol by 17.8% and triglycerides by 49%, which were remarkably higher in the control diabetic mice (Wang et al., 2015). Later, the same research group identified that the encapsulation of SPG within a coating of resistant starch (RS) could synergistically enhance its antidiabetic activity (Jiang et al., 2021). Despite the indigestible nature of RS, it could be metabolized by the gut microbiome, and thereafter it could reverse the high blood glucose level back to a normal level by stable gut microbiota organization and delayed gastric emptying. The RS-coated SPG controlled the hypolipidemic and hypoglycemic efficacies in the diabetic rat's liver by upregulating the gene expression of glycogen generation (*GS2* and *GYG1*), lipid oxidation (*Acox1*), glucose and cholesterol homeostasis (*Insig1* and *Insig2*), and lipogenesis suppression (*ACC*, *Fads1*) (Jiang et al., 2021). The abundance of *Proteobacteria* was relatively higher in the diabetic mice group that in turn enhanced gut inflammation and stimulated the commencement of type 2 diabetes. Notably, the RS-coated SPG reduced the occurrence of *Proteobacteria* in the gut, hampered the inflammation, and thereby downregulated the recurrence of diabetic conditions (Jiang et al., 2021). Taken together, it can be concluded that *Ganoderma* could serve as a potential alternative for the treatment of diabetes. However, more investigations are needed to explore other chemical constituents and their mechanical insights from *Ganoderma* in more detail.

7.4.3 Lipid Metabolic Disorder

Excessive body fat accumulation leads to the development of lipid metabolism disorder, a major health threat in community settings. Among different lipid metabolism disorders, hyperglycemia, hyperlipidemia, cardiovascular diseases, and nonalcoholic fatty liver maladies are widely explored across the globe. The increased number of triglycerides, total cholesterol, and low-density lipoprotein cholesterol with a declined level of high-density lipoprotein cholesterol in the blood are the major characteristics of hyperlipidemia, a major public health risk factor. The prevalence of hyperlipidemia leads to the development of other health hazards including atherosclerosis, peripheral artery malady, and coronary heart disease (Guo et al., 2020; Chalasani et al., 2018). Accumulating evidence has shown that dysbiosis in gut microflora could obstruct the hepatic cholesterol metabolism, leading to the rise of metabolic alterations. According to a few published reports, the chemical compositions derived from the medicinal mushroom *Ganoderma* could be used to modulate lipid metabolic disorders by impairing gut flora dysbiosis. In an interesting study by Guo et al. (2018) the polysaccharide from *G. lucidum* (GL95) not only maintained the normal body weight of the HFD-induced mice but also downregulated triglyceride and total cholesterol amount in hyperlipidemic rats compared with the control set. Intriguingly, the supplementation of GL95 for 8 weeks elevated the abundance of *Prevotella*, *Alloprevotella*, *Desulfovibrio*, *Alistipes*, *Butyricimonas*, *Sutterella*, and *Holdemania* and a *Bacteroidetes*/*Firmicutes* ratio that is significantly lower in the hyperlipidemic mice (Guo et al., 2018). Mechanistically, *Alistipes* abundance by GL95 supplementation markedly reduced the triglyceride and total cholesterol levels in serum and the liver. Furthermore, the prevalence of *Desulfovibrio* and *Alloprevotella* could elevate the production of SCFs that could sensitize the peripheral tissues toward insulin via GPCR 41 and GPCR 43 cascades after intestinal metabolization. Therefore, GL95 could ameliorate lipid metabolism disorders, specifically by reducing the total cholesterol and triglyceride levels by impairing gut microflora disturbances (Guo et al., 2018). In line with this evidence, GA, a triterpenoid isolated from the fruit body of *Ganoderma*, significantly reduced the total cholesterol and triglyceride level in serum in HFD-induced mice at 75 mg/kg. Apart from this, GA supplementation at a high dosage of 75 mg/kg could enhance the abundance of *Alistipes*, which enhances the energy metabolism within the host and stimulates the generation of antimicrobial peptides. Such resultant effects could negatively correlate with the colonization of

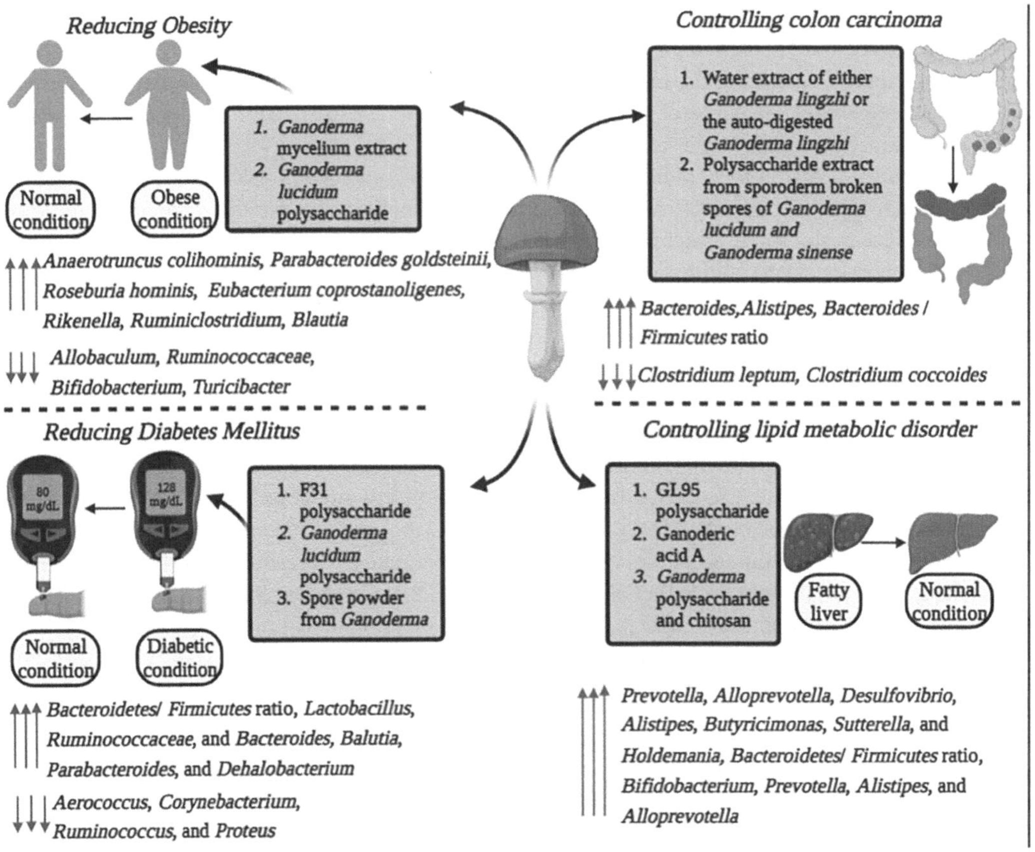

pathogenic organisms and thereby downregulate the lipid biochemical parameters such as triglyceride and total cholesterol levels in serum and the liver. Moreover, the abundance of *Bacteroides* by GA supplementation enhanced SCF formation and reduced the pH of the lumen, resulting in the disruption of pathogenic bacterial colonization (Guo et al., 2020). In another innovative report, Tong et al. (2020) determined that *Ganoderma* polysaccharide and chitosan (PC) synergistically reduced excessive lipid accumulation in the body by modulating the gut flora composition. As chitosan downregulates the absorption and digestion of visceral fat by impairing lipid metabolism and bile acid synthesis, its conjugation with *Ganoderma* polysaccharide could trigger the lowering of lipid levels inside the body (Wang et al., 2017; Tong et al., 2020). In corroboration with the previous reports, this PC complex could increase the abundance of *Bifidobacterium*, *Prevotella*, *Alistipes*, and *Alloprevotella* that could reduce the total cholesterol and triglyceride levels in serum and the liver of an HFD-stimulated golden hamster model (Tong et al., 2020). In summary, *Ganoderma* chemical constituents could reduce hyperlipidemia and other metabolic disorders, providing significant mechanical insight through the regulation of gut microbiota.

7.4.4 Colon Carcinoma

Colon carcinoma is the third most common cancer-associated mortality across the globe. A high-fat diet, obesity, excessive alcohol consumption, and an unhealthy lifestyle are the primary reasons for the malignant transformation of the normal colonic cells (Cappell, 2008). Studies have found that dysbiosis in intestinal microbial habitation could promote carcinogenesis with very poor outcomes. The disruption of the natural habitats within the colonic region could enhance the secretion of secondary bile acids such as lithocholic and deoxycholic acids that could stimulate colon carcinoma. Furthermore, the disappearance of gut flora lowers the synthesis of SCFs, which increases chronic inflammation, leading to colonic malignancy (Garrett, 2019). Accumulating reports have suggested that a healthy diet comprising phytochemicals could reduce the risk of colon cancer, providing an alternative strategy in their management. The polysaccharides and triterpenoids isolated from *Ganoderma* could hamper such intestinal flora dysbiosis and thereby prevent malignancy. Yang et al. (2017) determined that the supplementation of water extract of either the *G. lingzhi* or the autodigested *G. lingzhi* for 3 weeks markedly reduced the synthesis of secondary bile acids such as lithocholic and deoxycholic acids in the colon, lowering the risk of colon malignancy. The cytotoxic nature of deoxycholic acids transforms the normal colonic cells to malignant cells by DNA damage, activating NF-κβ signaling cascades wherein lithocholic acid acts as a toxic endobiotic to the colonic cells (Bernstein et al., 2005; Payne et al., 2007; Hofmann, 2004). Mechanistically, these two extracts significantly decreased the abundance of *Clostridium leptum* and *Clostridium coccoides* per gram of cecal digesta. These bacterial species are considered secondary bile acid producers, and their reduction attenuated the inflammation and malignant transformation of colonic cells (Yang et al., 2017). In another current study by Guo et al. (2021), the polysaccharide extract from sporoderm broken spores of *G. lucidum* impaired the colon carcinoma by interfering with the gut microbial dysbiosis through increasing SCF production and TLR4/MyD88, NF-κβ signaling machinery. Such an extract elevated the abundance of *Bacteroides* and *Alistipes*, increased SCF production, and negatively correlated with colon carcinoma progression. In a different study by Li et al. (2018) the polysaccharide extract from *G. lucidum* and *G. sinense* notably reduced the tumor size and inflammation of breast cancer cells by modulating the gut microflora. Such polysaccharide extracts increased the abundance of the *Bacteroides*/*Firmicutes* ratio that significantly correlates with previous observations. Such an increased *Bacteroides*/*Firmicutes* ratio could ameliorate breast cancer progression by enhancing SCF production and reducing the inflammatory response. Therefore, the intake of different polysaccharides and triterpenoid extraction from *Ganoderma* could serve as a potential alternative for cancer prevention. Further investigations are highly warranted to identify more components and to explore detailed mechanical insights before clinical trials can move further.

7.5 CONCLUSION

As the human gut microbiota harbors a highly diversified microbial population, which is significantly correlated with overall human health, any disturbance or disbalance of this enormously rich microflora, termed classically as "microbial dysbiosis," leads to several pathological conditions, including cancer. Recent reports suggest that *Ganoderma* possesses promising pharmacological potential in terms of treating several cellular and physiological conditions, including inflammation and carcinogenesis. *Ganoderma* does so by modulating the gut microbiota. Therefore, stringent research in deciphering its role in regulating the human gut microbiota and its potential as a probiotic drug is highly warranted.

REFERENCES

Adak, Atanu, and Mojibur R. Khan. 2019. An insight into gut microbiota and its functionalities. *Cellular and Molecular Life Sciences* 76 (3). Springer International Publishing: 473–93. doi:10.1007/s00018-018-2943-4.

Aron-Wisnewsky, Judith, Moritz V. Warmbrunn, Max Nieuwdorp, and Karine Clément. 2021. Metabolism and metabolic disorders and the microbiome: The intestinal microbiota associated with obesity, lipid metabolism, and metabolic health—pathophysiology and therapeutic strategies. *Gastroenterology* 160 (2). The American Gastroenterological Association: 573–99. doi:10.1053/j.gastro.2020.10.057.

Bernstein, H., C. Bernstein, C. M. Payne, K. Dvorakova, and H. Garewal. 2005. Bile acids as carcinogens in human gastrointestinal cancers. *Mutation Research — Reviews in Mutation Research* 589 (1): 47–65. doi:10.1016/j.mrrev.2004.08.001.

Cao, Ying Jia, Zi Rui Huang, Shi Ze You, Wei Ling Guo, Fang Zhang, Bin Liu, Xu Cong Lv, Zhan Xi Lin, and Peng Hu Liu. 2022. The protective effects of ganoderic acids from *Ganoderma lucidum* fruiting body on alcoholic liver injury and intestinal microflora disturbance in mice with excessive alcohol intake. *Foods* 11 (7). doi:10.3390/foods11070949.

Cappell, Mitchell S. 2008. Pathophysiology, clinical presentation, and management of colon cancer. *Gastroenterology Clinics of North America* 37 (1): 1–24. doi:10.1016/j.gtc.2007.12.002.

Chalasani, Naga, Zobair Younossi, Joel E. Lavine, Michael Charlton, Kenneth Cusi, Mary Rinella, Stephen A. Harrison, Elizabeth M. Brunt, and Arun J. Sanyal. 2018. The diagnosis and management of nonalcoholic fatty liver disease: Practice guidance from the american association for the study of liver diseases. *Hepatology* 67 (1): 328–57. doi:10.1002/hep.29367.

Chang, Chih Jung, Chuan Sheng Lin, Chia Chen Lu, Jan Martel, Yun Fei Ko, David M. Ojcius, Shun Fu Tseng, et al. 2015. *Ganoderma lucidum* reduces obesity in mice by modulating the composition of the gut microbiota. *Nature Communications* 6 (May). Nature Publishing Group: 1–17. doi:10.1038/ncomms8489.

Chen, Mingyi, Dan Xiao, Wen Liu, Yunfei Song, Baorong Zou, Lin Li, Pei Li, et al. 2020. Intake of *Ganoderma lucidum* polysaccharides reverses the disturbed gut microbiota and metabolism in type 2 diabetic rats. *International Journal of Biological Macromolecules* 155. Elsevier B.V: 890–902. doi:10.1016/j.ijbiomac.2019.11.047.

Chen, Nian Hong, Jian Wen Liu, and Jian Jiang Zhong. 2010. Ganoderic acid T inhibits tumor invasion *in vitro* and *in vivo* through inhibition of MMP expression. *Pharmacological Reports* 62 (1). Elsevier: 150–63. doi:10.1016/S1734-1140(10)70252-8.

Cör, Darija, Željko Knez, and Maša Knez Hrnčič. 2018. Antitumour, antimicrobial, antioxidant and antiacetylcholinesterase effect of *Ganoderma lucidum* terpenoids and polysaccharides: A review. *Molecules* 23 (3): 1–21. doi:10.3390/molecules23030649.

El-Sayed Moustafa, Julia S., and Philippe Froguel. 2013. From obesity genetics to the future of personalized obesity therapy. *Nature Reviews Endocrinology* 9 (7). Nature Publishing Group: 402–13. doi:10.1038/nrendo.2013.57.

Ferreira, Isabel C. F. R., Sandrina A. Heleno, Filipa S. Reis, Dejan Stojkovic, Maria João R. P. Queiroz, M. Helena Vasconcelos, and Marina Sokovic. 2015. Chemical features of *Ganoderma* polysaccharides with antioxidant, antitumor and antimicrobial activities. *Phytochemistry* 114. Elsevier Ltd: 38–55. doi:10.1016/j.phytochem.2014.10.011.

Garrett, Wendy S. 2019. The gut microbiota and colon cancer. *Science* 364 (6446): 1133–35. doi:10.1126/science.aaw2367.

Gomaa, Eman Zakaria. 2020. Human gut microbiota/microbiome in health and diseases: A review. *Antonie van Leeuwenhoek, International Journal of General and Molecular Microbiology* 113 (12). Springer International Publishing: 2019–40. doi:10.1007/s10482-020-01474-7.

Guo, Cuiling, Dandan Guo, Liu Fang, Tingting Sang, Jianjun Wu, Chengjie Guo, Yujie Wang, et al. 2021. *Ganoderma lucidum* polysaccharide modulates gut microbiota and immune cell function to inhibit inflammation and tumorigenesis in colon. *Carbohydrate Polymers* 267 (January). Elsevier Ltd: 118231. doi:10.1016/j.carbpol.2021.118231.

Guo, Wei Ling, Jian Bin Guo, Bin Yu Liu, Jin Qiang Lu, Min Chen, Bin Liu, Wei Dong Bai, Ping Fan Rao, Li Ni, and Xu Cong Lv. 2020. Ganoderic acid A from: *Ganoderma lucidum* ameliorates lipid metabolism and alters gut microbiota composition in hyperlipidemic mice fed a high-fat diet. *Food and Function* 11 (8): 6818–33. doi:10.1039/d0fo00436g.

Guo, Wei Ling, Yu Yang Pan, Lu Li, Tian Tian Li, Bin Liu, and Xu Cong Lv. 2018. Ethanol extract of *Ganoderma lucidum* ameliorates lipid metabolic disorders and modulates the gut microbiota composition in high-fat diet fed rats. *Food and Function* 9 (6): 3419–31. doi:10.1039/c8fo00836a.

Harrigan, Natasha. 2007. Risk factors for type 2 diabetes. *U.S. Pharmacist* 32 (10): 61–63.

Hofmann, Alan F. 2004. Detoxification of lithocholic acid, a toxic bile-acid: Relevance to drug hepatotoxicity. *Drug Metabolism Reviews* 36 (3–4): 703–22. doi:10.1081/DMR-200033475.

Hollister, Emily B., Chunxu Gao, and James Versalovic. 2012. Compositional and functional features of the gastrointestinal microbiome and their effects on human health. *Gastroenterology* 146 (6): 1449–58. doi:10.1053/j.gastro.2014.01.052. Compositional.

Hsu, Chin Lin, Yu Shan Yu, and Gow Chin Yen. 2008. Lucidenic acid B induces apoptosis in human leukemia cells via a mitochondria-mediated pathway. *Journal of Agricultural and Food Chemistry* 56 (11): 3973–80. doi:10.1021/jf800006u.

Hsu, Kai Di, and Kuan Chen Cheng. 2018. From nutraceutical to clinical trial: Frontiers in *Ganoderma* development. *Applied Microbiology and Biotechnology* 102 (21). Applied Microbiology and Biotechnology: 9037–51. doi:10.1007/s00253-018-9326-5.

Human Microbiome Project Consortium. 2012. Structure, function and diversity of the healthy human microbiome. *Nature* 486: 207–14.

Jiang, Yumei, Na Zhang, Yawen Zhou, Zhongkai Zhou, Yu Bai, Padraig Strappe, and Chris Blanchard. 2021. Manipulations of glucose/lipid metabolism and gut microbiota of resistant starch encapsulated *Ganoderma lucidum* spores in T2DM rats. *Food Science and Biotechnology* 30 (5). Springer Singapore: 755–64. doi:10.1007/s10068-021-00908-w.

Kino, K., A. Yamashita, K. Yamaoka, J. Watanabe, S. Tanaka, K. Ko, K. Shimizu, and H. Tsunoo. 1989. Isolation and characterization of a new immunomodulatory protein, Ling Zhi-8 (LZ-8), from *Ganoderma lucidium*. *Journal of Biological Chemistry* 264 (1): 472–78. doi:10.1016/s0021-9258(17)31282-6.

Kladar, Nebojsă V., Neda S. Gavarić, and Biljana N. Božin. 2016. *Ganoderma*: Insights into anticancer effects. *European Journal of Cancer Prevention* 25 (5): 462–71. doi:10.1097/CEJ.0000000000000204.

Li, Daotong, Pan Wang, Pengpu Wang, Xiaosong Hu, and Fang Chen. 2016. The gut microbiota: A treasure for human health. *Biotechnology Advances* 34 (7). Elsevier Inc.: 1210–24. doi:10.1016/j.biotechadv.2016.08.003.

Li, Li Feng, Hong Bing Liu, Quan Wei Zhang, Zhi Peng Li, Tin Long Wong, Hau Yee Fung, Ji Xia Zhang, Su Ping Bai, Ai Ping Lu, and Quan Bin Han. 2018. Comprehensive comparison of polysaccharides from *Ganoderma lucidum* and *G. sinense*: Chemical, antitumor, immunomodulating and gut-microbiota modulatory properties. *Scientific Reports* 8 (1). Springer US: 1–12. doi:10.1038/s41598-018-22885-7.

Li, Liu Dingji, Pei Wen Mao, Ke Di Shao, Xiao Hui Bai, and Xuan Wei Zhou. 2019. *Ganoderma* proteins and their potential applications in cosmetics. *Applied Microbiology and Biotechnology* 103 (23–24). Applied Microbiology and Biotechnology: 9239–50. doi:10.1007/s00253-019-10171-z.

Ma, Quantao, Yaqi Li, Pengfei Li, Min Wang, Jingkang Wang, Ziyan Tang, Ting Wang, Linglong Luo, Chunguo Wang, and Baosheng Zhao. 2019. Research progress in the relationship between type 2 diabetes mellitus and intestinal flora. *Biomedicine and Pharmacotherapy* 117 (May). Elsevier: 109138. doi:10.1016/j.biopha.2019.109138.

Mariat, D., O. Firmesse, F. Levenez, V. D. Guimarães, H. Sokol, J. Doré, G. Corthier, and J. P. Furet. 2009. The Firmicutes/Bacteroidetes ratio of the human microbiota changes with age. *BMC Microbiology* 9: 1–6. doi:10.1186/1471-2180-9-123.

Patterson, Elain E., Paul M. Ryan, John F. Cryan, Timothy G. Dinan, R. Paul Ross, Gerald F. Fitzgerald, and Cath Erine Stanton. 2016. Gut microbiota, obesity and diabetes. *Postgraduate Medical Journal* 92 (1087): 286–300. doi:10.1136/postgradmedj-2015-133285.

Payne, C. M., C. Weber, C. Crowley-Skillicorn, K. Dvorak, H. Bernstein, C. Bernstein, H. Holubec, B. Dvorakova, and H. Garewal. 2007. Deoxycholate induces mitochondrial oxidative stress and activates NF-κB through multiple mechanisms in Het-116 colon epithelial cells. *Carcinogenesis* 28 (1): 215–22. doi:10.1093/carcin/bgl139.

Ridaura, Vanessa K., Jeremiah J. Faith, Federico E. Rey, Jiye Cheng, Alexis E. Duncan, Andrew L. Kau, Nicholas W. Griffin, et al. 2013. Gut microbiota from twins discordant for obesity modulate metabolism in mice. *Science* 341 (6150). doi:10.1126/science.1241214.

Sang, Tingting, Chengjie Guo, Dandan Guo, Jianjun Wu, Yujie Wang, Ying Wang, Jiajun Chen, et al. 2021. Suppression of obesity and inflammation by polysaccharide from sporoderm-broken spore of *Ganoderma lucidum* via gut microbiota regulation. *Carbohydrate Polymers* 256 (June 2020). Elsevier Ltd: 117594. doi:10.1016/j.carbpol.2020.117594.

Shao, Weiming, Chun Xiao, Tianqiao Yong, Yifan Zhang, Huiping Hu, Ting Xie, Rongjie Liu, et al. 2022. A polysaccharide isolated from *Ganoderma lucidum* ameliorates hyperglycemia through modulating gut microbiota in type 2 diabetic mice. *International Journal of Biological Macromolecules* 197. Elsevier B.V.: 23–38. doi:10.1016/j.ijbiomac.2021.12.034.

Sittipo, Panida, Stefani Lobionda, Yun Kyung Lee, and Craig L. Maynard. 2018. Intestinal microbiota and the immune system in metabolic diseases. *Journal of Microbiology* 56 (3): 154–62. doi:10.1007/s12275-018-7548-y.

Tong, Ai Jun, Rong Kang Hu, Lin Xiu Wu, Xu Cong Lv, Xin Li, Li Na Zhao, and Bin Liu. 2020. *Ganoderma* polysaccharide and chitosan synergistically ameliorate lipid metabolic disorders and modulate gut microbiota composition in high fat diet-fed golden hamsters. *Journal of Food Biochemistry* 44 (1): 1–9. doi:10.1111/jfbc.13109.

Wang, Bin, Sicong Zhang, Xiaoya Wang, Shuo Yang, Qixing Jiang, Yanshun Xu, and Wenshui Xia. 2017. Transcriptome analysis of the effects of chitosan on the hyperlipidemia and oxidative stress in high-fat diet fed mice. *International Journal of Biological Macromolecules* 102. Elsevier B.V.: 104–10. doi:10.1016/j.ijbiomac.2017.03.187.

Wang, Fang, Zhongkai Zhou, Xiaochong Ren, Yuyang Wang, Rui Yang, Jinhua Luo, and Padraig Strappe. 2015. Effect of *Ganoderma lucidum* spores intervention on glucose and lipid metabolism gene expression profiles in type 2 diabetic rats. *Lipids in Health and Disease* 14 (1): 1–9. doi:10.1186/s12944-015-0045-y.

Yang, Yongshou, Dwi Eva Nirmagustina, Thanutchaporn Kumrungsee, Yukako Okazaki, Hiroyuki Tomotake, and Norihisa Kato. 2017. Feeding of the water extract from *Ganoderma* lingzhi to rats modulates secondary bile acids, intestinal microflora, mucins, and propionate important to colon cancer. *Bioscience, Biotechnology and Biochemistry* 81 (9). Taylor & Francis: 1796–1804. doi:10.1080/09168451.2017.1343117.

Zhong, Jiayi, Liu Fang, Rong Chen, Jing Xu, Dandan Guo, Chengjie Guo, Cuiling Guo, Jiajun Chen, Chaojie Chen, and Xingya Wang. 2021. Polysaccharides from sporoderm-removed spores of *Ganoderma lucidum* induce apoptosis in human gastric cancer cells via disruption of autophagic flux. *Oncology Letters* 21 (5): 1–12. doi:10.3892/ol.2021.12686.

8 Structural Elucidation and Medicinal Attributes of Secondary Metabolites from *Ganoderma*

Predrag Petrović[1] and Jovana Vunduk[2]
1 University of Belgrade, Belgrade, Serbia
2 Institute of General and Physical Chemistry, Belgrade, Serbia and Ekofungi Ltd., Belgrade, Serbia

8.1 INTRODUCTION

In the life of a human, male sexual development starts in the mother's womb. Much later, the same male will enter puberty and develop facial and chest hair, pelvic build, upper body muscular build, and increased body hair (Richards and Hawley, 2011). From the point of survival, all these changes are not necessary; however, almost every human being of both sexes goes through puberty. So why does that happen? As pointed out by Paciulli and Cromer (2018), the listed changes known as the secondary sex characteristics are not directly related to sexual reproduction and survival, but they serve to increase chances of mating. This further implies a higher possibility of horizontal gene transfer and thus survival. In plants, the role of increasing a plant's survival and reproductive fitness belongs to the secondary metabolites. They came out as adaptive characteristics selected during evolution (Wink, 2003). When it comes to microorganisms, Firn and Jones (2000) presented and reconciled two opposing views. First, the "negative" one gives no crucial fitness and survival role to secondary metabolites. The opposite view goes in line with the ones presented earlier for other organisms; such molecules are biologically active and endow the producer with increased fitness. The unifying hypothesis proposed that one single molecule can rarely possess strong biological activity, so the organism generates an array of similar chemical structures. They are families of compounds that share the same metabolic pathway. Many of them will not increase biological fitness, but the mechanism of generating secondary metabolites might serve as a powerful evolutionary playground. This is especially important in the kingdom of fungi, whose members populate every ecological niche, meaning a constantly changing environment. The most prolific producers of secondary metabolites are ascomycetes and basidiomycetes during their vegetative and sexual phases (Anke, 2020). In comparison with plant-derived secondary metabolites, higher fungi-derived secondary metabolites are underexplored (Chen and Liu, 2017). However, the interest in bioprospecting higher fungi as a resource of drugs of natural origin is constantly increasing (Zhong and Xiao, 2009). This is backed up by ever-growing evidence of a vast array of pharmacological activities of mushrooms (Vunduk et al., 2015; Petrović et al., 2016; Petrović and Vunduk, 2022). The research in this field is mainly focused on the isolation and characterization as well as on the biological activity of secondary metabolites and much less on biosynthesis regulation, purification, and commercial production.

The *Ganoderma* genus is among the most well-known for its bioactivity (Klaus et al., 2016) with more than 400 secondary metabolites isolated from several species (Baby, Johnson and Govindan, 2015). Highly oxygenated triterpenoids are the most characteristic secondary metabolites of *Ganoderma* spp. (Anke, 2020). Besides their proven biological activity, they are also responsible

 DOI: 10.1201/9781003354789-8

for the specific color of this mushroom, and some of them contribute to its bitter taste (Wang et al., 2020; Vunduk and Veljović, 2021; Andrejč, Knez and Marevci, 2022). However, the same genus is prized for the biological activity of its polysaccharides (Wang et al., 2020). They are proven immunomodulators and anticancer agents (Xu et al., 2011). Secondary metabolites, like meroterpenoids, on the other hand, possess some other biological activities like antibacterial, antiviral activity, liver protection, etc. (Niedermeyer et al., 2013; Zhang et al., 2014; Wang et al., 2020; Zhu et al., 2022).

8.2 PHARMACOLOGICAL ACTIVITY OF *GANODERMA* TRITERPENOIDS

Triterpenoids from *Ganoderma* species share an overall similar structure, but there is a great diversity in regard to their degree of oxygenation, presence, and positioning of double C=C bonds, as well as presence of the substituents, which may have great influence on their biological activity. As they are lanosterol derivatives, most of the characterized compounds have a tetracyclic structure, with a branch of varying length at C17 of the ring D. The best known are compounds that have a branch with a terminal carboxylic acid (ganoderic, ganoderenic, lucidenic acids, etc.). Apart from acids, ganoderic alcohols and aldehydes are also known, having a terminal alcohol or aldehyde group instead of carboxylic group (Min et al., 2000; Trigos and Medellin, 2011; Liang et al., 2019; Galappaththi et al., 2022. Compounds with an unusual *seco* backbone (opened rings) or with presence of seven-membered rings/lactones in the backbone have been isolated in recent times (Kleinwächter et al., 2001; El Dine et al., 2008a). The first of these highly oxygenated triterpenoids, ganoderic acids A and B, were isolated 40 years ago by Kubota et al. (1982) in their attempt to identify bitter compounds from *G. lucidum*. Since then, nearly 500 triterpenoids were isolated and characterized from various species of the *Ganoderma* genus (Galappaththi et al., 2022).

These compounds are found in the fruiting bodies (including spores), and some are produced by the mycelium of *Ganoderma* species. However, individual compounds are often found in minute quantities, which makes research difficult, along with the potential exploitation for medical purposes. Min et al. (1998) showed that triterpenoid content may vary greatly in fruiting bodies of *G. lucidum*; although they only had a limited number of standards (7 acids and 6 alcohols), they found that examined specimens of fruiting bodies contained 1–3.6 mg/g of triterpenoid acids and 244–362 μg/g of triterpenoid alcohols. The antlered form of *G. lucidum* was found to be exceptionally high in triterpenoid acid content (4.7 mg/g). Spores, on the other hand, were shown to be the richest in triterpenoids, with 24.6 mg/g of acids and 1.1 mg of alcohols; spores, however, represent only a small portion of the fruiting bodies. Ganoderic acid A was in all cases the most abundant triterpenoid.

8.2.1 Antitumor Activity

After numerous studies on *Ganoderma* species extracts (especially alcohol extracts), which demonstrated its potent cytotoxicity, the search began for its active ingredients, and it was soon found that triterpene-enriched fractions were those responsible for the activity (Liu et al., 2002; Jiang et al., 2006; Müller et al., 2006; Lu et al., 2004; Li et al., 2017; Lin et al., 2003; Thyagarajan et al., 2010). Since then, numerous triterpenoids were isolated from various species of *Ganoderma* and proven to act as antitumor agents, through various mechanisms, the most important of which is their cytotoxicity.

By now, there are a number of studies which aimed to uncover the mechanism of the cytotoxic activity of ganoderic acids and their analogues. It is well documented that they lead to programmed cell death—apoptosis—by interfering with various cell messaging pathways. Apoptosis may be induced by external or internal signals (extrinsic and intrinsic apoptotic pathways), but it may also be triggered by immune cells, specifically CD4+ T lymphocytes. The extrinsic pathway is triggered by activation of so-called death receptors; one of the best known is probably TNFR1, a receptor that is activated by tumor necrosis factor α (TNFα). The intrinsic pathway is triggered by a variety of different factors, which include toxins; viruses; stress conditions resulting from hyperthermia, hypoxia, or radiation; and reactive molecular species such as free radicals. Also, the intrinsic pathway may be induced by a lack of apoptosis-preventing extracellular signals. Different pathways result in activation of different caspases—proteases that cleave and activate other factors in the programmed cell death pathways, but both apoptotic pathways result in activation of caspases, which induce the executional stage of cell death (caspase-3, caspase-6, and caspase-7) by activating endonucleases that cause disintegration of the nucleus. Mitochondrial membrane dysfunction is another important, well-studied event in apoptosis, leading to release of cytochrome *c* into cytosol and activation of caspases.

Apoptosis is a normal physiological process, when controlled, allowing for remodeling and growth of tissues, as well as removal of dysfunctional or damaged cells. However, tumor cells may become resistant to apoptosis by downregulating or expressing nonfunctional death receptors or by overexpression of antiapoptotic protein Bcl-2 or downregulating proapoptotic protein Bax (Elmore, 2007).

The mechanism of cytotoxic activity of ganoderic acids in cancer cells may actually interfere with their apoptosis-evading strategies and apparently potentially involves numerous signaling pathways. Studies on ganoderic acids and their antitumor activity, however, did not focus only on their direct cytotoxicity but also explored other potential antitumor mechanisms these compounds may possess. Ganoderic acids were shown to exhibit tumor invasion inhibitory activity in both in vitro and in vivo models. Metastasis is a problem of special concern when treating cancer, although not all cancer types are equally invasive. Chen et al. (2010) investigated the mode of action of ganoderic acid T against invasive behavior of human colon carcinoma cell line (HCT-116 cells). Several enzymes and factors are recognized as very important for the ability of tumors to disseminate and expand through the healthy tissues; metalloproteinases (MMPs) hydrolyze proteins of the extracellular matrix; urokinase-type plasminogen activator, a serine protease uPA, through activation of plasmin, leads to secretion of angiogenic growth factors (vascular endothelial growth factor [VEGF] and basic fibroblast growth factor bFGF]); and nitric oxide synthase (NOS) leads to NO synthesis, an endothelial growth factor that causes local vasodilatation. The expression of all these factors is regulated by the transcription factor NF-κB. In vitro screening showed that ganoderic acid T promoted aggregation of HCT-116 cells—cancer cells that are otherwise loosely attached to one another and can easily detach and migrate. It also inhibited adhesion of cancer cells to extracellular matrix components. The wound healing assay also confirmed that ganoderic acid T inhibits migration of these highly invasive cells (and also lung cancer cells, 95 D cells, as previously shown by Tang et al. [2006]). In all three assays, the activity was comparable to that of the antineoplastic drug doxorubicin. Ganoderic acid T was found to inhibit the activity of NF-κB by disabling its translocation to the nucleus, presumably by inhibiting degradation of the κB inhibitor (IκB), which under normal circumstances is bound to NF-κB in the cytoplasm, making it inactive until certain stimuli cause its lysis. This consequently causes downregulation of enzymes that are the under control of NF-κB—MMPs, uPA, and NOS—and lowers the ability of cancer cells to migrate. The same study further explored the ability of ganoderic acid T to inhibit tumor invasion in vivo, using an animal model with C57B/6 mice implanted with Lewis lung carcinoma (LLC) cells. It was found that ganoderic acid T was able to inhibit tumor growth (measured by weight) in a dose-dependent matter. Also, it inhibited tumor metastasis in lungs as well; at the highest dosage regimen of 28 mg/kg/day, the growth inhibition was ~60% and the number of metastatic changes was reduced by ~80%. Still, cisplatin was shown to be far more superior in both cases. Analysis of a tumor mass taken from mice showed that ganoderic acid T caused downregulation of two types of MMPs, collagenases MMP2 and MMP9, in vivo as well.

In a similar study, Chen et al. (2008) found that another triterpenoid isolated from the cultivated mycelium of *G. lucidum*, ganoderic acid Me, exhibits anti-invasion activity of highly metastatic lung cancer cell line 95-D cells. The wound healing assay showed that ganoderic acid Me was also capable of inhibiting migration and extracellular matrix adhesion of the tumor cells and promoted homotypic cell aggregation. Ganoderic acid Me suppressed gene expression of metalloproteinases MMP2 and MMP9 at the messenger RNA (mRNA) and protein level, which seems to be of great importance for the anti-invasive activity. Ganoderic acid Me was found to be highly toxic to 95-D cancer cells and exhibited low cytotoxicity on normal cells; it was confirmed that it causes cell cycle arrest in the G1 phase as well (Chen and Zhong, 2009). Further research aimed to investigate the role of antitumor protein p53 in the mechanism of its activity, and it was shown that p53 knockout cell lines did exhibit different behavior when incubated with ganoderic acid Me, which caused an S phase or G1/S transition arrest, indicating that ganoderic acid Me controlled cell cycle arrest at least in part through p53-mediated pathways. The same group also showed that ganoderic acid Me possesses significant cytotoxic activity against HCT-116 human colon carcinoma cells in a dose-dependent manner; they found that the compound increased p53 expression in a time-dependent manner. It did not alter levels of Bcl-2, but increased levels of Bax, leading to an increase in the Bax/Bcl-2 ratio and activation of further apoptotic steps. Ganoderic acid Me caused mitochondrial disfunction, which was evident by the decrease of the transmembrane potential of mitochondria and with the release of cytochrome *c* in the cytosol and consequent activation of caspase-3 (Zhou et al., 2011). Ganoderic acid Me was also a subject of an in vivo study in which C57BL/6 mice were inoculated with LLC cells. After implanting the tumor cells, mice were administered intraperitoneally either 7 or 28 mg/kg of ganoderic acid Me daily during a 10-day period. It was shown that it could inhibit tumor growth and prevent lung metastasis. Surprisingly, this effect was (at least partially) associated with the immune system stimulation by the triterpenoid compound—evidenced both by increased levels of proinflammatory cytokines and the activation of natural killer (NK) cells. Compared to the control group, the enhanced activity of NK cells from the mice spleen was noticed in the group treated with ganoderic acid Me, which rose from ~30% in the control group to ~50% in the group on the higher dosage regime. NF-κB expression in spleen tissue was also connected to ganoderic acid Me treatment and consequent immune responses. The concentrations of interleukin (IL)-2 and interferon (IFN)-γ in the mice blood serum were also increased in a dose-dependent manner. IL-2 levels tripled, while IFN-γ levels almost doubled. This increase was also confirmed at the mRNA level, providing further evidence that ganoderic acids may also act as immunostimulatory agents (Wang et al., 2007).

Ganoderic acid A induced apoptosis in human leukemia cell line Nalm-6 and was shown to be relatively selectively toxic to these cells at certain concentrations. It was also confirmed that ganoderic acid A had an effect on levels of certain microRNAs (miRNAs); miRNAs are oligonucleotides that can act either as oncogenes or tumor suppressors; two types of miRNAs, miR-17-5p and miR-181b, are known oncomirs in Nalm-6 cells, and ganoderic acid A causes downregulation of both (Mortazavie et al., 2022). Yang et al. (2018) investigated the activity of ganoderic acid A towards breast cancer tumor cells MDA-MB-231 and found that it exhibited cytotoxic activity by inhibiting the Janus kinase 2/signal transducer and activator of transcription 3 (JAK2/STAT3, respectively) signaling pathway, which regulates expression of certain cytokines, but also takes part in regulation of cell functioning and pro- and antiapoptotic proteins involved in the intrinsic apoptotic pathway. This study found that ganoderic acid A inhibited MDA-MB-231 cell viability, invasive capacity, and induced apoptosis, all in a dose-dependent manner, with the best results obtained using a relatively high concentration of 0.4 mM. An increase in concentration of reactive oxygen species (ROS) was also evident in the cells, and their production rose from 27% in the control to 76% when cells were incubated with 0.4 mM ganoderic acid A. Proteins involved in cell cycle regulation were also affected, with cyclin D1 expression suppressed, while p21 and p27 levels were significantly increased, leading to cell cycle arrest in the G0/G1 phase. Finally, JAK2 and STAT3 phosphorylation was inhibited by the treatment with ganoderic acid A, so the levels of antiapoptotic Mcl-1 and Bcl-xL were decreased, while proapoptotic proteins connected to intrinsic/mitochondrial apoptosis were increased (Bak, Bax, and cytochrome *c*). Ganoderic acid A exhibited similar effects on the hepatocellular cancer cell line HepG2, causing cell proliferation inhibition, suppression of invasion, cell cycle arrest in the G0/G1 phase, and apoptosis in a dose- and time-dependent manner. At a molecular level, this study confirmed that ganoderic acid A suppresses the expression of cyclin D1 (and therefore cell cycle arrest before the S phase) and causes elevation of p21 and activation of caspase-3 (Wang et al., 2017).

Jiang et al. (2008) tested the ability of three different ganoderic acids—ganoderic acid A, F, and H—to inhibit the growth of breast cancer cells, MDA-MB-231 cells. The authors targeted the activity of two factors which are found to be overly active in invasive breast cancer—activator protein-1 (AP1) and NF-κB. Both proteins are involved, among others, in the regulation of invasive factors, such as uPA and its receptor, uPAR. Also, both AP1 and NF-κB are responsible for control of other genes involved in normal cell functioning and that are crucial for cancer cell proliferation, survival, and invasiveness (cyclin D1, Bcl proteins, VEGF, etc.). Ganoderic acids A and H showed inhibitory activity in terms of proliferation, colony formation, and invasive behavior, all in a dose- and time-dependent manner. At the highest concentration of 0.5 mM, both compounds reached the maximum growth-inhibitory activity after 72h, which was between 50% and 60%. Adhesion of cells was also inhibited by both compounds, and the adhesion was decreased to 35–45%, and cell invasion was reduced to 40% and 16% by ganoderic acid A and H, respectively, at the highest concentration of 0.5 mM. Both compounds inhibited constitutive activation of AP1 and NF-κB, which control expression of both uPA and uPAR, but it was found that ganoderic acid A and H only affected uPA expression by markedly reducing uPA excretion, but did not have any effect on uPAR levels, suggesting that they only lead to downregulation of uPA. Ganoderic acids A and H also suppressed expression of Cdk4, which is involved in cell cycle regulation. However, more interesting is that ganoderic acid F showed no activity whatsoever; these 3 compounds share an overall similar structure, and the only difference between ganoderic acids H and F is the presence of a hydroxyl group in the former and a keto group in the latter at the C3 position. The hydroxyl group is thus crucial for ganoderic acid's activity, suggesting its role in the compound's binding to the target receptor. However, Nguyen et al. (2015) found in their study that ganoderic acid F possessed prominent dose-dependent anti-angiogenesis activity in a human umbilical vein endothelial cell (HUVEC) model. The cells were shown to be viable at concentrations of the compound at which it showed the activity. The mechanism of action was shown to be via suppression of the proliferation of cells and also by inhibition of tube-like capillary structure formation (Figure 8.1).

Liu and Zhong (2011) also tried to analyze the structure-activity relationship of ganoderic acids by testing 2 positional isomers, ganoderic acids Mf and S, for their ability to induce apoptosis in human cervical carcinoma HeLa cells. These two triterpenoids differ in that ganoderic acid Mf has β-OAc and β-OH groups at C3 and C15, respectively, while ganoderic acid S has α-OH and β-OAc at C3 and C22, respectively. Both compounds showed time- and dose-dependent inhibition of cell growth, which was achieved through apoptosis induction and cell cycle progression arrest. Both compounds induced apoptosis by causing mitochondria membrane dysfunction, release of cytochrome c in the cytosol, and activation of caspase-3 and caspase-9. Proapoptotic Bax protein was upregulated, while antiapoptotic Bcl-2 was downregulated, leading to a significant increase in the Bax/Bcl-2 ratio. The compounds, however, differed significantly when it came to cell cycle arrest, with ganoderic acid S at various concentrations leading to an accumulation of cells in the S phase (30.76–46.23%) and ganoderic acid Mf causing accumulation of cells in the earlier, G1 phase (42.31–55.75%). The number of apoptotic cells increased with the increased concentrations of the compounds (0–97.7 μM), and ganoderic acid Mf showed both better selectivity towards cancer cells and potency to induce apoptosis. The authors noted a possibility that C3 substituent orientation may be crucial for different activity when it comes to cell cycle arrest, with C3α substituent configuration being axial, while C3β is equatorial to the planar ring, thus exhibiting quite different molecular geometry.

After determining that *G. amboniense* extracts are cytotoxic to cancer cell lines and that they are able to inhibit the activity of topoisomerases in vitro, Li et al. (2005) found that the most active compound of the extract was ganoderic acid X. They tested the isolated compound against human hepatoma cancer cell line HuH-7 cells. Ganoderic acid X caused apoptosis, with a decrease in Bcl-xL levels and apparent mitochondrial dysfunction, with increased cytochrome *c* levels in the cytosol and activation of caspase-3. The activity on topoisomerase was tested in an in vitro assay, assessing the ability of topoisomerase I and IIα to catalyze the conversion of supercoiled plasmid DNA into

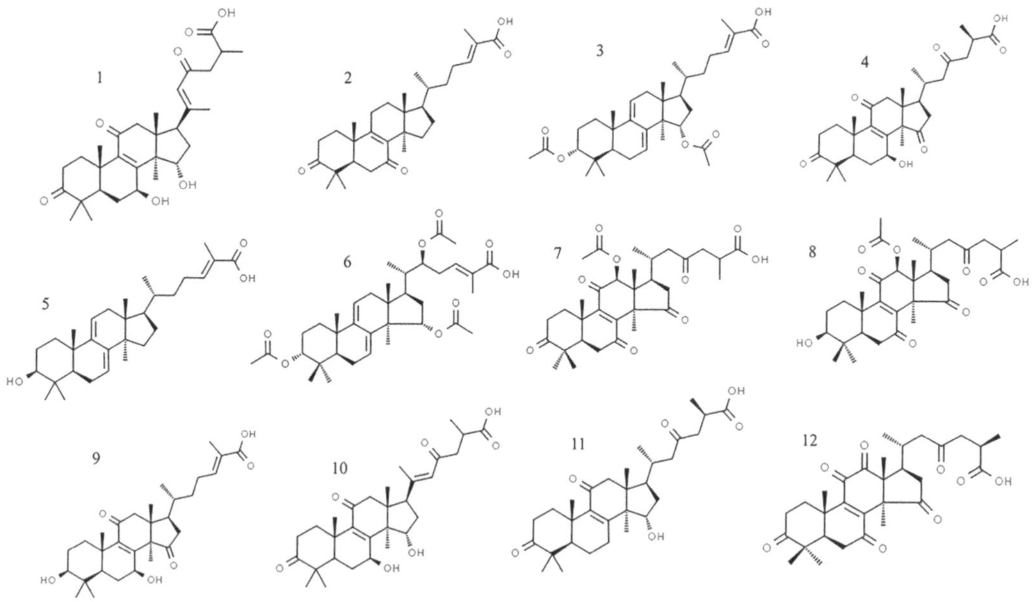

FIGURE 8.1 Structures of selected triterpenoid acids isolated from *Ganoderma* spp.: 1- ganoderic acid A, 2- ganoderic acid DM, 3- ganoderic acid Me, 4- ganoderic acid C1, 5- ganoderic acid Y, 6- ganoderic acid T, 7- ganoderic acid F, 8- ganoderic acid H, 9- ganoderic acid β, 10- ganoderic acid A, 11- ganolucidic acid A, 12- ganosporeric acid A.

a relaxed form in the presence of ganoderic acid X. While relaxation of the supercoiled plasmid was achieved in 30 mins (~90%) in normal conditions, ganoderic acid X caused a direct, dose-dependent inhibition of the enzymes' activity, being slightly more active towards topoisomerase IIα than topoisomerase I, with IC_{50} of 25 and 55 μM, respectively. The ability of topoisomerase IIα to decatenate a massive DNA network (kDNA) into minicircle DNA was inhibited by ganoderic acid X as well. Finally, in HuH-7 cell culture, the time-dependent inhibitory activity on DNA synthesis was observed; in the cells incubated with ganoderic acid X, during first 30 min, [3H]-thymidine incorporation was decreased about 33%, and after 3h it reached 66%. The inhibition of either topoisomerase may result in aberrant mitosis, leading to "mitosis catastrophe", an event which can trigger the apoptotic signaling pathways.

Ganoderic acid DM was marked as a promising natural-origin supplement for prostate cancer treatment due to its antiandrogenic activity. It was found to act as an inhibitor of 5α-reductase, an enzyme that catalyzes testosterone transformation into its more active form, dihydrotestosterone, and thus inhibiting growth of hormone-dependent prostate cancer; also, it can bind to and block the androgen receptor, which disrupts the normal dihydrotestosterone-mediated signaling pathway (Johnson et al. 2010). Ganoderic acid DM inhibits androgen-induced cell growth, affecting cancer cell proliferation, in a dose-dependent manner (Liu et al., 2009). When compared with the commercial steroid inhibitor (finasteride) and natural one (α-linolenic acid), ganoderic acid DM's potency was found to be in the middle (0.73 μM, 10.6 μM, and 116 μM, respectively). In addition, the same study examined the ability of ganoderic acid DM to inhibit osteoclastogenesis from the RAW264 cells without a cytotoxic effect. The observed effect is important in connection with later-stage prostate cancer development, which is metastasizing in the bones. This process is supported by the overproduction of osteoclasts (Waning, Mohammad and Guise, 2013). The authors concluded that ganoderic acid DM can be efficient therapy in both cases, androgen-dependent (proved by the test on lymph node carcinoma of the prostate cell line, LnCaP) and independent prostate cancer types (tested on PC-3 cell line).

The anticancer mechanism of ganoderic acid DM was proved in the case of breast cancer, which is, like prostate cancer, also hormone-sensitive (Wu et al., 2012). The effect was tested on two breast cancer cell lines, one was estrogen receptor (ER) positive (MCF-7) and another was ER-negative (MDA-MB-231). And although the proliferation of both types of cells was suppressed, ER-positive breast cancer cells were more sensitive to ganoderic acid DM. The mechanism behind the observed effect in MCF-7 cells was proven to be G1 cell cycle arrest, even at very low concentrations like 25 μM. The proteins whose downregulated expression was responsible for the G1 cell cycle arrest were identified as CDK2, CDK6, cyclin D1, and p-Rb, as well as the oncoprotein c-Myc. Furthermore, the authors demonstrated that ganoderic acid DM induces apoptosis, and the possible mechanism behind it might be the DNA damage registered after only 6 hours since ganoderic acid DM treatment of MCF-7 cells.

Hossain et al. (2012) explored the anticancer efficacy of ganoderic acid DM using several human melanoma cell lines (J3, HT-144, 1359-mel, and DM-331). It was reported that ganoderic acid DM decreased melanoma cells' viability and induced apoptosis, which was caspase-dependent. The apoptotic molecular mechanism was also investigated, and Western blot analysis showed a reduction in the level of survivin (an apoptotic inhibitor) expression as well as an increase of cytochrome *c* and Apaf-1. In addition, fine-tuning between autophagy and apoptosis was suggested as a possible mechanism of the ganoderic acid DM antitumor effect. At the same time, the mushroom's triterpene stimulated the immune response by promoting the CD4+ T cell's ability to recognize the tumor. However, this effect was only present when lower doses were used (20 μM), while the higher concentrations exhibited a cytotoxic effect. In an experiment in vivo, the mice bearing tumors were treated with ganoderic acid DM, which slowed tumor formation.

Xia et al. (2020) examined ganoderic acid DM efficacy in another type of cancer common in China, non–small-cell lung cancer (NSCLC). It was shown that ganoderic acid DM inhibits proliferation and induces apoptosis in two types of NSCLC cell lines (A549 and NCI-H460), which eventually leads to growth inhibition. Apoptosis stimulation was time- and dose-dependent, and it was more expressed in NCI-H460 than in A549 cells (31.2% and 40.3% at 20 μM, respectively). On the molecular level apoptosis was proved by the detection of a reduced level of Bcl-2, an apoptosis-related protein. Moreover, ganoderic acid DM induced autophagy in cancer cells, which might be an early step contributing to later apoptosis in test cells. Finally, it was shown that autophagic apoptosis was initiated by the inhibition of the PI3K/Akt/mTOR signaling pathway, which is known as being crucial in the modulation of the development and further growth of cancer in humans.

Das et al. (2019) conducted a pre-clinical study of a mixed ganoderic acid A and ganoderic acid DM effect in anaplastic meningioma treatment based on experiments in cell culture and a xenograft animal model. The results showed that this combination of ganoderic acids caused cell death via upregulation of NDRG2 protein expression while without having a cytotoxic effect toward normal neuron and human arachnoid cells. Expression of several proteins (MMP-9, p-P13K, p-AKT, p-mTOR, and Wnt-2) responsible for the cancer cell life cycle was suppressed, while apoptotic factors, like Bax, were promoted. In addition, it was shown that ganoderic acid A/DM downregulated the NDRG2 promoter mRNA level. When xenografted mice were treated with ganoderic acids for 14 days their tumor volume decreased significantly while the survival rate increased. Moreover, side effects like hepatic injury caused by the drug's cytotoxicity were also evaluated, and no changes in the liver alkaline phosphatase levels were observed for the applied dose of tested ganoderic acid A/DM.

Besides triterpenoid acids, triterpenoid alcohols were also shown to possess prominent antitumor ability; Min et al. (2000) tested 20 different triterpenoids isolated from *G. lucidum* for their cytotoxic ability; they included ganoderic acids A, B, C1, C2, C6, G, δ, ε, ζ, η, and θ; ganolucidic acids A and D; lucideric acid α and ganoderic alcohols; lucidumols A and B; ganoderiol F; ganodermanondiol; and ganodermanontriol. The activity against mouse fibrosarcoma (Meth-A) and LLC cells was more superior in the case of ganoderic alcohols, with ganoderic acids in many cases not being able to exhibit any cytotoxic ability even at the highest concentration used (20 μg/mL). Ganoderic alcohols, on the other hand, were especially effective against Meth-A cells, having ED_{50} values between 3.4 and 8.5 μg/mL; however, doxorubicin was still more than 100 times more effective in this study. Lucidumols C and D, isolated from *G. lingzhi*, were tested against several cancer cell lines (breast cancer MCF-7, hepatocellular carcinoma HepG2, cervical cancer HeLa and colon cancer Caco-2, and HCT-116, as well as noncancer colon epithelial cells CCD 841 and primary normal human dermal fibroblasts [NHDFs]); interestingly, lucidumol C showed relatively high toxicity towards all cell lines, while lucidumol D exhibited somewhat lower activity towards MCF-7, HepG2, and HeLa cells and significantly lower activity towards both colon cancer cell lines, while normal cell lines were practically not affected at the concentration range tested. The 2 compounds differ only in functional group at C11—with lucidumol C having a carbonyl group, while lucidumol D has a hydroxyl group—which implies that very small structural differences can strongly affect ganoderic alcohols' cytotoxic potency and/or affinity towards certain cell types (Satria et al., 2018). Ganodermanontriol is another relatively well-studied ganoderic alcohol with a slightly different mode of action, at least on colon cancer cells. It caused arrest of HCT-116 and HT-29 colon cancer cell proliferation (without significant effect on their viability), through interfering with ß-catenin signaling activity. This dual-functioning protein regulates cell adhesion to other cells, and also gene transcription, and is a part of the Wnt signaling pathway, which is involved in most cases of colorectal cancers. One of the genes regulated by ß-catenin is responsible for cyclin D1 transcription, which is involved in regulation of cell cycle progression, and consequently proliferation. High ß-catenin levels, caused by mutations of the protein itself or by other mutations in factors that regulate its levels, can consequently cause elevated cyclin D1 levels, which is seen in almost a third of all colorectal cancer cases. Ganodermanontriol inhibits ß-catenin's transcription activity, which leads to downregulation of cyclin D and cell cycle arrest, resulting in the suppression of tumor cell proliferation. Total levels of ß-catenin are, however, elevated after treatment with ganodermanontriol, but so are the levels of E-cadherin, an adhesion protein known to be an important tumor suppressor. This indicates that since ß-catenin's transcription activation ability is compromised, it actually migrates to the cell membrane, where it binds to E-cadherin, one of the functional proteins that recognize it and have the ability to bind to it. Most importantly, ganodermanontriol was tested in vivo, using a xenograft model with mice inoculated with HT-29 colon cancer cells in a dosage regimen of 3.0 mg/kg of body weight/day for 28 days intraperitoneally. The authors concluded that ganodermanontriol significantly inhibited tumor growth in the treated group compared to the control group, although the results were not consistent. The expression of cyclin D1 was downregulated in vivo, too, and there were no noticeable adverse reactions to ganodermanontriol administration. (Jedinak et al., 2011). Ganodermanontriol was further found to downregulate expression of cell cycle regulatory protein CDC20 in breast cancer cell lines MDA-MB-231, which was found to be overexpressed in tissue samples of breast cancer in human patients. Similar to ganoderic acid's mode of action, ganodermanontriol exhibited antiaggregation, antiadhesive, and antimigrating activity on breast cancer cells, which could be connected to downregulation of uPA (Jiang et al., 2011), whose role in tumor metastasis was previously discussed.

Ganodermanontriol can be derived from lanosterol through semisynthesis, which is of great importance, as obtaining bioactive triterpenoids solely from fruiting bodies or mycelium is a time-consuming and expensive process with relatively low yields. The method proposed by Kennedy et al. (2011) gives a 15.3% yield of ganodermanontriol and 3 of its stereoisomers; the naturally occurring form was still the most active against proliferation of estrogen-dependent and estrogen-independent breast cancer cell lines (MCF-7 and MDA-MB-231).

Ganoderol B was isolated from *G. lucidum* ethanol extract, using bioassay-guided isolation, as one of the compounds involved in the inhibitory activity of 5α-reductase by Liu et al. (2007). However, it was reported to be far less active than ganoderic acids TR and DM, as well as a-lanosta-7,9(11),24-triene-15a,26-dihydroxy-3-one, IC_{50} values of which were estimated by the same research group to be 8.5, 10.6, and 41.9 μM. Ganoderol B, on the other hand, inhibited only 37% of the enzyme's activity at 113 μM, and IC_{50} could not be reached, as the compound was not soluble enough at concentrations higher than 120 μM. Ganoderol B also showed significantly lower affinity for binding to the androgen receptor (AR), requiring a concentration of 15 μM for 50% of enzyme saturation, compared to dihydrotestosterone, which exhibited the same effect at only 0.018 μM. Lymph node carcinoma of the prostate cells (LNCaP) was not affected by ganoderol B when incubated at concentrations between 9.25 and 25 μM, but in the presence of androgens (testosterone or DHT), it inhibited cell growth in a dose-dependent manner, confirming that its mechanism of activity comes from its antiandrogen effect. Moreover, the expression of both AR and prostate-specific antigen (PSA) were downregulated at the mRNA level in the cancer cells. In an in vivo experiment, ganoderol B reduced testosterone-induced prostate growth by weight in rats, which was found to be reversible. Dosage regimens of 0.001 and 0.1 mg/kg/day inhibited prostate growth by 31% and 49%, respectively. *G. lucidum* ethanol extract, administered at 1 mg/kg/day, which contained 0.1% of ganoderol B, was found to inhibit prostate growth by 28%, suggesting that ganoderol B was one of the main antiandrogenic compounds responsible for its activity.

As semi-synthesis is often used in drug development to create compounds with better activity and safety profiles of physicochemical properties, Liu et al. (2012) designed semisynthetic derivatives of ganoderic acid T to test whether its cytotoxic ability can be improved. They synthesized 3 different esters, methyl-, ethyl- and propylganoderate T, as well as an amide derivative, "ganoderamide T". The compounds were tested on 3 cancer cell lines, HeLa, HepG2, and 95-D, and a normal cell line, MCF-10A. The amide derivative was shown to be the most effective against HeLa cells, with an IC_{50} value of just 4.1 μM, while exhibiting cytotoxicity against MCF-10A at much higher concentrations (IC_{50} = 100.8 μM). All compounds showed better activity than a standard used in this study, a DNA topoisomerase I inhibitor hydroxycamptothecin. Both ganoderic acid T and its amide derivative were further tested for the ability to inhibit DNA synthesis in bromodeoxyuridine assay and were found to decrease bromodeoxyuridine incorporation absorption in HeLa cells at 40 μM for 67% and 76%, respectively, while the decrease was significantly lower in the normal MCF-10A cells (25% and 18%, respectively). All 5 compounds were also found to arrest the cell cycle of HeLa cells at the G1 phase, causing accumulation of sub-G1 populations, and, again, the amide derivative was the most active of the compounds. MCF-10A cells, after treatment with either ganoderic acid T or its amide derivative, showed only a slight increase in the G1 phase population. The mechanism of apoptosis was confirmed to include mitochondrial membrane dysfunction and activation of caspase-3 and caspase-9. The study also confirmed that carboxylic acid moiety is not necessary for cytotoxic activity, as was also shown before in studies conducted with ganoderic alcohols.

Colossolactones were described from a Vietnamese *G. colossum* (Kleinwächter et al., 2001; El Dine, El Halawany, Ma et al., 2008). Some of these triterpenoids have a very unusual structure, being pentacyclic dilactones, with two cycloheptane rings in the backbone, one of which is formed by intramolecular cyclization and represents a lactone (ring A); the other δ-lactone is formed on the side chain of the ring D. Triterpenoids of such structure had been previously known from plants of the genus *Schisandra* (schisanlactones) and *Kidsura* (kadsulactone A, kadsudilactone, lancilactones). Colossolactone H was marked as one of the most cytotoxic among these compounds and was studied in more detail in regard to its antitumor mechanism of action in H1650 lung cancer cells. Using GeneChip analysis, 8568 genes were identified in the cells, among which 252 genes were significantly upregulated and 398 genes were significantly downregulated by the treatment with colossolactone H. Cell cycle progression gene downregulation, in combination with redox gene upregulation was responsible for the compound's cytotoxicity. Colossolactone H caused cell growth inhibition and a significant increase in ROS, causing DNA damage, as well as elevation of p53 protein levels. An in vivo study in athymic mice bearing xenograft tumors also showed that colossolactone H in combination with gefitinib was able to inhibit tumor growth (Chen et al., 2016) (Figure 8.2).

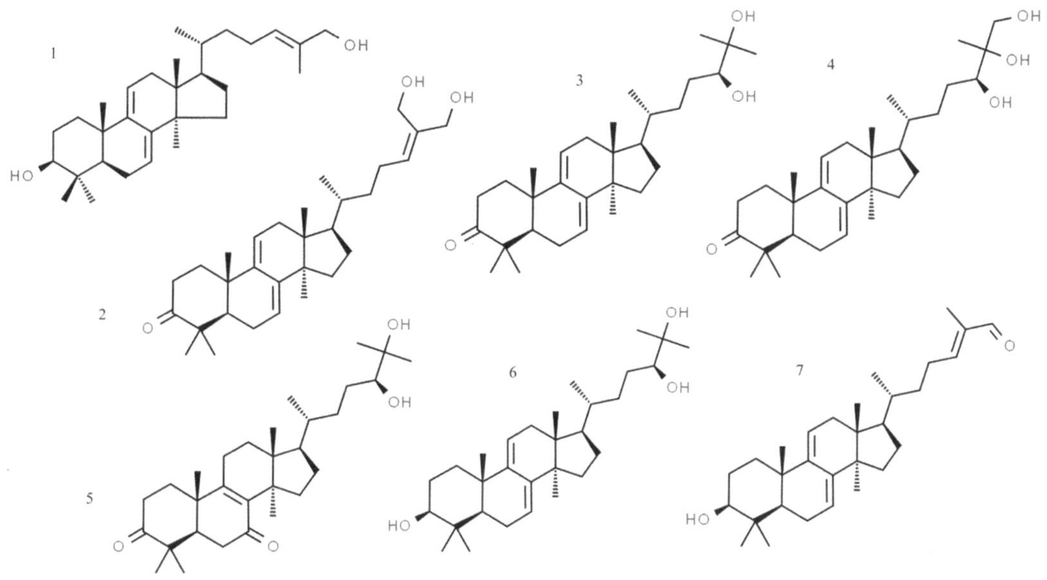

FIGURE 8.2 Structures of selected triterpenoid alcohols and an aldehyde isolated form *Ganoderma* spp.; 1- ganoderol B, 2- ganoderiol F, 3- ganodermanondiol, 4- ganodermanontriol, 5- lucidumol A, 6- lucidumol B, and 7- lucialdehyde A.

8.2.2 Antiviral Activity

After finding that methanol extracts of *G. lucidum* possess anti-HIV activity, fractionization of the extract with different solvents revealed that the $CHCl_3$-soluble fraction could inhibit HIV-1 protease. Soon, triterpenoids were identified to be the active components. Min et al. (1998) isolated 10 different triterpenoids from *G. lucidum* fruiting bodies and tested them for potential activity against HIV-1 protease. The compounds showed very different activity, with ganoderic acid β exhibiting IC_{50} at 20 μM, but with lucidumol A and ganoderiol F not being able to exhibit any activity at the maximum 500 μM tested. No QSAR (Quantitative structure–activity relationship) analysis was provided, and it is yet to be revealed how exactly these compounds interact with the enzyme. However, even the activity of ganoderic acid β was almost 100 times lower than that of the acetyl-pepsatin, which was used as a positive control in the assay.

At the same time, El-Mekkawy et al. (1998) were also investigating the anti-HIV activity of *Ganoderma* triterpenoids, although besides testing compounds for antiprotease activity, they examined their potential to inhibit an HIV-1–induced cytopathic effect in a leukemia T-cell line (MT-4 cells). Several compounds were tested for their antiprotease activity by both research groups, and the results obtained for ganoderic acids A, B, and C1 and ganoderiol F may be regarded as similar, with ganoderic acids B and C1 being moderately active and ganoderic acid A and ganoderiol F either showing weak activity or not showing activity at all in the range of the tested concentrations. However, the results obtained for ganodermanontriol differed drastically; while Min et al. (1998) found that it was among the most active compounds, with IC_{50} of 70 μM, El-Mekkawy et al. (1998) couldn't detect any activity even at 1 mM. Still, they found that ganodermanontriol and ganoderiol F were the only compounds among the 9 triterpenoids tested that were capable of inhibiting HIV-1–induced cytopathic effect in MT-4 cells (100%), and at a relatively low concentration of 7.8 μg/mL, although both compounds were exhibiting cytotoxicity on the cells at 15.6 μg/mL. None of the tested triterpenoids was able to inhibit HIV-1-RT, another essential HIV enzyme.

El Dine, El Halawany, Nakamura et al. (2008) tested 6 colossolactones and schisanlactone A against HIV-1 protease; colossolactones E, V, and VII, as well as schisanlactone A exhibited significant activity as enzyme inhibitors, with IC_{50} values ranging from 5 to 13.8 μg/mL. The authors noted that pentacyclic compounds with hydroxyl substitutes in the skeletal core exhibited lower activity, indicating that hydrophobicity of the core may be important for the activity. Among the tested compounds, 3 were of *seco* or opened structure (tri- and tetracyclic); for those compounds, the presence of a single double bond in the backbone ($\Delta^{8,9}$) seemed to be important for the activity, since compounds with two double bonds ($\Delta^{7,8}$, $\Delta^{9,11}$) were less active; the geometry of the compound is significantly altered by the changes in the positions and numbers of double bonds in the backbone.

Zheng and Chen (2017) tested 5 triterpenoids derived from *G. lucidum*, ganoderic acids A and B, ganoderol B, ganodermanondiol, and ganodermanontriol for their ability to inhibit Epstein-Barr virus (EBV) infection, which is a risk factor for lymphoid and epithelial malignancies, including Hodgkin and Burkitt lymphoma, nasopharyngeal and gastric carcinoma, etc. EBV primarily infects B lymphocytes, but epithelial cells and other immune cells may also carry the virus. The mechanism of malignant transformation induction by EBV in cells involves activation of telomerases, enzymes not normally active in somatic cells, which are involved in telomere length maintenance. Telomeres are DNA structures located at the chromosome ends and are crucial for DNA stabilization; every time replication occurs, telomeres are shortened, which prevents endless cell division. However, activation of telomerases disturbs this replication regulation system, causing cell "immortalization" and progression towards malignant transformation (Dolcetti et al., 2014; Kamranvar and Masucci, 2017). The triterpenoids were thus tested on Raji cells, lymphoblast-like cells derived from Burkitt lymphoma and B95-8 cells, and EBV-transformed tamarin cells, both carrying the EBV genome. The compounds did not exhibit cytotoxicity upon the cell lines, but significantly decreased EBV early antigen (Raju cells) and capsid antigen (B95-8 cells) activation. At a lower concentration of 3.2 nM, all the compounds caused moderate inhibition of both antigens, but ganoderic acids A and B were more active in lowering levels of early antigen-positive cells than other compounds, to about 60%. At higher concentrations of 16 nM, all compounds caused a very pronounced decrease of both early and capsid antigen-positive cells when compared to control, with ganoderic acids A and B again being superior in suppressing EBV early antigen activation and reducing early antigen-positive cells to 19% and 17%, respectively. The compounds also showed inhibitory activity on telomerase, lowering the enzyme's activity in nasopharyngeal cancer cells NPC 5-8 F to 80% on average; however, a docking molecular study showed that ganoderic acid A is able to bind to the telomerase ligand site.

Zhang et al. (2014) examined the antiviral activity of ganoderic alcohol, lanosta-7,9(11),24-trien-3-one,15,26-dihydroxy, and ganoderic acid Y against enterovirus 71, an agent responsible for hand, foot, and mouth disease in children, which can cause fatal complications and for which there is still no approved antiviral therapy. Both compounds exhibited antiviral activity, but did not show cytotoxicity against rhabdomyosarcoma (RD) cells at the same concentrations. RD cells were then infected with enterovirus 71 and treated with triterpenoids or ribavirin, an antiviral agent used as a positive control. While all compounds' antiviral activity increased with a concentration increase to 4 μg/mL, at higher concentrations of 20 and 100 μg/mL, ribavirin's activity decreased while the triterpenoids' activity still remained high and was superior to that of ribavirin. Both triterpenoids displayed antiviral activity of about 80%, while ribavirin's maximum activity was significantly lower, at about 50%. In two other assays used to determine the compounds' mode of action, either RD cells or virions were pre-incubated with the compounds and then mixed to initiate infection. In both cases, the viral inhibition rate exhibited by triterpenoids was very high at the concentration of 20 μg/mL and was between 80% and 90%. However, ribavirin's peak inhibition was lower than 30% in both assays. A molecular docking study showed that there is a possible means of interaction between the two compounds and the virion, via "F site", a hydrophobic pocket of the viral capsid protein, which may prevent uncoating of enterovirus 71 and thus inhibit its infective ability. Moreover, it was found that viral RNA levels were reduced when infected cells were treated with triterpenoids, significantly reduced when cells were pretreated with the compounds, and inhibited by ~80% when viral particles were first incubated with the compounds at the concentration of 20 μg/mL, indicating that viral RNA replication was inhibited via interaction of the compounds with the viral capsid, as predicted by docking analysis.

Bharadwaj et al. (2019) had a different, in silico approach when looking for potent anti– Dengue virus compounds among *Ganoderma* triterpenoids. They performed a molecular docking analysis in order to find compounds that can bind to and potentially inhibit DENV NS2B-NS3 protease, a Dengue virus enzyme recognized as a main target for antiviral therapy. Among 22 "virtually tested" triterpenoids, 4 were found to be the most active, having a better docking score than a reference inhibitor 1,8-dihydroxy-4,5-dinitroanthraquinone (–5.377 kcal/mol); these were ganosporeric acid A (–5.830 kcal/mol), ganoderic acid C2 (–5.948 kcal/mol), lucidumol A (–5.993 kcal/mol), and ganodermanontriol (–6.291 kcal/mol). Ganodermanontriol and ganoderic acid C2 were further tested in an in vitro Dengue virus inhibition assay, but only ganodermanontriol showed some inhibitory activity, lowering the viral titer for ~25 and ~40% at the concentrations of 25 and 50 μM, respectively.

Since inhibitors of neuraminidase represent the main antiviral agents used in the treatment of influenza, *Ganoderma* triterpenoids were also tested against neuraminidase originating from various strains of influenza virus: H1N1, H3N2, H5N1, and H7N9. Among 31 triterpenoids, which included ganoderic and ganoderenic acids, as well as ganoderols, ganoderiol F, ganodermanondiol, ganodermanontriol, and lucialdehydes, all tested at the concentration of 200 μM, 2 compounds emerged as potent inhibitors of both H1N1 and H5N1 neuraminidase, ganoderic acid T-Q and TR, with inhibition rates of 81.7% and 87.4% against H1N1 neuraminidase and 94.4% and 96.5% against H5N1 neuraminidase, respectively. Two other types of neuraminidases were also significantly inhibited by these compounds, but in a range between 50% and 60%. In comparison, the well-known ganoderic acid A exhibited only 5% and 31.4% inhibition activity against H1N1 and H5N1 neuraminidases, respectively. All triterpenoids showed better activity against neuraminidase type 1 (N1 variants). The study further concentrated on N1 types of neuraminidases, and IC_{50} values were estimated in an in vitro assay; ganoderic acids T-Q and TR were confirmed to be the most active compounds, with IC_{50} values falling in the range between 1.2 and 10.9 μM, while for most of the other compounds, it was found to be higher than the maximum concentration used in the assay (200 μM). However, their cytotoxicity, which was also estimated against the breast cancer cell line MCF7 cells, was also relatively high, with CC_{50} values of ganoderic acids T-Q and TR being 28.2 and 91.6 μM, respectively. Ganoderol B on the other hand, showed no cytotoxicity at 200 μM, but exhibited IC_{50} against H5N1 at 35.5 μM, meaning that through chemical modifications, it is possible to achieve better antiviral activity with minimal cytotoxicity and that ganoderic acids T-Q and TR are still a great molecular scaffold for the future design of neuraminidase inhibitors. Docking studies deduced that the backbone shared by these compounds—two double bonds ($\Delta^{7,8}$, $\Delta^{9,11}$) in the tetracyclic ring and a carboxylic acid moiety at the branch—were connected with the better activity, compared to two other types of backbones—two double bonds ($\Delta^{7,8}$, $\Delta^{9,11}$) in the tetracyclic ring and lack of a carboxylic acid group at the branch (alcohol or aldehyde group instead) or one double bond ($\Delta^{8,9}$) and a carboxylic acid moiety at the branch (Zhu et al., 2015).

8.2.3 Ganoderic Acids as Potential Agents in the Treatment of Metabolic Disorders

Although the research on the different biological activities of *Ganoderma* secondary metabolites started some decades ago, their possible use in diabetes prevention via the anti-α-glucosidase and aldose reductase effect started to be reported about a decade ago. Among 17 mushroom species, methanol and ethanol extracts of *G. lucidum* proved to be the most potent inhibitors of aldose reductase, an enzyme that can contribute to the accumulation of sorbitol, further leading to diabetic complications like cardiovascular ones (Fatmawati et al., 2008; Jannapureddy et al., 2021). The authors from the former study strongly suggested that the observed activity should be attributed to triterpenoids. Furthermore, Fatmawati et al. (2008) performed the in vivo test on galactitol accumulation in the galactosemic rat model. Again, *G. lucidum* extract was an effective reducer of galactitol in rats' lens. In the next study, Fatmawati et al. (2010) isolated and identified the compound responsible for aldose reductase inhibitory activity, ganoderic acid Df, proving that the activity is due to a carboxyl group of the compound's side chain. Moreover, the same group (Fatmawati et al., 2010) examined the $CHCl_3$ extract of cultivated *G. lucidum* fruit body in vitro and found it to show inhibitory activity on α-glucosidase. The isolated active compound ganoderol B had around 5 times lower IC_{50} than a positive control, acarbose. In addition, another species, *G. lingzhi*, turned out to be a good source of lanostane-type triterpenoids with α-glucosidase inhibitory activity (Fatmawati et al., 2013). As reported in the same research paper, there are several structural requirements necessary for the expression of anti-α-glucosidase activity. The OH substituent at C-11 as well as the carboxylic group in the side chain is necessary for the recognition of α-glucosidase inhibitory activity, while the OH substituent at C-3, the double-bond moiety at C-24 and C-25, and the double moiety at C-20 and C-22 in the side chain improve α-glucosidase inhibitory activity. Another group of authors performed ethanol extraction of another cultivated *Ganoderma* species, *G. leucocontextum*, and isolated 26 triterpenes, all of the lanostane type (16 new ones), tested them all on α-glucosidase activity, and found that 3 of them have very potent inhibitory activity in vitro (using α-glucosidase from baker's yeast and small intestinal mucosa from rats) in comparison with acarbose (Wang, Bao et al., 2015). As the authors described, the observed effect is due to farnesyl hydroquinone moiety in ganoleucoins M, N, and P. Binh et al. (2018) identified both new and several known lanostane triterpenoids from methanol extracts of the fruit bodies of two *Ganoderma* species, *G. lucidum* and the less researched *G. multipileum*. The extracts of both species exhibited α-glucosidase inhibition, and 3β-lanosta-7,9(11)-dien-3-ol was the most effective with the lowest IC_{50} value, 198.8 μM, in comparison with the acarbose standard whose activity was 712.4 μM. Ren (2019) tested in vitro inhibitory activity of ganoderic acid A on α-amylase and α-glucosidase enzymes and observed significantly stronger activity of this compound in comparison with acarbose. Furthermore, ganoderic acid was administered to rats, which appear to have better utilization of glucose when supplemented with this triterpene. The same supplementation significantly reduced blood glucose levels and the incidence of diabetes.

Piling up the evidence of antidiabetic activity of *Ganoderma* triterpenes resulted in an interest in their extraction optimization. Ryu et al. (2021) combined ultrasonication and heating and by response surface methodology, optimized the extraction procedure. The highest antidiabetic activity was achieved when frieze-dried *G. lucidum* fruit bodies were extracted at temperatures between 66.8° and 70°C for more than 2.8 h.

Promising in vitro results for antidiabetic activity based on specific enzyme inhibition were further extended to in vivo models, like rats. In the study by Wang, Zhou et al. (2015), diabetic male rats were fed *G. lucidum* shell-broken spores for 4 weeks. The spores of this species are known to be rich in triterpenoids (Ma et al., 2011). As reported, the spore supplement (1 g per day) lowered blood glucose levels by promoting glycogen synthesis and inhibiting gluconeogenesis.

Hypercholesterolemia is recognized as one of the main causes of blood vessel deterioration and consequent cardiac-related complications such as thrombosis and increased risk of heart attack. Lowering high blood cholesterol levels—specifically a type of cholesterol protein carriers called low density lipoprotein (LDL)—is therefore crucial for the prevention of cardiovascular diseases. Targeting enzymes in the cholesterol anabolic pathway is by far the most efficient way to lower endogenic cholesterol. The enzyme that was shown to be rate-limiting in the cholesterol biosynthesis process in hepatocytes, 3-hydroxy-3-methylglutaryl co-enzyme A reductase (HMG-CoA reductase), catalyzes the conversion of HMG-CoA to mevalonic acid (MVA). Inhibition of HMG-CoA reductase is the mechanism of action of the statins, which are among most prescribed drugs (Atlı et al., 2015). Lovastatin, originally isolated from micromycetes such as *Aspergillus terreus* and *Monascus purpureus* (Gunde-Cimerman, Friedrich et al., 1993), was later confirmed to be produced by the oyster mushroom (*Pleurotus ostreatus*) (Gunde-Cimerman, Plemenitaš et al., 1993) and other macromycetes (Petrović, Ivanovic et al., 2019; Petrović and Vunduk, 2022). Since triterpenes share a similar structure with cholesterol, testing these compounds for their potential cholesterol synthesis inhibitory activity was a next logical step. Wang, Bao et al. (2015) and Zhang et al. (2018) isolated several triterpenoids from *G. leucocontextum*, native to the Tibetan plateau, which possessed unique structures and were named ganoleucoins (although the authors also used the term "ganoleuconins", which was presumably a typing error). Ganoleucoins J, K, and L were particularly interesting, as they possessed 3-hydroxy-3-methylglutaryl (HMG) moiety in the position C4. Ganoleucoins M,

N, O, and P, on the other hand, represented conjugates of triterpenoids and farnesyl hydroquinone groups, known as ganomycins. Indeed, all compounds with HMG moiety showed inhibitory activity towards HMG-CoA reductase in vitro; ganoleucoin J, which represents ganoleucoin D substituted with a HMG group at C4, exhibited almost 4 times better inhibitory activity than ganoleucoin D (IC_{50} =26.4 μM for ganoleucoin J and IC_{50} = 97.5 μM for ganoleucoin D), leading to a conclusion that HMG moiety increases a compound's affinity towards the enzyme. Two other compounds with HMG moiety had even more pronounced activity, although ganoderic acids DM and Y, as well as ganoderiol J (which does not possess such a group) had similar and in some cases even better activity. Perhaps introducing HMG moiety in the structure of these compounds can increase their activity even further. In a subsequent study, more similar compounds isolated from the same *Ganoderma* species were tested for their potential anti–HMG-CoA reductase activity. Ganoleucoins T and Y represent structural analogues but differ in that ganoleucoin Y contains the HMG group at C4. Both compounds, however, showed very similar activity, meaning that HMG group did not have any influence on it. The authors performed docking studies and found that both compounds can bind in the same pocket on the enzyme as atorvastatin, but different amino acid residues are involved. The majority of the tested triterpenoids didn't show any activity in the range of tested concentrations (≤100 μM), but several compounds (lucidenic acid E, ganolucidic acid η, and ganoderenic acid K) exhibited in vitro activity that was comparable to that of atorvastatin. These compounds have different structures, and a QSAR analysis is needed to understand the structure-activity relationship. More research by independent groups is needed before reaching any definitive conclusions.

Su et al. (2020) isolated a total of 22 triterpenoid compounds from *G. applanatum* and evaluated their antiadipogenic effect, measuring their ability to inhibit lipid accumulation in 3T3-L1 adipocytes in an in vitro assay. Two compounds, ganoapplanoids K and Q, were found to possess the best inhibitory activity towards adipogenesis. The compounds were first shown not to exert cytotoxic effects towards cells at the concentrations up to 50 and 200 μM, respectively, so cells were incubated with the compounds in the range of concentrations between 2.5 and 40 μM. Both ganoapplanoids K and Q showed dose-dependent activity and at the highest concentration reduced the lipid content to ~60 and ~55%, respectively. The compounds also decreased significantly the content of intracellular cholesterol and, especially triglycerides, in a dose-dependent manner. The same research group further isolated more triterpenoids from *G. applanatum* as a part of screening for more potential antiadipogenic compounds; ganodapplanoic acid I was found to be the most active compound, with the safest profile on 3T3-L1 preadipocyte, with no observable cytotoxicity at concentrations as high as 200 μM. It reduced lipid contents in the cells to 68.8%, compared to control, at the concentration of 20 μM. However, increased compound concentrations did not cause any significant further reduction in cell lipid content. However, peroxisome proliferator-activated receptor γ (PPARγ), CAAT/enhancer-binding protein β (C/EBPβ), and fatty acid synthase (FAS) expression were all downregulated by ganodapplanoic acid I, meaning that the compound may inhibit differentiation of 3T3-L1 preadipocytes and decrease the adipogenesis through regulation of expression of these factors (Su et al., 2021) (Figure 8.3).

8.2.4 Hepatoprotective Ability of *Ganoderma* Triterpenoids

Extracts of *Ganoderma* species have been known to possess strong antioxidant activity, which is backed up by numerous publications (Mustafin et al., 2022; Yan et al., 2019; Mohsin et al., 2011). The antioxidant activity of these mushrooms can be beneficial in some conditions, such as heart toxicity in acute alcohol poisoning (Wong et al., 2004) or liver injury (He et al., 2008; Thuy et al., 2022). Moreover, *Ganoderma* tinctures have been used to treat liver dysfunction in death cap (*Amanita phalloides*) poisonings, with clinically proven beneficial effects on recovery (Zhang, Tang et al., 2022). Triterpene-enriched extracts were shown to possess both antioxidant (Smina et al., 2011) and hepatoprotective abilities (Zhao et al., 2019), which prompted more detailed research on *Ganoderma* triterpenoids in regard to their ability to alleviate the oxidative damage of hepatocytes.

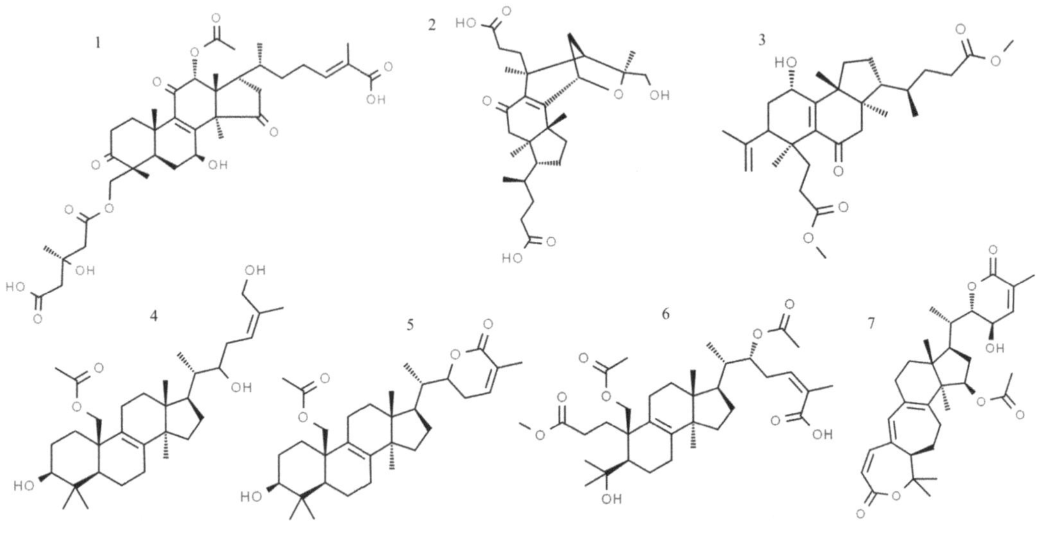

FIGURE 8.3 Structures of some unusual triterpenoids isolated from *Ganoderma* spp.: 1- ganoleucoin J, 2- fornicatin A, 3- fornicatin F, 4- colossolactone A, 5- colossolactone B, 6- colossolactone V, and 7- colossolactone VIII.

The hepatoprotective effects of triterpenoids isolated from *G. theaecolum* were tested in vitro using a DL-galactosamine–induced HL-7702 cell damage model. Out of 10 compounds tested, 3 maintained a high survival rate of the hepatocytes (≥75%), at 10 μM, compared to control (43%). The effect on hepatocyte survival was superior to that of bicyclol, a drug used in China to treat viral hepatitis. The mechanism of action of ganoderic acids and their analogues is not yet investigated (Liu et al., 2014). Zhang, Gao et al. (2022) also surveyed the potential hepatoprotective effect of triterpenoids isolated from *G. lucidum* mycelial mat, but they used the H_2O_2-induced oxidative injury model in HepG2 cells. Some of the compounds tested showed a positive impact on the cell survival rate, comparable to that of *N*-acetylglucosamine, which is frequently used to treat liver damage. In this study the authors analyzed the structure-activity relationship of the compounds and drew a tentative conclusion that triterpenoids with a carboxyl group at C-26 (ganoderic acids) containing 8(9)-en-7-one moiety had greater potential as hepatoprotective agents, as the majority of the active compounds shared this structure; these compounds reduced liver alanine aminotransferase (ALT), aspartate aminotransferase (AST), and lactate dehydrogenase (LDH) activity and also increased the concentration of glutathione, which is an important intracellular antioxidant. Peng et al. (2014) performed similar research with some unusual triterpenoids isolated from *G. cochlear*, using the same H_2O_2-induced oxidative injury model in HepG2 cells. Fornicatins A, D, and F, 3,4-seco-triterpenoids, showed moderate ability to lower ALT levels in treated hepatocytes, at nontoxic concentrations, with fornicatin F showing the greatest decrease of ALT of 23.3% at 59.9 μM and also showing the lowest cytotoxicity towards the cells. At a similar concentration (65.7 μM), fredelin, a pentacyclic triterpenoid with a single keto group at C3, achieved even an higher ALT decrease of about 30%. Fredelin also exhibited the highest activity in regard to AST level decrease, which was 41%.

An in vivo study on mice was performed to see whether ganoderic acid A may have positive effects on nonalcoholic steatohepatitis. Mice were fed a high-fat, high-cholesterol diet, and ganoderic acid A was orally administered to 2 groups, at 2 different dosage regimens, 25 or 50 mg/kg/day, for 3 months. It was revealed that ganoderic acid A exhibited beneficial effects, compared to the group that did not receive it; the body weight of the mice treated with ganoderic acid A was reduced, and fat accumulation, steatosis, inflammation, and fibrosis in the liver were significantly inhibited. The hepatoprotective effect against the liver injury was further confirmed by lower levels of ALT and AST in the serum of treated animals compared to the model group, which was not treated with ganoderic acid A. Other biochemical parameters connected to high-fat diet were also affected, with total triglycerides, total cholesterol, and LDL levels being reversed by the treatment. Ganoderic acid A also reduced number of Kupffer cells and macrophage infiltration in s of high-fat-fed mice, as well as serum inflammation parameters and proinflammatory cytokines (TNF-α, IL-1β, and IL-6) levels, which were increased in the model group. The hepatoprotective effects of ganoderic acid A were associated with its ability to alleviate oxidative stress damage caused by a high-fat diet in the liver. The concentration of malondialdehyde, an oxidative stress marker, was markedly increased in the model group, while levels of an antioxidant enzyme, superoxide dismutase, were deceased. Ganoderic acid A treatment, however, lowered malondialdehyde levels and increased superoxide dismutase levels. The high-fat diet also induced expression of endoplasmic reticulum stress genes, which was reversed by the treatment with ganoderic acid A (Zhu et al., 2022).

8.2.5 *Ganoderma* Triterpenoids in the Treatment of Inflammatory Conditions

The effects that ganoderic acids exert on immune cells may be beneficial in the treatment of various diseases where the immune system is involved. Liu et al. (2015) tested a mixture of aqueous extracts of *G. lucidum* and two medicinal plants, *Sophora flavescens* and *Glycyrrhiza uralensis*, named ASHMI, in the treatment of asthma, and it was shown to be both clinically efficient and safe for patients. An in vitro study in a murine asthma model found that this preparation could suppress both Th2- and TNF-α–mediated inflammation; however, only *G. lucidum* was found to be responsible for inhibition of TNF-α production by murine macrophages. This is of great importance, as non-Th2–mediated asthma is resistant to corticoid treatment. Further investigation found that triterpenoid-enriched fraction, and not polysaccharide fraction of the *G. lucidum* extract, was responsible for the activity, and ganoderic acids were suspected to be the key active compounds. Among 15 isolated triterpenoids, several compounds exhibited inhibitory activity on TNF-α production to some degree (ganoderic acids A, J, K, and U), but only 1 though, ganoderic acid C1, exhibited significant inhibitory activity on lipopolysaccharide (LPS)–induced TNF-α production in murine macrophages and also in peripheral blood mononuclear cells isolated from asthma patients. This modulation of TNF-α production was not due to cell apoptosis, as cells were found to be viable at the concentrations that showed the anti–TNF-α activity, but due to interference with cell signaling pathways. The reduction of TNF-α was connected to inhibition of the NF-κB signaling pathway, through which LPS induces TNF-α production and also AP-1 and MAPK pathways to some degree. Interestingly, ganoderic acid C1 did not suppress TNF-α production in nonstimulated macrophages and actually caused a slight increase in TNF-α levels, similar to the mode of action of β-glucans. This suggests that ganoderic acids may rather act as immune modulators and not as strict immune suppressants. This also implicates that ganoderic acids may be useful in the treatment of other autoimmune diseases in which overproduction of TNF-α plays the key role, like in rheumatoid arthritis or Crohn disease.

8.2.6 Ganoderic Acid DM as a Potential Agent in the Treatment of Osteoporosis

Expectedly, ganoderic acid DM's efficiency in osteoclastogenesis inhibition proved in the case of prostate cancer was tested in prospects of osteoporosis treatment. Miyamoto et al. (2009) examined ganoderic acid DM effects in ovariectomized rats. The study reported that ethanol extract of *G. lucidum* prevented bone resorption; however, the effect was not dose-dependent. The authors explained that in this case, lower concentrations were more effective due to the estrogen-like effect. At the same time, higher concentrations inhibited cellular molecules in charge of cell signaling, growth, and death. Furthermore, the observed inhibition of osteoclastic differentiation was examined in the bone marrow cells and RAW 264 cell D-clone. In both cases, ganoderic acid DM inhibited the process of osteoclastogenesis by suppressing the expression of c-Fos, which further downregulated the transcription factor NFATc1 expression by RANKL. Finally, DC-STAMP expression was inhibited, causing the reduction of cell fusion. Moreover, ganoderic acid DM suppressed cathepsin

K gene expression responsible for osteoclastic bone resorption. Overall, ganoderic acid DM proved to be a bone-protective compound of *G. lucidum*.

8.2.7 *Ganoderma* Triterpenoids in Cosmetic Products—Melanogenesis Inhibitors

Hyperpigmentation of the skin may arise from various conditions, and though it's normally benign, it may cause a serious impact on the quality of a patient's life. Melasma is the most common hyperpigmentation disorder that affects mostly women and is associated with both genetic and external factors, with exposure to the sun being the most important (Nicolaidou and Katsambas, 2014). Postinflammatory hyperpigmentation is another common disorder affecting mostly people with a darker skin tone, but hyperpigmentation may also arise from systemic disorders or may be induced by certain drugs (Goswami and Sharma, 2020), or it can be connected to skin aging (Kim et al., 2022). In all of these conditions there is a hyperproduction of melanin, pigment that is synthesized from L-DOPA by the enzyme tyrosinase, which is why this enzyme is the main target in the treatment of the hyperpigmentation conditions. Melanogenesis inhibitors are widely used in topical products and include hydroquinone arbutin, kojic acid, and niacinamide, although many plant extracts are used as well (Goswami and Sharma, 2020). Mushrooms were also found to contain compounds with tyrosinase-inhibitory activity (Petrović, Vunduk et al., 2019), and there are actually products for hyperpigmentation treatment that are based on *G. lucidum* extracts, and their efficacy has prompted the research for their active ingredients. Ganodermanondiol was found to possess inhibitory activity on melanin production on the cellular level. The compound was tested for its ability to inhibit melanogenesis in B16F10 melanoma cells. Ganodermanondiol inhibited melanin production in a dose-dependent manner, and it was shown that this inhibition was through downregulation of several proteins involved in melanogenesis—tyrosinase, tyrosinase-related proteins 1 and 2, and most importantly microphthalmia-associated transcription factor (MITF). MITF suppression could at least partially be associated with inhibition of MAPKs and the cAMP-dependent signaling pathway that is also seen in B16F10 cells after treatment with ganodermanondiol (Kim et al., 2016) (Table 8.1).

Compound	Source	Activity
3β-lanosta-7,9(11)-dien-3-ol	*G. lucidum*	Antihyperglicemic (α-glucosidase inhibition) (Binh et al., 2018)
Colossolactone V Colossolactone VII Colossolactone E	*G. colossus*	Anti-HIV activity (El Dine et al., 2008)
Colossolactone H	*G. colossus*	Cytotoxic, tumor growth inhibition *in vivo* (Chen et al., 2016)
Fornicatin A Fornicatin D Fornicatin F Fredelin	*G. cochlear*	Hepatoprotective activity *in vitro* (Peng et al., 2014)
Ganodapplanoic acid I	*G. applantum*	Antiadipogenic (lipid accumulation inhibition *in vitro*) (Su et al., 2021)
Ganoapplanoid K Ganoapplanoid Q	*G. applantum*	Antiadipogenic (lipid accumulation inhibition *in vitro*) (Su et al., 2020)
Ganoderenic acid K	*G. leucocontextum*	HMG-CoA reductase inhibitory activity (Zhang et al., 2018)
Ganoderic acid A	*G. lucidum*	Cytotoxic (Mortazavie et al., 2022, Yang et al., 2018, Wang et al., 2017, Das et al., 2019), antitumor *in vivo* (Das et al., 2019) anti-invasive (Yang et al., 2018, Wang et al., 2017, Jiang et al., 2008), anti-EBV (Zheng & Chen, 2017), anti-diabetic (α-glucosidase inhibition) (Ren, 2019), hepatoprotective activity *in vivo* (Zhu et al., 2022)
Ganoderic acid B	*G. lucidum*	Antiviral (anti-EBV) (Zheng & Chen, 2017)
Ganoderic acid C1	*G. lucidum*	Anti-inflammatory (Liu et al., 2015)
Ganoderic acid Df	*G. lucidum*	Antihyperglicemic (α-glucosidase inhibition) (Fatmawati et al., 2010)

(Continued)

Compound	Source	Activity
Ganoderic acid DM	*G. lucidum*, *G. leucocontextum*	Cytotoxic (Wu et al. 2012), antitumor *in vivo* (Das et al., 2019) 5α-reductase inhibition, androgen receptor inhibition (Johnson et al. 2010), HMG-CoA reductase inhibition (Wang, Bao et al., 2015), bone-protective (Miyamoto et al., 2009)
Ganoderic acid F	*G. lucidum*	Antiangiogenic (Nguyen et al., 2015)
Ganoderic acid H	*G. lucidum*	Cytotoxic, anti-invasive (Jiang et al., 2008)
Ganoderic acid Me	*G. lucidum* mycelium	Cytotoxic (Zhou et al., 2011, Chen & Zhong, 2009), anti-invasive (Chen et al., 2008), immunostimulation *in vivo* (Wang et al., 2007)
Ganoderic acid Mf	*G. lucidum* mycelium	Cytotoxic (Liu & Zhong, 2011)
Ganoderic acid S	*G. lucidum* mycelium	Cytotoxic (Liu & Zhong, 2011)
Ganoderic acid T	*G. lucidum* mycelium	Cytotoxic (Liu et al., 2012), anti-invasive *in vitro* and *in vivo* (Chen et al., 2010)
Ganoderic acid T amide ("ganoderamide T")	semisynthetic	Cytotoxic, more potent than ganoderic acid T (Liu et al., 2012)
Ganoderic acid T-Q Ganoderic acid TR	*G. lingzhi*	Antiviral (H1N1 and H5N1) (Zhu et al., 2015)
Ganoderic acid X	*G. amboniense*	Cytotoxic (Li et al., 2004)
Ganoderic acid Y	*G. lucidum*, *G. leucocontextum*	Antiviral (enterovirus 71) (Zhang et al., 2014), HMG-CoA reductase inhibitory activity (Wang, Bao et al., 2015)
Ganoderic acid β	*G. lucidum*	Anti-HIV (Min et al., 1998)
Ganoderiol F	*G. lucidum*	Anti-HIV (El-Mekkawy et al., 1998)
Ganoderiol J	*G. leucocontextum*	HMG-CoA reductase inhibitory activity (Wang, Bao et al., 2015)
Ganodermanondiol	*G. lucidum*	Cytotoxic (Min et al., 2000) anti-EBV (Zheng & Chen, 2017), melanogenesis inhibition (Kim et al., 2016)
Ganodermanontriol	*G. lucidum*	Cytotoxic (Min et al., 2000), Tumor growth inhibition *in vivo* (Jedinak et al., 2011), anti-invasive (Jiang et al., 2011), anti-HIV (Min et al., 1998, El-Mekkawy et al., 1998) anti-EBV (Zheng & Chen, 2017)
Ganoderol B	*G. lucidum*	Testosterone-induced prostate growth inhibition *in vivo* (Liu et al., 2007)., anti-EBV (Zheng & Chen, 2017), anti-viral (H5N1) (Zhu et al., 2015), antihyperglicemic (α-glucosidase inhibition) (Fatmawati et al., 2010)
Ganoleucoin J Ganoleucoin T Ganoleucoin Y	*G. leucocontextum*	HMG-CoA reductase inhibition (Wang, Bao et al., 2015, Zhang et al., 2018)
Ganoleucoin M Ganoleucoin N Ganoleucoin P	*G. leucocontextum*	Antihyperglicemic (α-glucosidase inhibition) (Wang, Bao et al., 2015)
Ganolucidic acid η	*G. leucocontextum*	HMG-CoA reductase inhibition (Zhang et al., 2018)
Lanosta-7,9(11),24-trien-3-one,15,26-dihydroxy	*G. lucidum*	Anti-viral (enterovirus 71) (Zhang et al., 2014)
Lucidenic acid E	*G. leucocontextum*	HMG-CoA reductase inhibitory activity (Zhang et al., 2018)
Lucidumol C Lucidumol D	*G. lingzhi*	Cytotoxic (Satria et al., 2018)
Schisanlactone A	*G. colossus*	Anti-HIV (El Dine et al., 2008)

Selected triterpenoids isolated from *Ganoderma* spp.; names of the species that compounds were isolated from are given as in the publications, however, *G. lucidum* was described from Europe, and it was relatively recently shown that species cultivated in East Asia as "*G. lucidum*" is a distinct species, named *G. sichuanense*, with *G. lingzhi* being a synonym that is also still in use (Thawthong et al., 2017)

8.3 Studies on Metabolism of *Ganoderma* Triterpenoids in In Vitro and In Vivo Models

Currently, knowledge about the metabolism of ganoderic acids and its analogues is very limited and confined to research in isolated cells or animals. Cao et al. (2017) performed the most detailed metabolic pathway analysis of ganoderic acid A in rats and identified 37 metabolites by analyzing plasma, bile, and urine after intravenous administration of the compound at a dosage of 20 mg/kg. The authors also performed an in vitro study using both rat and human liver microsomes. The in vivo study showed that ganoderic acid A goes through extensive metabolism and that it undergoes oxidation, reduction, and hydroxylation, as well as glucuronidation and sulfation, with a reduction product, ganoderic acid C2, being the most abundant. The main route of excretion of ganoderic acid A metabolites is via bile (as expected for terpenoids), with a total of 34 metabolites detected in it. These metabolites included both phase I metabolism products and phase II conjugates. In urine, on the other hand, 12 different metabolites were detected, all phase II biotransformation products. A total of 22 metabolites were found in plasma, which included both phase I and phase II metabolic products. The liver (hepatocytes) was found to be the main point of ganoderic acid A clearance, and an in vitro study confirmed that both in rat and human liver microsomes, CYP3A was involved in its metabolism. All metabolites detected in rat liver microsomes were also found in the in vivo samples, as well as those detected in human liver microsomes, indicating that the ganoderic acid A metabolic pathway is similar in humans.

Guo, Liu et al. (2013) performed a similar study on ganoderic acid C2 metabolism by analyzing plasma samples of rats after oral administration. They identified 10 metabolites, among which was ganoderic acid A, implying that these compounds may undergo reversible conversion from one to another in vivo; both of these compounds are known to be bioactive, which also means that the physiological effects of these compounds may be prolonged. Other metabolites that were identified as known compounds found in *Ganoderma* species included ganoderic acids B, G, AM1, C6, and η; ganolucidic acid B; and others.

Guo, Shen et al. (2013) studied the distribution and metabolism of ganoderic acid B in rats after oral administration and identified 14 of its metabolites in total, along with the unchanged ganoderic acid B, in plasma, bile, stomach, intestines, liver, lungs, and kidneys. Interestingly, ganoderic acid B 3-O- and 7-O- glucuronides, which are relatively hydrophilic metabolites, were isolated only from bile. The unchanged compound was found in all samples, except for bile. Most of the other metabolites represented compounds naturally occurring in *Ganoderma* species and included ganoderic acids A, AM1, C2, C6, D, E, G, and η.

Zhang et al. (2009) investigated the metabolism of ganoderiol F, a very potent cytotoxic triterpenoid; the authors wanted to test whether this compound is biotransformed by intestinal bacteria. They incubated ganoderiol F with bacteria isolated from rat feces, but also healthy human volunteers, and determined that in both cases bacteria transforms ganoderiol F to ganodermatriol, although the clinical significance of this finding is yet to be validated. The authors found both ganoderol F and ganodermatriol in fecal samples of rats after up to 24h, meaning that ganoderiol F has only limited absorption and that transformation to ganodermatriol happens in vivo as well. They could not be detected in either urine or bile samples. They also determined the LD_{50} for both ganoderiol F and ganoderic acid A and found that ganoderiol F is much more toxic, with an LD_{50} of 2.95 mg/kg compared to ganoderic acid A, with an LD_{50} of 178.57 mg/kg.

Ansari et al. (2022) studied the pharmacokinetics of triterpenoids from a triterpenoid-enriched fraction produced from an Indian variety of *G. lucidum*. They found that ganoderic acid H was the dominant compound in the fraction, at almost 24% w/w; other triterpenoids were detected as well and included ganoderic acids A, D, and N and tsugaric acid A. After oral administration of the preparation to rats, blood samples were collected and analyzed by Ultra-high performance liquid-chromatography–mass spectrometry (UPLC-MS), although the analytical standard was used only for ganoderic acid H, while identification of other compounds was based on available mass spectra database libraries. All of these triterpenoids were found to be bioavailable, leading to their identification in blood samples, except for tsugaric acid A, although this compound represents acetyl ester, which may undergo rapid deacetylation by esterases. For ganoderic acid H, it was found that it is readily absorbed, reaching peak concentration after 1h. The compound concentration could be detected in blood for up to 12h after administration. It was not detected in urine, as it is almost certainly excreted after chemical transformation. For ganoderic acid A, it was found that after per os administration to rats, it reaches maximum concentrations in blood relatively quickly, after only 18 min. The half-lives of its distribution ($t1/2\alpha$) and elimination phases ($t1/2\beta$) were found to be 12.4 and 451 min, respectively. On the other hand, for ganoderiol F, it was determined that it reaches maximum blood concentration after around 100 min, but with $t1/2\alpha$ and $t1/2\beta$ being much lower, 2.4 and 34.8 min, respectively. Both compounds were found to have low bioavailability of about 10%, which might be due to poor absorption but also due to a liver first-pass effect (Zhang et al., 2009).

8.4 Other Secondary Metabolites from *Ganoderma* spp.

Meroterpenoids are a structurally diverse group of naturally occurring compounds, having in common a partial synthesis from terpenoid pathways ("mero" = "partial") (Matsuda and Abe, 2016). In meroterpenoids, terpenoid structures are often combined with polyketides, alkaloids, phenols, and amino acids, and thus exhibit great structural diversity (Nazir et al., 2021). They include well-known compounds such as tocopherols, ubiquinone, and alkaloid vinblastine, as well as cannabinoids (Matsuda and Abe, 2016).

Although somehow "shadowed" by ganoderic acid and its analogues, there are more than 100 different meroterpenoids isolated from *Ganoderma* spp., most of which were shown to possess some kind of pharmacological activity (Peng and Qiu, 2018).

Ganomycins A and B were the first meroterpenoids isolated from a *Ganoderma* species (*G. pfeifferi*) back in 2000. They represent farnesyl hydroquinones and possess a carboxylic group on the C3 of the farnesyl moiety. Ganomycins exhibited antimicrobial activity towards both gram-positive and gram-negative bacteria, with no preferences towards one or another, but they did not show any activity against yeasts. The best activity was achieved against *Micrococcus flavus*, 2.5 μg/mL, which was still about 10 times lower activity to that of ampicillin (Mothana et al., 2000). Ganomycins were also shown to possess HMG-CoA reductase inhibitory activity, which was comparable to that of atorvastatin in vitro (Chen et al., 2017). More farnesylated hydroquinones were isolated from *G. capense* and named ganocapensins. Two of the compounds possessed 13- and 14-memebred ether rings, and all ganocapensins exhibited significant antioxidant activity in an in vitro DPPH (2,2-diphenyl-1-picrylhydrazyl) radical scavenging assay Peng et al. (2016) (Figure 8.4).

Polycyclic phenolic meroterpenoids named cochlearols A and B were the first to be characterized from *G. cochlear.* These compounds were tested for their renoprotective ability and were shown to interfere with the TGF-β/Smads signaling pathway in TGF-β1–induced rat renal proximal tubular cells. The TGF-β/Smads pathway was marked as a key pathway which leads to a fibrotic cascade (Dou et al., 2014). Wang et al. (2019) characterized more cochlearols and tested them as potential inhibitors of TGF-β1–induced rat interstitial fibroblast cells (NRKe49F) proliferation; some of the compounds were confirmed to be active. Cochlearoids are also known from *G. cochlear* and represent phenolic polycyclic compounds, which also show inhibitory activity on hyperproduction of fibronectin. The authors dismissed the possibility that cytotoxicity was the reason for the compounds' activity since the compounds were shown not to cause cell death in the MTT (3-(4,5-dimethylthiazol-2-yl)-2,5-diphenyltetrazolium bromide) test at higher concentrations than those needed for fibronectin production inhibition (Wang et al., 2016). Cochlearoids L and M were further isolated and characterized and were tested for their cytotoxicity and ability to inhibit bromodomain-containing protein 4 (BRD4); colchlearoid L was indeed found to possess weak cytotoxic activity against 3 human cell lines (A549, K562, Huh-7) and weak-to-moderate BRD4 inhibitory activity (Qin et al., 2018a). Cochlearoids N-P are similar in structure to cochlearoids

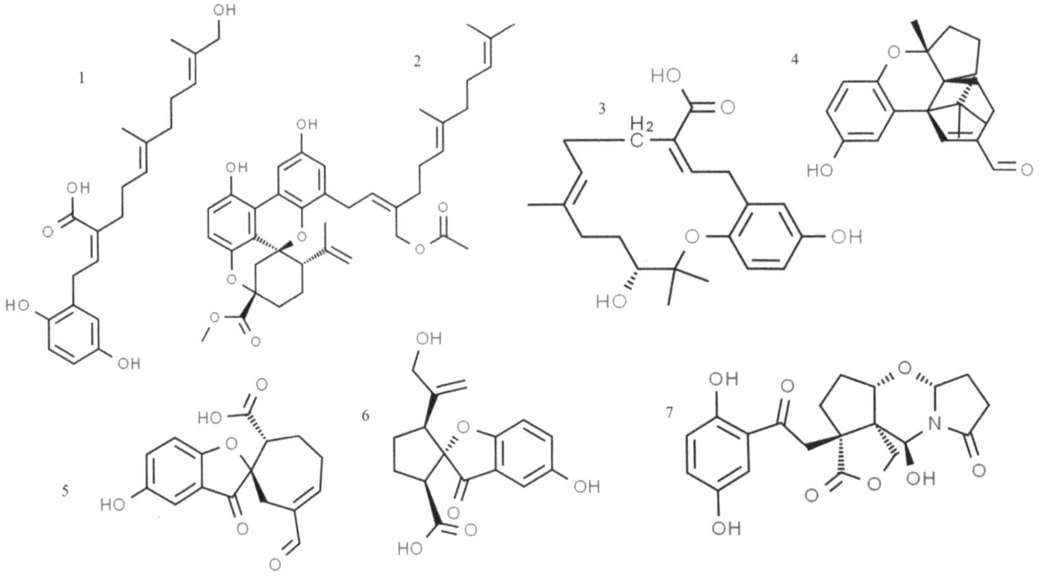

FIGURE 8.4 Structure of selected meroterpenoids isolated from *Ganoderma* spp.: 1- ganomycin A, 2- cochlearoid A, 3- petchiether B, 4- cochlearol B, 5- spiroapplanatumine H, 6- spiroapplanatumine M, and 7- sinensilactam A.

A-E and were also tested for their BRD4 inhibitory activity and cytotoxicity, as well as for their antibacterial activity against *Staphylococcus aureus*; however, although some of the enantiomers showed activity, it was less pronounced than that of the control reagents (Qin et al., 2018b). The same research group characterized more meroterpenoids from *G. cochlear*, with benzopyran and benzofuran motifs, named ganodercins. Both enantiomers of ganodercin V, (+)-ganodercin S, and cochlearol F were shown in the same study to be able to reduce the expression of collagen I and thus exhibit renoprotective activity (Qin et al., 2022).

Sinensilactam A represents a pair of enantiomeres and was isolated from *G. sinense* (as "*G. sinensis*"), and unlike other meroterpenoids known from *Ganoderma* species, it represents an alkaloid, having a nitrogen atom in its polycyclic structure. It was proposed that it is a hybrid of shikimic acid, mevalonic acid, and amino acid pathways. It was tested to see whether it can be active against renal fibrosis. Both enantiomers of sinensilactam A were shown to be selective inhibitors of Smad3 phosphorylation, and (-)-enantiomer was more potent (Luo et al., 2015). Cao et al. (2016) isolated another 7 meroterpenoids from *G. sinense*, 6 of which were previously unknown and were named zizhines A-F. They all represented 5-phenyl-5H-furan-2-one but were inactive as inhibitors of Smad3 phosphorylation.

Petchiethers are interesting macrocyclic ethers with a 14-member (petchiether A) or 15-member ring (petchiether B) that resemble macrolides. They are isolated from *G. petchii*, hence the name. Like other meroterpenoids, they were shown to inhibit fibronectin secretion from TGF-β1–induced rat kidney tubular epithelial cells (Li et al., 2016). Dai et al. (2020) characterized polycyclic lactones from the same species, named pethiilactones; (-)-enantiomeres of petchiilactone A and C were found to be able to induce transformation of umbilical cord mesenchymal stem cells to keratinocyte-like cells, which may be helpful in wound healing.

Ganotheaecoloids represent terminal cyclohexane-type meroterpenoids characterized from the fruiting bodies of *G. theaecolum*. One of these compounds, ganotheaecoloid J, was found to have relatively good inhibitory activity on cyclooxygenase-2 (COX-2), which is involved in inflammation

(Luo et al., 2018), although the authors did not compare their activity to the known inhibitors, such as celecoxib.

Meroapplanins, isolated from *G. applanatum*, are another group of alkaloid meroterpenoids containing tetrahydropyridine moiety and are structurally similar to some degree to sinensilactams. The compounds were tested in a H_2O_2-induced PC12 cell damage model to assess their potential neuroprotective properties, and only one enantiomer of Meroapplanin C showed some ability to increase the cell survival (Peng et al., 2020).

There are over 20 characterized spiroapplanatumines from *G. applanatum*, which possess a spiro[benzofuran-2(3H), 1'-cyclohpetane] or spiro[benzofuran-cyclopenatne] structure, and two of them (G and K) were found to be potent inhibitors of JAK3 kinase, which is important in some autoimmune diseases (Luo, Di et al., 2016). The same research group further isolated two compounds from *G. applanatum*, one with a unique spiro[benzofuran-2,2'bicyclo[3.2.2]nonane] ring and another with an unusual dioxacyclopenta[cd]inden structure. These were named applanatumols A and B, respectively, and were assessed for the potential activity against renal fibrosis using rat proximal tubular epithelial cells. The compounds were active to some degree, and the activity differed for different applanatumol B enantiomers (Luo, Wei et al., 2016).

Yan et al. (2015) isolated 12 meroterpenoid compounds from *G. lingzhi*, among which 4 compounds also exhibited a unique spiro[benzofuran-2,10-cyclopentane] structure. They were named spirolingzhines A–D and tested for their potential neuroregenerative activity. Most of the compounds promoted rat neural stem cell (NSC) proliferation in vitro, but spirolingzhine A showed the most prominent activity, comparable to that of a known NSC-proliferation stimulator, forskolin, and was marked as a candidate for future in vivo studies.

Besides terpenoid compounds, Kawagishi et al. (1993) isolated interesting nucleosides, two epimers of 5'-deoxy-5'-(methylsulfinyl)adenosine, from a butanol extract of *G. lucidum* fruiting bodies, which exhibited platelet-aggregation inhibitory activity. Interestingly, the compounds had been previously synthesized by Kuhn and Jahn (1965). They tested compounds for their ability to lower blood pressure in cats, and only 1 epimer was shown to be active. When Kawagishi et al. (1993) tested the same compounds in the antiplatelet aggregation assay, again, only 1 epimer showed inhibitory activity at 50 µg/mL—the same epimer that possessed blood-pressure lowering ability in vivo. The activity was comparable to that of adenosine, used as a positive control, when platelet aggregation was induced by either adenosine diphosphate (ADP) or a thromboxane mimetic, prostaglandin endoperoxide 15-hydroxy-11α,9α-(epoxymethano) prosta-5,13-dienoic acid. Other sulfur-containing organic compounds are also known as antiplatelet aggregation agents, and differences in the activity of sulfoxide derivatives epimers is well known, due to sulfoxide asymmetry (Figure 8.5).

8.3 CONCLUSION

Species of the genus *Ganoderma* are the source of many unique compounds which show a vast array of biological activity. The most studied are highly oxygenated triterpenoid compounds known as ganoderic acids and their analogues. They have been a subject of research for 40 years, mostly because of their cytotoxic/antitumor activity. However, both ganoderic acids and their analogues, as well as other secondary metabolites produced by *Ganoderma* species, mostly meroterpenoids, have been shown to possess antiviral, antibacterial, antihypercholesterolemic, antihyperglicemic, anti-inflammatory, hepatoprotective, and other pharmacological properties. Despite intensive research, there is still a lack of clinical data regarding the compounds' efficacy and safety in vivo. One of the main drawbacks of extensive use of ganoderic acids and other *Ganoderma* metabolites is that they are present in relatively low concentrations in fruiting bodies; it is why the possibility to obtain these compounds from mycelium obtained using submerged cultivation is the subject of many recent studies, as well as optimization of the production process.

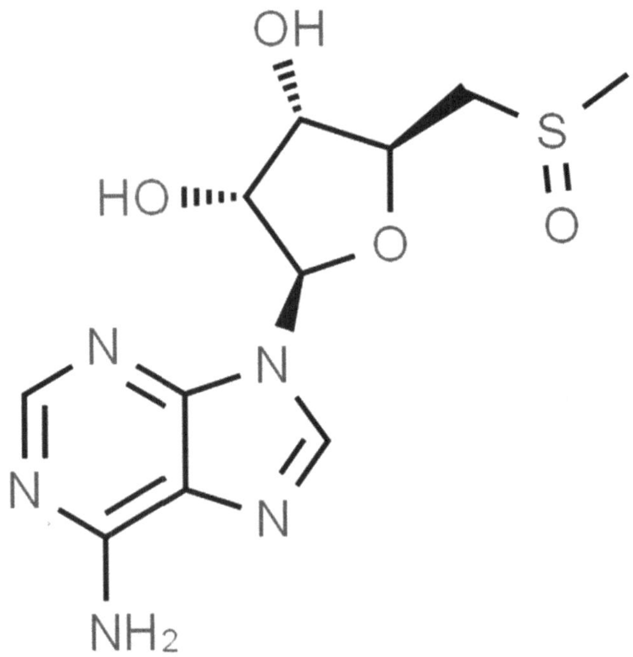

FIGURE 8.5 5'-Deoxy-5'-(methylsulfinyl)adenosine, nucleoside isolated from *G. lucidum*.

ACKNOWLEDGMENT

This work was supported by the Ministry of Education, Science and Technological Development of the Republic of Serbia (Contract No. 451–03–68/2022–14/200135 & Contract No. 451–03–68/2022–14/200051). All chemical structures were taken from ChemSpider (www.chemspider.com).

REFERENCES

Anke, T. 2020. Secondary metabolites from mushrooms. *J. Antibiot.*, 73(10): 655–6.

Ansari, M. H. R., Khan, W., Parveen, R., Saher, S., and Ahmad, S. 2022. Pharmacokinetic, metabolomic, and stability assessment of Ganoderic acid H based triterpenoid enriched fraction of *Ganoderma lucidum* P. Karst. *Metabolites*, 12(2) (January): 97. www.mdpi.com/2218-1989/12/2/97.

Atli, B., Yamaç, M., Yildiz, Z., and Isikhuemnen, O. S. 2015. Enhanced production of lovastatin by *Omphalotus olearius* (DC.) Singer in solid state fermentation. *Rev. Iberoam. Micol.*, 32(4): 247–51.

Baby, S., Johnson, A. J., and Govindan, B. 2015. Secondary metabolites from *Ganoderma*. *Phytochemistry*, 114: 66–101.

Bharadwaj, S., Lee, K. E., Dwivedi, V. D., Yadava, U., Panwar, A., Lucas, S. J., Pandey, A., and Kang, S. G. 2019. Discovery of *Ganoderma lucidum* triterpenoids as potential inhibitors against dengue virus NS2B-NS3 protease. *Sci. Rep.*, 9(1) (December): 19059. www.nature.com/articles/s41598-019-55723-5.

Binh, P. T., Thao, N. P., Luyen, N. T., Trang, D. T., Binh, P. T., Nguyen, N. P., Hiep, N. T., Dang, N. H., Tri, T. M., and Dat, N. T. 2018. Lanostane-type triterpenoids from *Ganoderma lucidum* and *G. multipileum* fruiting bodies. *Nat. Prod. Commun.*, 13(11): 1441–4.

Cao, F., Feng, L., Ye, L., Wang, L., Xiao, B., Tao, X., and Chang, Q. 2017. Ganoderic acid A metabolites and their metabolic kinetics. *Front. Pharmacol.*, 8 (March): 101. www.frontiersin.org/articles/10.3389/fphar.2017.00101/full.

Cao, W., Luo, Q., Cheng, Y., and Wang, S. 2016. Meroterpenoid enantiomers from *Ganoderma sinensis*. *Fitoterapia*, 110: 110–15.

Chen, B., Tian, J., Zhang, J., Wang, K., Liu, L., Yang, B., Bao, L., and Liu, H. 2017. Triterpenes and meroterpenes from *Ganoderma lucidum* with inhibitory activity against HMGs reductase, aldose reductase and α-glucosidase. *Fitoterapia*, 120: 6–16.

Chen, H., and Liu, J. 2017. Secondary metabolites from higher fungi. In *Progress in the Chemistry of Organic Natural Products*, ed. A. D. Kinghorn, H. Falk, S. Gibbons and J. Kobayashi, 106: 1–201. Cham: Springer International Publishing.

Chen, N., Liu, J., and Zhong, J. 2008. Ganoderic acid Me inhibits tumor invasion through down-regulating matrix metalloproteinases 2/9 gene expression. *J. Pharm. Sci.*, 108(2): 212–16.

Chen, N., Liu, J., and Zhong, J. 2010. Ganoderic acid T inhibits tumor invasion *in vitro* and *in vivo* through inhibition of MMP expression. *Pharmacol. Rep.*, 62(1): 150–63.

Chen, N., and Zhong, J. 2009. Ganoderic acid Me induces G1 arrest in wild-type p53 human tumor cells while G1/S transition arrest in p53-null cells. *Process Biochem.*, 44(8): 928–33.

Chen, S., Chang, C., Chen, T., Chang, Y., and Lin, S. 2016. Colossolactone H, a new *Ganoderma* triterpenoid exhibits cytotoxicity and potentiates drug efficacy of gefitinib in lung cancer. *Fitoterapia*, 114: 81–91.

Cör Andrejč, D., Knez, Ž., and Knez Marevci, M. 2022. Antioxidant, antibacterial, antitumor, antifungal, antiviral, anti-inflammatory, and neuro-protective activity of *Ganoderma lucidum*: An overview. *Front. Pharmacol.*, 13 (July): 934982. www.frontiersin.org/articles/10.3389/fphar.2022.934982/full.

Dai, W., Zhu, Y., Qin, F., Chang, J., Zeng, Y., Wang, J., Zhang, Y., and Cheng, Y. 2020. Skeletal meroterpenoids from *Ganoderma petchii* mushrooms that potentially stimulate umbilical cord mesenchymal stem cells. *Bioorg. Chem.*, 97 (April): 103675. www.sciencedirect.com/science/article/pii/S0045206819319972.

Das, A., Alshareef, M., Henderson, F., Martinez Santos, J. L., Vandergrift, W. A., Lindhorst, S. M., Varma, A. K., Infinger, L., Patel, S. J., and Cachia, D. 2019. Ganoderic acid A/DM-induced NDRG2 over-expression suppresses high-grade meningioma growth. *Clin. Transl. Oncol.*, 22(7): 1138–45.

Dolcetti, R., Giunco, S., Dal Col, J., Celeghin, A., Mastorci, K., and De Rossi, A. 2014. Epstein-Barr virus and telomerase: From cell immortalization to therapy. *Infect. Agents Cancer*, 9 (February): 8. https://infectagentscancer.biomedcentral.com/articles/10.1186/1750-9378-9-8.

Dou, M., Di, L., Zhou, L., Yan, Y., Wang, X., Zhou, F., Yang, Z., Li, R., Hou, F., and Cheng, Y. 2014. Cochlearols A and B, polycyclic meroterpenoids from the fungus *Ganoderma cochlear* that have rfenoprotective activities. *Org. Lett.*, 16(23): 6064–7.

El Dine, R. S., El Halawany, A. M., Ma, C., and Hattori, M. 2008. Anti-HIV-1 protease activity of lanostane triterpenes from the Vietnamese mushroom *Ganoderma colossum*. *J. Nat. Prod.*, 71(6): 1022–6.

El Dine, R. S., El Halawany, A. M., Nakamura, N., Ma, C., and Hattori, M. 2008. New lanostane triterpene lactones from the Vietnamese mushroom *Ganoderma colossum* (Fr.) C. F. Baker. *Chem. Pharm. Bull.*, 56(5): 642–6.

El-Mekkawy, S., Meselhy, M. R., Nakamura, N., Tezuka, Y., Hattori, M., Kakiuchi, N., Shimotohno, K., Kawahata, T., and Otake, T. 1998. Anti-HIV-1 and anti-HIV-1-protease substances from *Ganoderma lucidum*. *Phytochemistry*, 49(6): 1651–7.

Elmore, S. 2007. Apoptosis: A review of programmed cell death. *Toxicol. Pathol.*, 35(4): 495–516.

Fatmawati, S., Kondo, R., and Shimizu, K. 2013. Structure-activity relationships of lanostane-type triterpenoids from *Ganoderma lingzhi* as α-glucosidase inhibitors. *Bioorg. Med. Chem. Lett.*, 23(21): 5900–3.

Fatmawati, S., Kurashiki, K., Takeno, S., Kim, Y., Shimizu, K., Sato, M., Imaizumi, K., Takahashi, K., Kamiya, S., Kaneko, S., and Kondo, R. 2008. The inhibitory effect on aldose reductase by an extract of *Ganoderma lucidum*. *Phytother. Res.*, 23(1): 28–32.

Fatmawati, S., Shimizu, K., and Kondo, R. 2010. Ganoderic acid Df, a new triterpenoid with aldose reductase inhibitory activity from the fruiting body of *Ganoderma lucidum*. *Fitoterapia*, 81(8): 1033–6.

Fatmawati, S., Shimizu, K., and Kondo, R. 2011. Ganoderol B: A potent α-glucosidase inhibitor isolated from the fruiting body of *Ganoderma lucidum*. *Phytomedicine*, 18(12): 1053–5.

Firn, R. D., and Jones, C. G. 2000. The evolution of secondary metabolism — a unifying model. *Mol. Microbiol.*, 37(5): 989–94.

Galappaththi, M. C., Patabendige, N. M., Premarathne, B. M., Hapuarachchi, K. K., Tibpromma, S., Dai, D., Suwannarach, N., Rapior, S., and Karunarathna, S. C. 2022. A review of *Ganoderma* triterpenoids and their bioactivities. *Biomolecules*, 13(1) (December): 24. www.mdpi.com/2218-273X/13/1/24.

Goswami, P., and Sharma, H. K. 2020. Skin hyperpigmentation disorders and use of herbal extracts: A review. *Curr. Trends Pharm. Res.*, 7 (July): 132. https://fjps.springeropen.com/articles/10.1186/s43094-021-00284-6.

Gunde-Cimerman, N., Friedrich, J., Cimerman, A., and Benički, N. 1993. Screening fungi for the production of an inhibitor of HMG CoA reductase: Production of mevinolin by the fungi of the genus *Pleurotus*. *FEMS Microbiol. Lett.*, 111(2–3): 203–6.

Gunde-Cimerman, N., Plemenitaš, A., and Cimerman, A. 1993. *Pleurotus* fungi produce mevinolin, an inhibitor of HMG CoA reductase. *FEMS Microbiol. Lett.*, 113(3): 333–7.

Guo, X., Liu, D., Ye, M., Han, J., Deng, S., Ma, X., Zhao, Y., Zhang, B., Shen, X., and Che, Q. 2013. Structural characterization of minor metabolites and pharmacokinetics of ganoderic acid C2 in rat plasma by HPLC coupled with electrospray ionization tandem mass spectrometry. *J. Pharm. Biomed. Anal.*, 75: 64–73.

Guo, X., Shen, X., Long, J., Han, J., and Che, Q. 2013. Structural identification of the metabolites of ganoderic acid B from *Ganoderma lucidum* in rats based on liquid chromatography coupled with electrospray ionization hybrid ion trap and time-of-flight mass spectrometry. *Biomed. Chromatogr.*, 27(9): 1177–87.

He, H., He, J., Sui, Y., Zhou, S., and Wang, J. 2008. The hepatoprotective effects of *Ganoderma lucidum* peptides against carbon tetrachloride-induced liver injury in mice. *J. Food Biochem.*, 32(5): 628–41.

Hossain, A., Radwan, F.F., Doonan, B.P., God, J.M., Zhang, L., Bell, P.D., and Haque, A. 2012. A possible cross-talk between autophagy and apoptosis in generating an immune response in melanoma. *Apoptosis*, 17(10): 1066–78.

Jannapureddy, S., Sharma, M., Yepuri, G., Schmidt, A.M., and Ramasamy, R. 2021. Aldose reductase: An emerging target for development of interventions for diabetic cardiovascular complications. *Front. Endocrinol.*, 12 (March): 636267. www.frontiersin.org/articles/10.3389/fendo.2021.636267/full.

Jedinak, A., Thyagarajan-Sahu, A., Jiang, J., and Sliva, D. 2011. Ganodermanontriol, a lanostanoid triterpene from *Ganoderma lucidum*, suppresses growth of colon cancer cells through ß-catenin signaling. *Int. J. Oncol.*, 38: 761–7.

Jiang, J., Grieb, B., Thyagarajan, A., and Sliva, D. 2008. Ganoderic acids suppress growth and invasive behavior of breast cancer cells by modulating AP-1 and NF-κb signaling. *Int. J. Mol. Med.*, 21(5): 577–84.

Jiang, J., Jedinak, A., and Sliva, D. 2011. Ganodermanontriol (GDNT) exerts its effect on growth and invasiveness of breast cancer cells through the down-regulation of CDC20 and uPA. *Biochem. Biophys. Res. Commun.*, 415(2): 325–9.

Jiang, J., Slivova, V., and Sliva, D. 2006. *Ganoderma lucidum* inhibits proliferation of human breast cancer cells by down-regulation of estrogen receptor and NF-κb signaling. *Int. J. Oncol.*, 29(3): 695–703.

Johnson, B., Doonan, B., Radwan, F., and Haque, A. 2010. Ganoderic acid DM: An alternative agent for the treatment of advanced prostate cancer. *Open Prost. Cancer J.*, 3(1): 78–85.

Kamranvar, S., and Masucci, M. 2017. Regulation of telomere homeostasis during Epstein-Barr virus infection and immortalization. *Viruses*, 9(8): 217.

Kawagishi, H., Fukuhara, F., Sazuka, M., Kawashima, A., Mitsubori, T., and Tomita, T. 1993.5′-Deoxy-5′-methylsulphinyladenosine, a platelet aggregation inhibitor from *Ganoderma lucidum. Phytochemistry*, 32(2): 239–41.

Kennedy, E.M., P'Pool, S.J., Jiang, J., Sliva, D., and Minto, R.E. 2011. Semisynthesis and biological evaluation of Ganodermanontriol and its stereoisomeric triols. *J. Nat. Prod.*, 74(11): 2332–7.

Kim, J. C., Kim, H., Kim, J., Kwon, O., Son, E., Lee, C., and Park, Y. 2016. Effects of Ganodermanondiol, a new melanogenesis inhibitor from the medicinal mushroom *Ganoderma lucidum. Int. J. Mol. Sci.*, 17(11) (October): 1798. www.mdpi.com/1422-0067/17/11/1798.

Kim, J.C., Park, T.J., and Kang, H.Y. 2022. Skin-aging pigmentation: Who is the real enemy? *Cells*, 11(16) (August): 2541. www.mdpi.com/2073-4409/11/16/2541.

Klaus, A., Kozarski, M., Vunduk, J., Petrovic, P., and Niksic, M. 2016. Antibacterial and antifungal potential of wild basidiomycete mushroom *Ganoderma applanatum. Lekovite sirovine*, 36: 37–46.

Kleinwächter, P., Anh, N., Kiet, T.T., Schlegel, B., Dahse, H., Härtl, A., and Gräfe, U. 2001. Colossolactones, new triterpenoid metabolites from a Vietnamese mushroom *Ganoderma colossum. J. Nat. Prod.*, 64(2): 236–9.

Kubota, T., Asaka, Y., Miura, I., and Mori, H. 1982. Structures of Ganoderic acid A and B, two new lanostane type bitter triterpenes from *Ganoderma lucidum* (FR.) Karst. *Helv. Chim. Acta*, 65(2): 611–19.

Kuhn, R., and Jahn, W. 1965. On the thio ether and S-oxide derivatives of adenosine. *Chem. Ber.*, 98(6): 1699–704.

Li, C., Chen, P., Chang, U., Kan, L., Fang, W., Tsai, K., and Lin, S. 2005. Ganoderic acid X, a lanostanoid triterpene, inhibits topoisomerases and induces apoptosis of cancer cells. *Life Sci.*, 77(3): 252–5.

Li, C., Luo, Q., Guo, P., Chen, L., and Cheng, Y. 2016. Petchiethers A and B, novel meroterpenoids with a 14- or 15-membered ring from *Ganoderma petchii. Phytochem. Lett.*, 18: 14–18.

Li, K., Na, K., Sang, T., Wu, K., Wang, Y., and Wang, X. 2017. The ethanol extracts of sporoderm-broken spores of *Ganoderma lucidum* inhibit colorectal cancer *in vitro* and *in vivo*. *Oncol. Rep.*, 38(5): 2803–13.

Liang, C., Tian, D., Liu, Y., Li, H., Zhu, J., Li, M., Xin, M., and Xia, J. 2019. Review of the molecular mechanisms of *Ganoderma lucidum* triterpenoids: Ganoderic acids A, C2, D, F, DM, X and Y. *Eur. J. Med. Chem.*, 174: 130–41.

Lin, S., Li, C., Lee, S., and Kan, L. 2003. Triterpene-enriched extracts from *Ganoderma lucidum* inhibit growth of hepatoma cells via suppressing protein kinase C, activating mitogen-activated protein kinases and G2-phase cell cycle arrest. *Life Sci.*, 72(21): 2381–90.

Liu, C., Yang, N., Song, Y., Wang, L., Zi, J., Zhang, S., Dunkin, D., Busse, P., Weir, D., Tversky, J., Miller, R. L., Goldfarb, J., Zhan, J., and Li, X. 2015. Ganoderic acid C1 isolated from the anti-asthma formula, ASHMI™ suppresses TNF-α production by mouse macrophages and peripheral blood mononuclear cells from asthma patients. *Int. Immunopharmacol.*, 27(2): 224–31.

Liu, J., Shimizu, K., Konishi, F., Kumamoto, S., and Kondo, R. 2007. The anti-androgen effect of ganoderol B (I) isolated from the fruiting body of *Ganoderma lucidum*. *Bioorg. Med. Chem.*, 15(14): 4966–72.

Liu, J., Shiono, J., Shimizu, K., Kukita, A., Kukita, T., and Kondo, R. 2009. Ganoderic acid DM: Anti-androgenic osteoclastogenesis inhibitor. *Bioorganic Med. Chem. Lett.*, 19(8): 2154–7.

Liu, L., Chen, H., Liu, C., Wang, H., Kang, J., Li, Y., and Chen, R. 2014. Triterpenoids of *Ganoderma theaecolum* and their hepatoprotective activities. *Fitoterapia*, 98, 254–9.

Liu, R., Li, Y., and Zhong, J. 2012. Cytotoxic and pro-apoptotic effects of novel ganoderic acid derivatives on human cervical cancer cells *in vitro*. *Eur. J. Pharmacol.*, 681(1–3): 23–33.

Liu, R., and Zhong, J. 2011. Ganoderic acid Mf and S induce mitochondria mediated apoptosis in human cervical carcinoma Hela cells. *Phytomedicine*, 18(5): 349–55.

Liu, X., Yuan, J., Chung, C., and Chen, X. 2002. Antitumor activity of the sporoderm-broken germinating spores of *Ganoderma lucidum*. *Cancer Lett.*, 182(2): 155–61.

Lu, Q., Jin, Y., Zhang, Q., Zhang, Z., Heber, D., Go, V.L., Li, F.P., and Rao, J.Y. 2004. *Ganoderma lucidum* extracts inhibit growth and induce actin polymerization in bladder cancer cells *in vitro*. *Cancer Lett.*, 216(1): 9–20.

Luo, Q., Di, L., Yang, X., and Cheng, Y. 2016. Applanatumols A and B, meroterpenoids with unprecedented skeletons from *Ganoderma applanatum*. *RSC Adv.*, 6(51): 45963–7.

Luo, Q., Tian, L., Di, L., Yan, Y., Wei, X., Wang, X., and Cheng, Y. 2015. (±)-Sinensilactam A, a pair of rare hybrid metabolites with Smad3 phosphorylation inhibition from *Ganoderma sinensis*. *Org. Lett.*, 17(6): 1565–8.

Luo, Q., Tu, Z., Yang, Z., and Cheng, Y. 2018. Renoprotective meroterpenoids from the fungus *Ganoderma cochlear*. *Fitoterapia*, 125: 273–80.

Luo, Q., Wei, X., Yang, J., Luo, J., Liang, R., Tu, Z., and Cheng, Y. 2016. Spiro meroterpenoids from *Ganoderma applanatum*. *J. Nat. Prod.*, 80(1): 61–70.

Ma, B., Ren, W., Zhou, Y., Ma, J., Ruan, Y., and Wen, C. 2011. Triterpenoids from the spores of *Ganoderma lucidum*. *N. Am. J. Med. Sci.*, 3(11): 495–8.

Matsuda, Y., and Abe, I. 2016. Biosynthesis of fungal meroterpenoids. *Nat. Prod. Rep.*, 33(1): 26–53.

Min, B., Gao, J., Nakamura, N., and Hattori, M. 2000. Triterpenes from the spores of *Ganoderma lucidum* and their cytotoxicity against Meth-A and LLC tumor cells. *Chem. Pharm. Bull.*, 48(7): 1026–33.

Min, B., Nakamura, N., Miyashiro, H., Bae, K., and Hattori, M. 1998. Triterpenes from the spores of *Ganoderma lucidum* and their inhibitory activity against HIV-1 protease. *Chem. Pharm. Bull.*, 46(10): 1607–12.

Miyamoto, I., Liu, J., Shimizu, K., Sato, M., Kukita, A., Kukita, T., and Kondo, R. 2009. Regulation of osteoclastogenesis by ganoderic acid DM isolated from *Ganoderma lucidum*. *Eur. J. Pharmacol.*, 602(1): 1–7.

Mohsin, M., Negi, P.S., and Ahmed, Z. 2011. Determination of the antioxidant activity and polyphenol contents of wild Lingzhi or Reishi medicinal mushroom, *Ganoderma lucidum* (W.Curt. Fr.) P. Karst. (Higher basidiomycetes) from central Himalayan hills of India. *Int. J. Med. Mushrooms*, 13(6): 535–44.

Mortazavie, F., Taheri, S., Tandel, P., Zare, F., and Tamaddon, G. 2022. The effect of Ganoderic acid A on Mir-17–5p and Mir-181b expression level and apoptosis induction in human leukemia nalm-6 cells. *Iran. J. Ped. Hematol. Oncol.*, 12(3): 152–63.

Mothana, R.A., Jansen, R., Jülich, W., and Lindequist, U. 2000. Ganomycins A and B, new antimicrobial farnesyl hydroquinones from the basidiomycete *Ganoderma pfeifferi*. *J. Nat. Prod.*, 63(3): 416–18.

Müller, C.I., Kumagai, T., O'Kelly, J., Seeram, N.P., Heber, D., and Koeffler, H.P. 2006. *Ganoderma lucidum* causes apoptosis in leukemia, lymphoma and multiple myeloma cells. *Leuk. Res.*, 30(7): 841–8.

Mustafin, K., Bisko, N., Blieva, R., Al-Maali, G., Krupodorova, T., Narmuratova, Z., Saduyeva, Z., and Zhakipbekova, A. 2022. Antioxidant and antimicrobial potential of *Ganoderma lucidum* and *Trametes versicolor*. *Turkish J. Biochem.*, 47(4): 483–9.

Nazir, M., Saleem, M., Tousif, M.I., Anwar, M.A., Surup, F., Ali, I., Wang, D., Mamadalieva, N.Z., Alshammari, E., Ashour, M.L., Ashour, A.M., Ahmed, I., Elizbit, Green, I.R., and Hussain, H. 2021. Meroterpenoids: A comprehensive update insight on structural diversity and biology. *Biomolecules*, 11(7) (June): 957. www.mdpi.com/2218-273X/11/7/957.

Nguyen, V. T., Tung, N. T., Cuong, T. D., Hung, T. M., Kim, J. A., Woo, M. H., Choi, J. S., Lee, J., and Min, B. S. 2015. Cytotoxic and anti-angiogenic effects of lanostane triterpenoids from *Ganoderma lucidum*. *Phytochem. Lett.*, 12: 69–74.

Nicolaidou, E., and Katsambas, A. D. 2014. Pigmentation disorders: Hyperpigmentation and hypopigmentation. *Clin. Dermatol.*, 32(1): 66–72.

Niedermeyer, T. H., Jira, T., Lalk, M., and Lindequist, U. 2013. Isolation of farnesylhydroquinones from the basidiomycete *Ganoderma pfeifferi*. *Nat. Prod. Bioprospect.*, 3(4): 137–40.

Paciulli, R. M., and Cromer, C. M. 2018. Secondary sex characteristics. In *Encyclopedia of Animal Cognition and Behaviour*, ed. J. Vonk and T. Shackelford: 1–9. Cham: Springer International Publishing.

Peng, X., Li, L., Wang, X., Zhu, G., Li, Z., and Qiu, M. 2016. Antioxidant farnesylated hydroquinones from *Ganoderma capense*. *Fitoterapia*, 111: 18–23.

Peng, X., Liu, J., Wang, C., Li, X., Shu, Y., Zhou, L., and Qiu, M. 2014. Hepatoprotective effects of triterpenoids from *Ganoderma cochlear*. *J. Nat. Prod.*, 77(4): 737–43.

Peng, X., and Qiu, M. 2018. Meroterpenoids from *Ganoderma* species: A review of last five years. *Nat. Prod. Bioprospect.*, 8(3): 137–49.

Peng, X., Shi, Q., Yang, J., Su, H., Zhou, L., and Qiu, M. 2020. Meroapplanins A-E: Five meroterpenoids with a 2,3,4,5-tetrahydropyridine motif from *Ganoderma applanatum*. *J. Org. Chem.*, 85(11): 7446–51.

Petrovic, P., Ivanovic, K., Jovanovic, A., Simovic, M., Milutinovic, V., Kozarski, M., Petkovic, M., Cvetkovic, A., Klaus, A., and Bugarski, B. 2019. The impact of puffball autolysis on selected chemical and biological properties: Puffball extracts as potential ingredients of skin-care products. *Arch. Biol. Sci.*, 71(4): 721–33.

Petrović, P., and Vunduk, J. 2022. Nature and chemistry of bioactive components of wild edible mushrooms. In *Wild Mushrooms: Characteristics, Nutrition, and Processing*, ed. S. B. Dhull, A. Bains, P. Chawla and P. K. Sadh: 211–57. Boca Raton: CRC Press.

Petrović, P., Vunduk, J., Klaus, A., Carević, M., Petković, M., Vuković, N., Cvetković, A., Žižak, Ž., and Bugarski, B. 2019. From mycelium to spores: A whole circle of biological potency of mosaic puffball. *S. Afr. J. Bot.*, 123: 152–60.

Petrović, P., Vunduk, J., Klaus, A., Kozarski, M., Nikšić, M., Žižak, Ž., Vuković, N., Šekularac, G., Drmanić, S., and Bugarski, B. 2016. Biological potential of puffballs: A comparative analysis. *J. Funct. Foods*, 21: 36–49.

Qin, F., Wang, D., Xu, T., Zhang, B., and Cheng, Y. 2022. Meroterpenoids containing benzopyran or benzofuran motif from *Ganoderma cochlear*. *Phytochemistry*, 199 (July): 113184. www.sciencedirect.com/science/article/pii/S0031942222001005?via%3Dihub.

Qin, F., Yan, Y., Tu, Z., and Cheng, Y. 2018a. (±) Cochlearoids N-P: Three pairs of phenolic meroterpenoids from the fungus *Ganoderma cochlear* and their bioactivities. *J. Asian Nat. Prod. Res.*, 21(6): 542–50.

Qin, F., Yan, Y., Tu, Z., and Cheng, Y. 2018b. Cochlearoids L and M: Two new meroterpenoids from the fungus *Ganoderma cochlear*. *Nat. Prod. Commun.*, 13(3): 1934578X1801300.

Ren, L. 2019. Protective effect of ganoderic acid against the streptozotocin induced diabetes, inflammation, hyperlipidemia and microbiota imbalance in diabetic rats. *Saudi J. Biol. Sci.*, 26(8): 1961–72.

Richards, J. E., and Hawley, R. S. 2011. Sex determination: How genes determine a developmental choice. In *The human genome*, third edition: 273–98. Cambridge: Academic Press.

Ryu, D. H., Cho, J. Y., Sadiq, N. B., Kim, J., Lee, B., Hamayun, M., Lee, T. S., Kim, H. S., Park, S. H., Nho, C. W., and Kim, H. 2021. Optimization of antioxidant, anti-diabetic, and anti-inflammatory activities and ganoderic acid content of differentially dried *Ganoderma lucidum* using response surface methodology. *Food Chem.*, 335 (January): 127645. www.sciencedirect.com/science/article/pii/S0308814620315077.

Satria, D., Amen, Y., Niwa, Y., Ashour, A., Allam, A. E., and Shimizu, K. 2018. Lucidumol D, a new lanostane-type triterpene from fruiting bodies of Reishi (*Ganoderma lingzhi*). *Nat. Prod. Res.*, 33(2): 189–95.

Smina, T., Mathew, J., Janardhanan, K., and Devasagayam, T. 2011. Antioxidant activity and toxicity profile of total triterpenes isolated from *Ganoderma lucidum* (Fr.) P. Karst occurring in South India. *Environ. Toxicol. Pharmacol.*, 32(3): 438–46.

Su, H., Wang, Q., Zhou, L., Peng, X., Xiong, W., and Qiu, M. 2020. Highly oxygenated lanostane triterpenoids from *Ganoderma applanatum* as a class of agents for inhibiting lipid accumulation in adipocytes. *Bioorg. Chem.*, 104 (November): 104263. www.sciencedirect.com/science/article/pii/S0045206820315613?via%3Dihub.

Su, H., Wang, Q., Zhou, L., Peng, X., Xiong, W., and Qiu, M. 2021. Functional triterpenoids from medicinal fungi *Ganoderma applanatum*: A continuous search for antiadipogenic agents. *Bioorg. Chem.*, 112 (July): 104977. www.sciencedirect.com/science/article/pii/S0045206821003540.

Tang, W., Liu, J., Zhao, W., Wei, D., and Zhong, J. 2006. Ganoderic acid T from *Ganoderma lucidum* mycelia induces mitochondria mediated apoptosis in lung cancer cells. *Life Sci.*, 80(3): 205–11.

Thawthong, A., Hapuarachchi, K. K., Wen, T., Raspé, O., Thongklang, N., Kang, J., and Hyde, K. D. 2017. *Ganoderma sichuanense* (Ganodermataceae, Polyporales) new to Thailand. *MycoKeys*, 22: 27–43.
Thuy, N.H., Diem, V.T., Thuong, T.T., Anh, T.T., Nhut, T.M., Dat, T.V., and Trinh, H.N. 2022. *In vitro* antioxidant activity and *in vivo* hepatoprotective effects of ethanolic extracts from wall-broken *Ganoderma lucidum* spores. *Open Access Maced. J. Med. Sci.*, 10(A): 1450–5.
Thyagarajan, A., Jedinak, A., Nguyen, H., Terry, C., Baldridge, L.A., Jiang, J., and Sliva, D. 2010. Triterpenes from *Ganoderma lucidum* induce autophagy in colon cancer through the inhibition of p38 mitogen-activated kinase (p38 MAPK). *Nutr. Cancer*, 62(5): 630–40.
Trigos, A., and Medellin, J.S. 2011. Biologically active metabolites of the genus *Ganoderma*: Three decades of myco-chemistry research. *Rev. Mex. Micol.*, 34: 63–83.
Vunduk, J., Klaus, A., Kozarski, M., Petrovic, P., Zizak, Z., Niksic, M., and Van Griensven, L.J. 2015. Did the iceman know better? Screening of the medicinal properties of the birch polypore medicinal mushroom, *Piptoporus betulinus* (higher basidiomycetes). *Int. J. Med. Mushrooms.*, 17(12): 1113–25.
Vunduk, J., and Veljović, S. 2021. Macrofungi in the production of alcoholic beverages: Beer, wine, and spirits. In *Advances in Macrofungi: Industrial Avenues and Prospects*, ed. K. R. Sridhar and S. K. Deshmukh: 1–34. Boca Raton: CRC Press.
Wang, F., Zhou, Z., Ren, X., Wang, Y., Yang, R., Luo, J., and Strappe, P. 2015. Effect of *Ganoderma lucidum* spores intervention on glucose and lipid metabolism gene expression profiles in type 2 diabetic rats. *Lipids Health Dis.*, 14 (May): 49. https://lipidworld.biomedcentral.com/articles/10.1186/s12944-015-0045-y.
Wang, G., Zhao, J., Liu, J., Huang, Y., Zhong, J., and Tang, W. 2007. Enhancement of IL-2 and IFN-γ expression and NK cells activity involved in the anti-tumor effect of ganoderic acid me *in vivo*. *Int. Immunopharmacol.*, 7(6): 864–70.
Wang, K., Bao, L., Xiong, W., Ma, K., Han, J., Wang, W., Yin, W., and Liu, H. 2015. lanostane triterpenes from the Tibetan medicinal mushroom *Ganoderma leucocontextum* and their inhibitory effects on HMG-CoA reductase and α-glucosidase. *J. Nat. Prod.*, 78(8): 1977–89.
Wang, L., Li, J., Zhang, J., Li, Z., Liu, H., and Wang, Y. 2020. Traditional uses, chemical components and pharmacological activities of the genus *Ganoderma* P. Karst.: A review. *RSC Adv.*, 10(69): 42084–97.
Wang, X., Sun, D., Tai, J., and Wang, L. 2017. Ganoderic acid A inhibits proliferation and invasion, and promotes apoptosis in human hepatocellular carcinoma cells. *Mol. Med. Rep.*, 16(4): 3894–900.
Wang, X., Wu, Z., Di, L., Zhou, F., Yan, Y., and Cheng, Y. (2019. Renoprotective meroterpenoids from the fungus *Ganoderma cochlear*. *Fitoterapia*, 132: 88–93.
Wang, X., Zhou, F., Dou, M., Yan, Y., Wang, S., Di, L., and Cheng, Y. 2016. Cochlearoids F-K: Phenolic meroterpenoids from the fungus *Ganoderma cochlear* and their renoprotective activity. *Bioorg. Med. Chem. Lett.*, 26(22): 5507–12.
Waning, D., Mohammad, K., and Guise, T. 2013. Cancer-associated osteoclast differentiation takes a good look in the miR(NA)ror. *Cancer Cell*, 24(4): 407–9.
Wink, M. 2003. Evolution of secondary metabolites from an ecological and molecular phylogenetic perspective. *Phytochemistry*, 64(1): 3–19.
Wong, K., Chao, H., Chan, P., Chang, L., and Liu, C. 2004. Antioxidant activity of *Ganoderma lucidum* in acute ethanol-induced heart toxicity. *Phytother. Res.*, 18(12): 1024–6.
Wu, G., Lu, J., Guo, J., Li, Y., Tan, W., Dang, Y., Zhong, Z., Xu, Z., Chen, X., and Wang, Y. 2012. Ganoderic acid DM, a natural triterpenoid, induces DNA damage, G1 cell cycle arrest and apoptosis in human breast cancer cells. *Fitoterapia*, 83(2): 408–14.
Xia, J., Dai, L., Wang, L., and Zhu, J. 2020. Ganoderic acid DM induces autophagic apoptosis in non-small cell lung cancer cells by inhibiting the PI3K/Akt/mTOR activity. *Chem. Biol. Interact.*, 316 (January): 108932. www.sciencedirect.com/science/article/pii/S0009279719313894?via%3Dihub.
Xu, Z., Chen, X., Zhong, Z., Chen, L., and Wang, Y. 2011. *Ganoderma lucidum* polysaccharides: Immunomodulation and potential anti-tumor activities. *Am. J. Chinese Med.*, 39(1): 15–27.
Yan, Y., Wang, X., Luo, Q., Jiang, L., Yang, C., Hou, B., Zuo, Z., Chen, Y., and Cheng, Y. 2015. Metabolites from the mushroom *Ganoderma lingzhi* as stimulators of neural stem cell proliferation. *Phytochemistry*, 114: 155–62.
Yan, Z., Li, S., Yu, H., and Xin, X. 2019. Extraction and antioxidant activity of *Ganoderma lucidum* polysaccharides using electrolyzed oxidizing water. *IOP Conf. Ser.: Earth Environ. Sci.*, 252(2) (April): 022074. https://iopscience.iop.org/article/10.1088/1755-1315/252/2/022074/pdf.
Yang, Y., Zhou, H., Liu, W., Wu, J., Yue, X., Wang, J., Quan, L., Liu, H., Guo, L., Wang, Z., Lian, X., and Zhang, Q. 2018. Ganoderic acid A exerts antitumor activity against MDA-MB-231 human breast cancer cells by inhibiting the Janus kinase 2/signal transducer and activator of transcription 3 signaling pathway. *Oncol. Lett.*, 16(5): 6515–21.

Zhang, J., Ma, K., Han, J., Wang, K., Chen, H., Bao, L., Liu, L., Xiong, W., Zhang, Y., Huang, Y., and Liu, H. 2018. Eight new triterpenoids with inhibitory activity against HMG-CoA reductase from the medical mushroom *Ganoderma leucocontextum* collected in Tibetan Plateau. *Fitoterapia*, 130: 79–88.

Zhang, Q., Zuo, F., Nakamura, N., Ma, C., and Hattori, M. 2009. Metabolism and pharmacokinetics in rats of ganoderiol F, a highly cytotoxic and antitumor triterpene from *Ganoderma lucidum*. *J. Nat. Med.*, 63(3): 304–10.

Zhang, W., Tao, J., Yang, X., Yang, Z., Zhang, L., Liu, H., Wu, K., and Wu, J. 2014. Antiviral effects of two *Ganoderma lucidum* triterpenoids against enterovirus 71 infection. *Biochem. Biophys. Res. Commun.*, 449(3): 307–12.

Zhang, X., Gao, X., Long, G., Yang, Y., Chen, G., Hou, G., Huo, X., Jia, J., Wang, A., and Hu, G. 2022. Lanostane-type triterpenoids from the mycelial mat of *Ganoderma lucidum* and their hepatoprotective activities. *Phytochemistry*, 198 (June): 113131. www.sciencedirect.com/science/article/pii/S0031942222000474?via%3Dihub.

Zhang, Y., Tang, X., Jiang, T., Zhang, F., Zeng, M., Zhou, D., Xu, Q., Zhang, W., and Jahanzaib, R. 2022. The potential activity of Lingzhi or Reishi medicinal mushroom, *Ganoderma lucidum* (Agaricomycetes), to alleviate liver injury in adults with acute mushroom poisoning: A retrospective study. *Int. J. Med. Mushrooms*., 24(5): 57–72.

Zhao, C., Fan, J., Liu, Y., Guo, W., Cao, H., Xiao, J., Wang, Y., and Liu, B. 2019. Hepatoprotective activity of *Ganoderma lucidum* triterpenoids in alcohol-induced liver injury in mice, an itraq-based proteomic analysis. *Food Chem*., 271: 148–56.

Zheng, D., and Chen, L. 2017. Triterpenoids from *Ganoderma lucidum* inhibit the activation of EBV antigens as telomerase inhibitors. *Exp. Ther. Med*., 14(4): 3273–8.

Zhong, J., and Xiao, J. 2009. Secondary metabolites from higher fungi: Discovery, bioactivity, and bioproduction. *Adv. Biochem. Eng. Biotechnol*., 113: 79–150.

Zhou, L., Shi, P., Chen, N., and Zhong, J. 2011. Ganoderic acid Me induces apoptosis through mitochondria dysfunctions in human colon carcinoma cells. *Process Biochem*., 46(1): 219–25.

Zhu, J., Ding, J., Li, S., and Jin, J. 2022. Ganoderic acid A ameliorates non-alcoholic streatohepatitis (NASH) induced by high-fat high-cholesterol diet in mice. *Exp. Ther. Med*., 23(4) (February): 308. www.spandidos-publications.com/10.3892/etm.2022.11237.

Zhu, Q., Bang, T. H., Ohnuki, K., Sawai, T., Sawai, K., and Shimizu, K. 2015. Inhibition of neuraminidase by *Ganoderma* triterpenoids and implications for neuraminidase inhibitor design. *Sci. Rep*., 5 (August): 13194. www.nature.com/articles/srep13194.

9 Anti-Inflammatory and Anti-Arthritis Properties of *Ganoderma*

Kunal Kumar Saha[1], *Anik Barman*[2], *and Narayan Chandra Mandal*[1*]
1 Visva Bharati University, Santiniketan, India
2 Bose Institute Kolkata, Kolkata, India

9.1 INTRODUCTION

Inflammation is the immune system's response to harmful stimuli such as infections, damaged cells, toxic chemicals, or irradiation. It simultaneously eliminates damaging impulses and initiates the healing process (Medzhitov, 2010; Ferrero-Miliani et al., 2007). Therefore, inflammation is a defensive process that is essential for maintaining normal health (Nathan and Ding, 2010). Typically, cellular and molecular activities and interaction during acute inflammatory reactions effectively reduce the risk of harm or infection. The acute inflammation is reduced and homeostasis in tissues is restored as a result of this mitigation mechanism. Yet untreated acute inflammation may progress to chronic inflammation, resulting in the onset of a variety of chronic inflammatory illnesses (Zhou et al., 2016). Redness, swelling, heat, discomfort, and loss of tissue function are signs of inflammation at the tissue level and appear due to localized immunological, vascular, and inflammatory cell responses to infection or damage (Takeuchi and Akira, 2010). Vascular permeability alterations, leukocyte recruitment and accumulation, and release of inflammatory mediators are significant microcirculatory processes that take place throughout the inflammatory phase (Chertov et al., 2000). Inflammation may be induced by pathogenic agents that harm tissue, such as infection, tissue damage, or myocardial infarction. Inflammation may have infectious or non-infectious causes (Chen et al., 2018). When a tissue is damaged, the body initiates a chemical signaling cascade that promotes healing of the damaged tissue. These cues direct leukocytes to relocate from the bloodstream to injury sites. Inflammation is initiated by these activated leukocytes' cytokine production (Jabbour et al., 2009).

In several instances, inflammation has been found to be associated with another clinical condition known as arthritis. It was the Greeks who first used the phrase "disease of the joints" to describe what we now call arthritis. It causes discomfort and structural damage to joints, and it may be either acute or chronic (Ma et al., 2009). Pain in a joint, regardless of its cause, is called arthralgia, which is different from arthritis (joint inflammation being a possible cause). Despite the fact that Neandertals and ancient Egyptians had arthritis in common, the name osteoarthritis was not really created until 1886 by Dr. John K. Spencer. It is estimated that there are over a hundred distinct forms of arthritis, the most prevalent of which is osteoarthritis, often known as degenerative arthritis or non-inflammatory arthritis. Inflammatory arthritis can be brought on by a number of factors, including autoimmune processes (such as rheumatoid arthritis, psoriatic arthritis, ankylosing spondylitis, etc.), crystal deposition–generated inflammation (such as gout, pseudogout, basic calcium phosphate illness), and infections (such as septic arthritis, Lyme's disease, etc.).

Every year a large number of people are affected either by the inflammatory or arthritic diseases and are thereby forced to adopt a painful as well as stressful life. Both these clinical conditions

DOI: 10.1201/9781003354789-9

cannot be cured permanently, and patients often have to take non-steroidal anti-inflammatory drugs (NSAIDs). Most NSAIDs are reported to have no selectivity for different isoforms of cyclo-oxygenase (COX) enzymes, a potential target for developing anti-inflammatory principals. Thus, this causes a wide range of side effects on human health which include gastrointestinal bleeding, stroke, and renal failure (Qin et al., 2020). In an alternative approach, the use of different traditional medicinal practices, especially medicinally important *Ganoderma* spp., have been found to have a significant satisfactory effect against different inflammatory as well as arthritic diseases. *Ganoderma* spp. are a group of saprophytic or facultative parasitic Basidiomycete mushroom chiefly growing on dead or decaying woods in high humid, subtropical, and temperate regions of Asia, Europe, and North and South America. Since ancient times, different species of *Ganoderma*, especially species like *Ganoderma lucidum*, have had a massive impact as traditional medicinal practice in Asia, particularly in China. So far, approximately 400 different bioactive compounds have been identified from different species of *Ganoderma*. Antioxidant, anti-inflammatory, anti-arthritic, anti-diabetic, anti-cancerous, anti-obesity, antibacterial, antiviral, immunomodulatory, etc., are only a few names of the therapeutic effects they have (Ahmad et al., 2021; Cör et al., 2018).

Here in this chapter, we will briefly discuss the different types of inflammatory and arthritic diseases and the associated immunopathogenesis processes. These findings indicate key areas to focus on while developing any anti-inflammatory and anti-arthritis medication. *Ganoderma* spp. specifically is being studied for its anti-inflammatory and anti-arthritis effects, and we will critically examine the findings reported to date. Herein, we will mainly focus on the chemical nature of the active constituents along with their extraction strategy and mode of action against different inflammatory or arthritic conditions. Additionally, different approaches to improve the quality of the *Ganoderma*-derived active constituents will also be discussed.

9.2 INFLAMMATORY DISEASES AND ANTI-INFLAMMATORY POTENTIAL OF *GANODERMA* SPP.

Infections, cellular damage, and chemical exposure are just a few of the many potential triggers for the immune system's inflammatory response. Acute or chronic inflammatory responses may cause tissue damage or infection in a variety of organs, including the heart, pancreas, liver, kidney, lung, brain, digestive tract, and reproductive system. Infectious and non-infectious stimuli as well as cell damage trigger inflammatory signaling pathways and activate inflammatory cells (Chen et al., 2018). Inflammatory disorders that are immune-mediated are frequent and clinically varied. In spite of their incurability, the therapeutic inventory for immune-mediated inflammatory diseases has evolved during the last two decades. We have transitioned from the widespread use of broad-spectrum immune modulators to the routine use of medicines with precise specificity as a result of advancements in monoclonal and molecular biotechnology, as well as more recently, highly focused medicinal chemistry. Here, we discuss the types of inflammatory diseases with characteristics of immunological and physiological conditions.

They all have a common mechanism that may be summed up as follows. Inflammatory response mechanisms depend on the specificity of the original stimulus and its location in the body. The recognition of damaging stimuli by cell surface pattern receptors triggers the activation of inflammatory pathways, production of inflammatory markers, and recruitment of inflammatory cells. Unfortunately, there are currently no treatments available for the prevalent and clinically varied group of illnesses known as immune-mediated inflammatory diseases (IMIDs) (Kuek et al., 2007). Many IMIDs are regularly treated with biological therapy; the principal routes and diseases that have been effectively addressed are shown in Figure 9.1. Included among these biological pharmaceutical targets are cytokines and their receptors as well as specific subsets of cells, with the goal of altering trafficking, changing the activation state of cells, or depleting cells. A wide range of immunological characteristics may provide different therapeutic benefits. Hence, if described, for instance, based on T-cell subset dominance, appropriate cognate treatments have been effectively

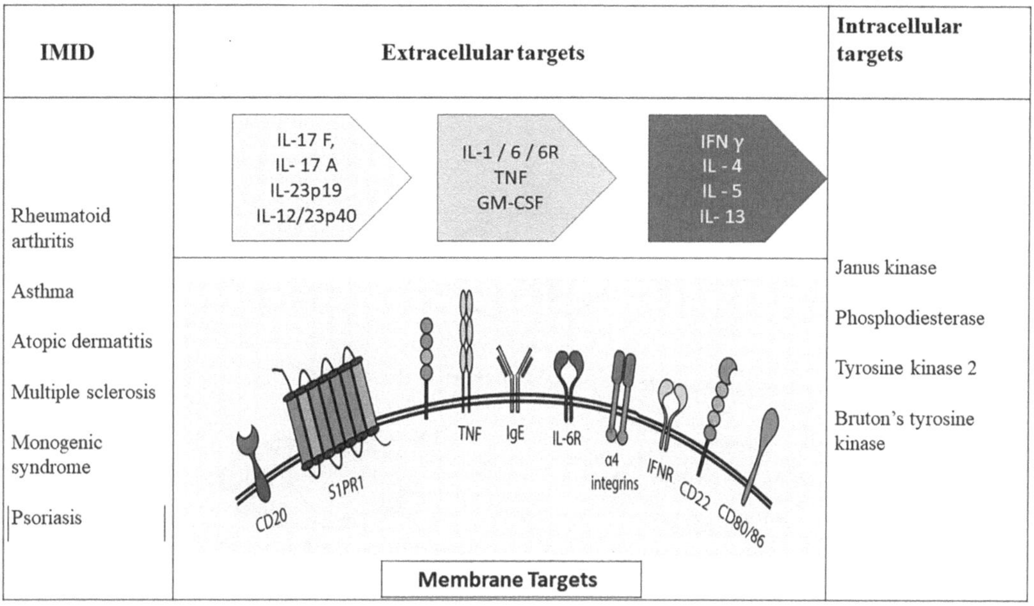

FIGURE 9.1 The image represents potential extracellular, intracellular and membrane bound targets for controlling IMID.

applied to type 1–, type 2–, and type 17–associated conditions (McInnes and Gravallese, 2021). $CD4^+$ T cells are often regarded as the major architects of the immune response, despite the fact that other cell types secrete immunoregulatory cytokines. T helper cells are often subdivided into the functionally heterogeneous Th 1 and Th 2 fractions according to the cytokines they secrete (Figure 9.2). Th 1 cells, which primarily promote inflammation, have been linked to the immunopathogenesis of IMIDs. In this situation, it is believed that Th 2 cells have anti-inflammatory or protective properties (Lucey et al., 1996). Modulation of T-cell activity has emerged as a desirable target for intervention because of its crucial involvement in IMIDs. Unfortunately, the exact origin of rheumatoid arthritis (RA) is still unclear. It is thought, however, that some antigen, either exogenous or endogenous, triggers an abnormal immune response in those who are genetically susceptible to developing the disease (Gregersen et al., 1987). When activated by antigen-presenting cells and co-stimulatory pathways, Th 1 cells enter the synovium and stimulate the production of inflammatory cytokines and damaging proteinases. Despite the fact that several cell types and mediators are involved in the pathophysiology of RA, tumor necrosis factor (TNF)-α, interleukin (IL)-1, and IL-6 are recognized as the principal inflammatory mediators (Choy and Panayi, 2001). Clinical studies have shown conclusively that anti–TNF-α and anti–IL-1 antibodies may lessen illness manifestations. However, IL-4 and IL-10, two anti-inflammatory cytokines, may provide resistance to RA (Sugiyama et al., 1995). For instance, it has been demonstrated in vitro that the Th 2 cell–produced IL-4 inhibits the activation of Th 1 cells, resulting in a reduction in the generation of IL-1 and TNF-α (Figure 9.2) (Van Roon et al., 1995). A lot is still unknown regarding the causes of RA; however, B lymphocytes have been hypothesized to have a role. Rheumatoid factor and other pathogenic autoantibodies are produced (Edwards et al., 1999), Perhaps well-described mechanisms include the loss of B-cell tolerance. Additional mouse studies have shown that the presence of B cells (Takemura et al., 2001) is crucial for T-cell activation within the rheumatoid synovium.

Anti-inflammatory medications are being developed with the purpose of inhibiting the production of these inflammatory mediators. Prominent examples include inflammatory mediators like

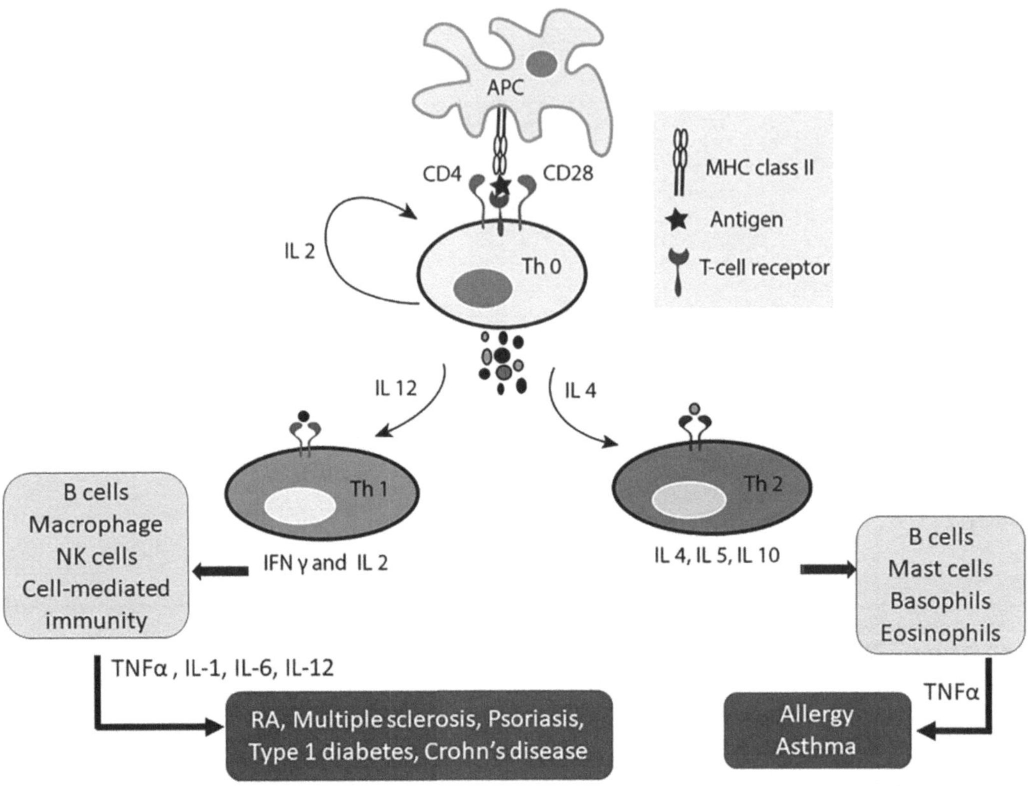

FIGURE 9.2 The image represents crosstalk between antigen present cells and T cell receptors resulting production of several interleukins and interferons. The consequence in turn causes rheumatoid arthritis, multiple sclerosis, psoriasis, type 1 diabetes, Crohn's disease, allergy and asthma.

eicosanoids, which are produced from arachidonic acid by one of three metabolic routes involving epoxygenase (COX), or Lipoxygenase (LOX). Specifically, the COX2 and COX1 isoforms are often the focus of efforts to create new NSAIDs (Qin et al., 2020; Rašeta et al., 2021). Similarly formyl peptide receptors (FPRs), a group of chemoattractant receptors, are highly present in macrophages. FPRs are responsible for regulating inflammation by increased activity of phagocytosis of microphages, superoxide production, increased activity of neutrophils, etc. Thus FPRs serve as perfect targets for the development of anti-inflammatory drugs (Peng et al., 2021). Following this principle several active constituents with varying degree of anti-inflammatory activities have been extracted from medicinally important *Ganoderma* spp. (Table 9.1).

9.3 ARTHRITIC DISEASES AND ANTI-ARTHRITIC POTENTIAL OF *GANODERMA* SPP.

The autoimmune disease RA is characterized by chronic pain and inflammation. Inflammation comes from the Latin word arthritis and the Greek word arthron, both of which imply joint and are the origins of the word arthritis. Inflammation of one or more joints occurs with arthritis, which itself comes in more than a hundred distinct varieties (Chhem et al., 1994). Osteoarthritis (OA), often known as degenerative joint disease, is by far the most common form of arthritis. The following are autoimmune disorders including arthritic conditions and others (Table 9.2).

Numerous other medical diseases can have arthritis as a secondary ailment. It is also possible to have undifferentiated arthritis, which does not fall into any recognized clinical illness classification

TABLE 9.1
List of Different Species of *Ganoderma* and Their Anti-Inflammatory Properties

Sr. No.	Source	Active Constituents and Chemical Nature	Extraction/Isolation Methodology	IC_{50} Value/Effective Concentration (EC) and Mode of Action	References
1.	*Ganoderma atrum* (powder of dried fruit body)	Eight ganoderic acids were identified in ESAC.	ESAC was extracted and purified using saturated $NaHCO_3$ solution, followed by acidification and re-extraction using chloroform. GC-MS analysis of ESAC was further carried out.	EC: 50–200 mg kg^{-1}. Inhibiting phagocytosis, NO and reducing IL-1β, and also by upregulation of MR and IL-10 among LPS induced macrophages like RAW264.7	Yao et al., 2022
2.	*Ganoderma lucidum*	Two sterols, ergosterol peroxide and ergosterol, were identified in GLS.	Crude methanolic extract of GLS extracted using alcohol/salt aqueous two-phase system (ATPS). GLS profile was further characterized by high-performance liquid chromatography–evaporative light scattering detector (HPLC-ELSD).	EC: 0.5, 2.5 and 5 μg ml^{-1}. Generating anti-inflammatory response by inhabiting phosphorylation of p65 and p38, thus, blocking the entire pathway for NF-κB and MAPK, respectively, among LPS-treated RAW264.7	Xu et al., 2021
3.	*Ganoderma lucidum*	FIP–glu. (glycoprotein with N-glycosylation)	Recombinant FIP-glu proteins were purified using Ni-NTA column chromatography and further analyzed using UPLC/Q-TOF-MS analysis	EC: 4 μg ml^{-1}. Inhibiting MAPK pathways by blocking phosphorylation of p38 in LPS-treated RAW264.7	Li, Chen et al., 2021; Li, Chang et al., 2019
4.	*Ganoderma lucidum* (heat dried)	Ganoderic acids (GAs) (oxygenated tetracyclic triterpenoids)	Dried crude ethanolic extract was prepared from heat-dried fruit bodies. Crude extract was further analyzed using UPLC/Q-TOF-MS analysis.	EC: 10 μg ml^{-1}. Inhibiting inflammatory-related cytokines, IL-23/p19, CCL20, and S100A7 at mRNA level in human immortalized keratinocyte line, HaCaT cells.	Ryu et al., 2021
5.	*Ganoderma applanatum*	Applandimeric acids D (4) (meroterpenoid dimer)	Ethanolic extract was initially fractionated using different column chromatography (macroporus resin → silica gel→ Rp–C18→ Sephadex LH-20 in order). Following this compound 4 was isolated using P-TLC and semi-preparative HPLC. Additionally, structural and chemical configuration was determined using coupling constant, NMR, and ECD.	IC_{50}: 7.93μM. Inhibited formyl peptide receptor 2 (FPR2) in human promyeloblast leukemia HL–60 cells.	Peng et al., 2021
6.	*Ganoderma cochlear*	Dispirocochlearoids A–C (1–3) (meroterpenoid)	All the compounds in isolated form are racemic mixtures through HPLC. Naturality and chemical nature of the compounds were further elucidated by HRESIMS, ^{13}C NMR with DEPT spectral data.	IC_{50}: 386 nM. Anti-inflammatory, inhibiting COX-2 inhibitor in LPS-induced lung injury among ALI mice.	Qin et al., 2020

(*Continued*)

TABLE 9.1 *(Continued)*
List of Different Species of *Ganoderma* and Their Anti-Inflammatory Properties

Sr. No.	Source	Active Constituents and Chemical Nature	Extraction/Isolation Methodology	IC_{50} Value/Effective Concentration (EC) and Mode of Action	References
7.	*Ganoderma lucidum* (powder of dried fruit body)	Ganoluciduone A (octonorlanostane), ganoluciduone B (lanostane triterpenoids)	Methanolic extract was initially fractionated through different column chromatography (silica gel→ Sephadex LH–20→ Octadecylsilyl). Following this P-TLC and semi-preparative HPLC were used to screen out 15 lanostanoid compounds. Among these, the two previously undescribed lanostinoids were characterized in detail through HRESIMS, ^{1}H and ^{13}C-DEPT NMR, and X-ray crystallographic analysis.	EC: 12.5 μM. Inhibiting NO generation among RAW264.7 macrophages treated with LPS.	Su et al., 2020
8.	*Ganoderma pfeifferi, G. resinaceum, G. lucidum and G. applanatum* (dried, powdered fruit bodies)	Protocatechuic acid (phenolics)	Ethanolic as well as chloroform extracts were prepared and phenolics from that crude extract (finally dissolved in DMSO) were quantified using LC-MS.	IC_{50}: 0.4–7.51 mg ml^{-1} for ethanolic extracts and 0.39–0.70 mg ml^{-1} for chloroform extracts. Inhibiting activity of 12–HHT, PGE2, and TXB2, some major metabolites of COX-1. It also inhibited 12-LOX by blocking 12-HETE.	Rašeta et al., 2021
9.	*Ganoderma atrum* (dried fruit bodies)	*Ganoderma atrum* polysaccharides (PSG–1).	Polysaccharides were precipitated from initial aqueous crude extract with 80% ethanol. Polysaccharide extract was further deproteinized following Sevag method. PSG-1 was purified using column chromatography with Superdex-G 200.	EC: 160 μg ml^{-1}. Anti-inflammatory response by reducing levels of LPS-induced ROS, pro-inflammatory cytokines, and COX-2. It also prevented activation of MAPK pathway both in LPS-treated inflammatory macrophages and also in intestinal-like Caco-2/RAW264.7 co-culture model for inflammation.	Hu et al., 2020a
10.	*Ganoderma lucidum* (dried and chopped fruit body)	Ganodermanontriol (triterpene)	Chloroform extract was passed through silica gel chromatography and subsequently triterpenes were isolated using silica gel chromatography and UPLC-MS analysis.	EC: 5, 10, and 20 mg kg^{-1}. Inhibiting expressions and interactions of TLR4 and MyD88, inhibiting phosphorylation of p38, ErK1/2 and JNK, and inhibiting translocation from nucleus and binding of NF-κB to DNA. Ultimately caused inactivation of pathways like NF-κB and MAPK among LPS/D-GalN-treated female BALB/C mice.	Hu et al., 2020b

11.	*Ganoderma cochlear*	Compound 18 (meroterpenoid)	95% ethanol was used to prepare the crude extract, which was further extracted using *n*-butanol and ethyl acetate. The ethyl acetate extract was then passed through different chromatography (macroporus resin → silica gel→ Rp–C18→ Sephadex LH–20) and following P-TLC and semi-preparative HPLC compound 18 was purified to 95%. HRESIMS and ^{13}C NMR were used to determine the molecular formula of C18.	Compound 18 binds with FPRs and causes conformational change in them, which blocked the WKYMVm signaling pathway. This reduces superoxide production and cell chemotaxis, which reduces inflammation among RBL-FPR2 cells (rat basophils leukemia cells [RBL-2H3] transfected with human FPR-2).	Wang et al., 2020
12.	*Ganoderma lucidum*	*Ganoderma lucidum* polysaccharides (GLPS)	GLPS were precipitated from aqueous extract using alcohol precipitation method and finally purified through HPLC.	EC: 5, 100, and 150 mg kg^{-1}. Inhibiting NLRP3, suppressing free radical lipid peroxidation and synthesis of NO and CYP2E1 in carbon tetra chloride–induced liver injury in Kunming mice.	Chen et al., 2019
13.	*Ganoderma lucidum*	Lucidone D (LUC)—terpenoid	Isolated from ethanolic extract of *G. lucidum* fruit bodies.	EC: 10, 20, and 40 μM. Anti-inflammatory effect by decreasing the synthesis of NO and COX-2 among LPS-induced RAW264.7 cells.	Feng and Wang, 2019
14.	*Ganoderma lucidum*	Ganosidone A (1) and its other derivatives (compounds 2–9). (lanostane triterpenoids)	Methanolic extract of fruit bodies were prepared, and after being reextracted with ethyl acetate, it was subjected to silica gel chromatography and LH-20 Sephadex gel column. Fractions from gel column were then subjected to HPLC to isolate the compounds. The structure was determined by HRESIMS and 1D and 2D NMR study.	EC: 50 μM. For compounds 4 and 7, this showed anti-inflammatory effect by decreasing NO production in LPS-treated RAW264.7 cells.	Koo et al., 2021
15.	*Ganoderma lucidum* (air dried powdered fruit bodies)	Lucidumins A–D (meroterpenoids) and lucidimine E and compound 10 (alkaloids)	Crude extract was prepared using 95% methanol, which was then reextracted with ethyl acetate. Following this the extract was purified first through a series of chromatography techniques (macroporus resin → silica gel→ Rp–C18→ Sephadex LH–20), and finally through PTLC and semi-preparative HPLC. Additional structural elucidation of purified compounds was done using HRESIMS and NMR.	IC_{50}: 4.68–15.49 μM, where compound 10 is most potent (EC_{50}: 2.49 μM). Anti-inflammatory effect by decreasing NO levels in LPS-treated macrophage RAW264.7 cells.	Lu et al., 2019
16.	*Ganoderma sinense* (air-dried powdered fruit bodies)	Ganocalidophins A–C (1–3) and compounds 6 (ergosterols)	A mixture of chloroform and methanol was used to prepare the crude extract. The crude extract was fractionalized using ethyl acetate. Compounds were isolated by subjecting these fractions into a combination of silica gel column chromatography, MPLC (ODS), and HPLC. Finally different spectroscopic techniques and X-ray diffraction (XRD) studies were employed for structural elucidation.	IC_{50}: 17.7–32.4 μM. Anti–inflammatory effect by inhibiting NO production in LPS-treated macrophage RAW264.7 cells.	Mei et al., 2019

(Continued)

TABLE 9.1 *(Continued)*
List of Different Species of *Ganoderma* and Their Anti-Inflammatory Properties

Sr. No.	Source	Active Constituents and Chemical Nature	Extraction/Isolation Methodology	IC_{50} Value/Effective Concentration (EC) and Mode of Action	References
17.	*Ganoderma resinaceum*	Compounds 5, 6, 22, 23, and 25–28 (ergostane type C_{28} steroids)	90% aqueous methanolic crude extract prepared at 60°C was further reextracted using petroleum ether and ethyl acetate. This extract was further purified following silica gel and RP18 column chromatography, P-TLC, and semi-preparative HPLC. Comprehensive spectroscopic analysis and XRD studies revealed structural details.	IC_{50}: 3.24–35.19 μM. Inhibited NO production in LPS-induced mouse mononuclear macrophages RAW264.7.	Shi et al., 2019
18.	*Ganoderma duripora* (dried fruit body powder)	Ganoduriporols A and B (fernesyl phenolic compounds)	Crude extract was prepared using 95% ethanol. After dried and resuspended into water, the crude extract was further treated with ethyl acetate. Following this the ethyl acetate extract was applied to different chromatography techniques (macroporus resin → silica gel→ Sephadex LH–20). Finally semi-preparative HPLC was performed to obtain purified ganoduriporols. HRESIMS and NMR were used to determine the structure and chemical formula of the compounds.	EC: 8 μM. Suppressed COX2, MAPK, and NF-κB signaling pathways and inhibited the production of IL-6, IL-1β, and TNF-μ in LPS-stimulated murine macrophage RAW264.7 cells.	Liu et al., 2018
19.	*Ganoderma lucidum*	GLPss58 (sulfated form of GLP20 a β-D-glucan polysaccharide)	Purified GLP20 was chemically modified using sulfation agent cholrosulfonic acid/pyridine method. The mixture was then dialyzed to remove pyridine, salts, and other degradation products. Finally 80% ethanol was used to concentrate and precipitate the dialysate. Then sulfated derivative GLPss58 was obtained by redissolving the dialysate into distilled water followed by freeze drying.	IC_{50}: 13.5 μg ml^{-1} Able to inhibit inflammation in HPBLs and in MSLs from C57BL/6mice by blocking inflammatory mediators like L-selectin, complement systems, and different cytokines.	Zhang et al., 2018
20.	*Ganoderma lucidum* (dried aqueous extract)	Ganoderic acid C1 (GAC1)—triterpene	Dried aqueous extract was redissolved in methylene chloride. The methylene chloride fraction was subjected to repeated silica gel, preparative HPLC, and Sephadex LH-20 column chromatography in order to obtained purified GAC1.	EC: 10 μg ml^{-1}. Suppressed the inflammatory response by inhibiting inflammatory cytokines TNF-α, IFN-α, and IL-17A in LPS-treated murine macrophage RAW264.7 cells, which can also be helpful for individuals with Crohn's disease.	Liu et al., 2015
21.	*Ganoderma lucidum* (dried mycelial powder)	Polysaccharide peptides (PsP)	NA	EC: 50, 100, and 150 mg kg^{-1}. Effective suppression of foam cells (produced in chronic inflammation) and Il-6, a inflammatory cytokine among atherosclerotic rats.	Sargowo et al., 2015

22.	*Ganoderma lucidum*	Crude extract	Crude extract was prepared by boiling the fruit bodies in hot water following filter sterilization.	EC: 20 mg kg–1. Anti-inflammatory effect by inhibiting TNF-α and IL-8. These along with anti-oxidant activities are protective against cerebral ischemia/reperfusion injury in rat.	Zhang et al., 2014
23.	*Ganoderma lucidum*	Butyl lucidenateD2 (GT–2), E2 P, N and Q; ganoderiol F, methyl ganodenate J. (lanostane triterpenes)	NA	GT2 inhibited the production of NO, COX-2, TNF-α, and IL-6. GT2 also induced heme oxygenase (HO)-1, which generates strong anti-inflammatory responses in LPS-treated, LPS-stimulated murine macrophage RAW264.7.	Choi et al., 2014
24.	*Ganoderma lucidum*	*Ganoderma lucidum* ethanol extract (EGL)	Freeze dried and milled fruit bodies were extracted using 25% ethanol at room temperature for 10 hours. Following this the extract was filter sterilized.	EC: 0.5 μg ml^{-1} EGL inhibited the proliferation of TNF-α, NO, and IL-6; additionally it suppressed TLR signaling pathway and NF-κB and COX-2 to generate anti-inflammatory effect in LPS-induced murine BV2 microglia. Hence also have a neuroprotective function.	Yoon et al., 2013
25.	*Ganoderma lucidum* (dried fruit body at 40–50ºC)	*Ganoderma lucidum* polysaccharides (GLP)	Powdered mushroom was defatted using petroleum ether, extracted with double distilled water, and filtered. 95% ethanol was added in concentrated aqueous extract and kept at 4ºC for 48 hours to precipitate the polysaccharides, which were further washed in ethanol and dried. The dried extract was then redissolved in deionized water and subjected to DEAE cellulose column chromatography. Then all the anthrone-positive fractions were repeatedly washed using Sevag solution. Finally polysaccharides were re-precipitated using 95% chilled ethanol.	EC: 100 mg kg^{-1} Carrageenan-treated (acute) and formalin-treated (chronic) inflammation comparable to diclofenac against EAC scell line in mice.	Joseph et al., 2011

TABLE 9.2
Comparison of the Main Types of Arthritis and a Feature Demonstration

Types of Arthritis	Feature Demonstration (Ultrasound Based)
RA	High synovial vascularity, persistent synovitis, tenosynovitis, and erosive changes.
OA	Synovial changes within joints, osteophytosis, bony erosions, soft tissue pathologies, and bakers cyst's and bursitis.
Infectious arthritis	Joint effusions.
Psoriatic arthritis	Characterized by joint effusions, synovial tissue, erosions, and hyperemia.
Spondyloarthritis	Calcifications and bony erosions and tendon thickening.
Pseudogout/calcium pyrophosphate deposition disease	Punctated fibrocartilage and thin band parallel to hyaline cartilage, also soft tissue calcification.
GA	Tophi, erosions, synovial hypertrophy, and fluid collections.

(Jessar and Hollander, 1955). Joint RA inflammation is intimately linked to immune cell infiltration, synovium hyperplasia, and excessive pro-inflammatory cytokine release, which cause cartilage degradation and bone erosion. Monocytes/macrophages and T cells are two crucial cellular elements found in the joint synovium of RA patients, which contains a range of immune cellular types. In the RA synovium, monocytes and macrophages can draw in T cells and encourage their development into inflammatory phenotypes. Similar to this, certain T-cell subtypes can draw in monocytes and macrophages, encourage the development of osteoblasts, and produce inflammatory cytokines. This chapter will explain how interactions between T cells, monocytes, and macrophages facilitate the onset of RA, opening up fresh ideas on the disease's pathophysiology and the creation of targeted treatments (Roberts et al., 2015; Fonseca et al., 2002).

Joint degeneration and deformity are common outcomes of RA, a common clinical phenotype defined by non-infectious chronic polyarticular swelling, most often affecting smaller joints (Smolen et al., 2016). Nevertheless, persistent periarticular synovitis is associated with the deterioration and erosion of joints, and this is true independent of the immunopathogenesis; as a result, the same clinical picture manifests itself regardless of the immunological trigger (Komatsu and Takayanagi, 2012). The clinical RA phenotype may be connected to a variety of immunologically different circumstances according to this theory. A rigorous paradigm for the assessment, diagnosis, and prognosis of patients who arrive with polyarthritis indicative of early RA is needed in the translational scenario, particularly in the present, when the primary function of innate immunity or autoinflammation is well understood. The pathogenesis of RA involves a large number of immune cells, especially those of the innate immune system viz. cells of the innate immune system, such as NK cells, as well as cells of the adaptive immune system, such as T lymphocytes and B lymphocytes (Giannini et al., 2020). Cell types of the immune system include macrophages, DCs, NK cells, and B cells. The development of RA also involves some non-immune cells, fibroblasts, and endothelial cells. These cellular elements of the joint synovium, such as macrophages and fibroblasts (Tu et al., 2018), T cells and DC cells (Wehr et al., 2019), T cells and NK cells (Shegarfi et al., 2012), etc., interact in a complex manner. T cells (Toh and Miossec, 2007) and macrophages are two of them that are known to be important cellular players in RA.

Although the roles of T cells and monocytes/macrophages in RA have long been studied, little research has been done on how these cells interact. Monocyte and T-cell colocalization has been seen in the synovium of RA (Fonseca et al., 2002), suggesting that T cell–monocyte/macrophage interactions may take place at the site of inflammation. Given the crucial roles that T cells and macrophages play in RA, it may be important to take into account how they interact because it may also be a key element in the emergence of this autoimmune illness (Roberts et al., 2015). In order to comprehend the molecular etiology of RA, it is crucial to demonstrate the precise interaction between T cells and macrophages. Inflammation of the synovial lining is a characteristic feature of the chronic inflammatory conditions of RA (synovitis). The inflammation in a RA joint is caused by a number of causes, including immune cell infiltration, synovial hyperproliferation, and the overproduction of pro-inflammatory mediators such as TNF, interferon (IFN), IL-1, IL-6, and IL-17, which eventually cause damage to the cartilages and underlying bones. The RA joint has a wide variety of immune cells, including monocytes, macrophages, and $CD4^+$ T cells (both pro-inflammatory and regulatory). The ratio of effector to regulatory $CD4^+$ T cells and the interaction between $CD14^+$ myeloid cells and $CD4^+$ T cells may have profound effects on $CD4^+$ T-cell function. Pannus development and synovial hyperplasia, brought on by multiplying fibroblasts and invading immune cells, are typical RA characteristics. These events promote leukocyte migration, immune cell activation, inflammatory mediator generation, and proteinase production, all of which contribute to joint deterioration. Many different types of immune cells, including $CD4^+$ T cells, $CD8^+$ T cells, B cells, NK cells, T cells, mast cells, and myeloid cells, have been identified in the RA joint. Many soluble mediators secreted by these immune cells, including rheumatoid factors, anti-citrullinated peptide antibodies, TNF, IL-6, IL-1, and IL-17A, have been shown to have a role in the pathogenesis of autoimmune diseases (McInnes and Schett, 2007, 2011). Recent advances in biological therapy that have successfully targeted key inflammatory cytokines, immunological components, and immune cells (examples include CTLA 4-Ig, which blocks CD80/CD86-mediated stimulation, anti-IL-6R treatment, and TNF blocking) emphasize the role of the immune system in disease development (e.g., B-cell depletion). This knowledge further fueled the discoveries of several anti-arthritic constituents from various sources (McInnes and Schett, 2007, 2011).

However, as per the anti-arthritic properties of *Ganoderma* spp. are concerned, it has been observed that the previous investigation was more focused on several arthritis-associated clinical conditions in addition to several inflammatory cytokines, as mentioned earlier. For instance, hyperactivity of XO caused increased uric acid levels in blood, a condition known as hyperuricemia. Gouty arthritis (GA), caused by hyperuricemia, is very painful. Hyperactivity of receptors involved for the reabsorption of uric acid and purine, such as GLUT9, CNT2, and URAT1, may also lead to elevated uric acid levels in the blood. Different active constituents of *Ganoderma* spp. are reported to have varying inhibitory properties against these receptors as well as XO, thereby reducing the chances for GA (Table 9.3). Similarly, fibromyalgia, a rheumatic disorder characterized by musculoskeletal pain accompanied by fatigue, distress, irregular sleep, etc., is mostly prevalent among women. It has been reported that a daily intake of 3 to 6 g of micro milled powder of *G. lucidum* is a cost-effective alternative that improved the health conditions in women with fibromyalgia (Pazzi et al., 2020; Garcia-Gordillo et al., 2015).

9.4 APPROACHES TO IMPROVE ANTI-INFLAMMATORY AND ANTI-ARTHRITIS PRINCIPLES OF *GANODERMA*

As mentioned earlier, several compounds with prolific anti-inflammatory and anti-arthritis potential have been reported from *Ganoderma* spp. (Table 9.1 and Table 9.3). Nonetheless, a large portion of these compounds are also purified and characterized following different chromatographic and spectroscopic techniques. However, there is always a need to improve the efficacy of these active constituents. This never-ending endeavor is employed in the following approaches.

9.4.1 Post-Harvest Processing and Extraction Conditions

Post-harvest processing and extraction conditions (especially temperature and time) can be one of the deciding factors for target-oriented investigation of different therapeutic properties of *G. lucidum*. For instance, Ryu et al. (2021) reported that heat-dried *G. lucidum*, when extracted at 50°C for 3 hours, showed the most promising anti-inflammatory activity. In contrast to this, crude extract prepared from freeze-dried *G. lucidum* at 64.2°C–70°C for 1.2 hours maximized the release of

TABLE 9.3
List of Different Species of *Ganoderma* and Their Anti-Arthritis Properties

Sr. No.	Source	Active Constituents and Chemical Nature	Extraction/Isolation Methodology	IC_{50} Value/Effective Concentration (EC) and Mode of Action	References
1.	*Ganoderma lucidum*	Lyophilized powder of mycelia (GM) and extracellular polysaccharides (GP) from submerged fermentation.	Both GM and GP were harvested and lyophilized from seven-day-old submerged culture of *G. lucidum*.	EC: 200–400 mg kg^{-1} body weight. Inhibit the activity of XO and reduce the occurrence of gout in potassium oxonate (PO)–injected rats.	Huang et al., 2022
2.	*Ganoderma lucidum*	Ganomycin 1 (GMI) (meroterpenoid)	HPLC was performed to obtain GMI from basidiocarps of *G. lucidum*. The purity of the GMI was further analyzed and ensured by ^{1}H and ^{13}C NMR.	EC: 30 μM of GMI. GMI inhibited the receptor activator of nuclear factor-κB (RANKL)–induced osteoclast formation in mouse bone marrow–derived macrophages (BMMs) and RAW364.7 cells. GMI not only inhibited phosphorylation of extracellular signal–regulated kinase (ERK), c-Jun N-terminal kinase (JNK), and p38 MAPKs but also suppressed the expression of c-Fos and NFATc1 transcription factors. It therefore reduced the chances for rheumatoid arthritis.	Tran et al., 2019
3.	*Ganoderma lucidum*	Ganoderic acid A (GAA) (triterpene)	NA	EC: 20, 40 mg kg^{-1}. GAA showed a determinate effect in joint inflammation among collagen- (type II) and FCA-treated arthritic rats. GAA improves overall health conditions by reducing oxidative stress, serum levels of different proinflammatory cytokines like Janus kinase, arthritis index, and joint erosion.	Cao et al., 2020
4.	*Ganoderma applanatum*	Ethanolic extract (GLE) and water extract (GLW)	Ethanolic extract was prepared from fruit bodies of *G. applanatum at* 65°C and filtered. On the other hand, water-based extract was prepared at 85°C and lyophilized.	EC: 30, 60, 120 mg kg^{-1}. Reduced the level of serum uric acid (SUA) in hyperuricemic rats. GLE and GLW upregulated the expression of organic anion transporter 1 (OAT1), which is responsible for excretion of uric acid. On the other hand, both the extracts inhibit GLUT9 and CNT2, thereby preventing the reabsorption of uric acid and purine. Additionally, GLE also reduced the expression of URAT1, although it was not able to inhibit the activity of XO.	Yong et al., 2018
5.	*Ganoderma applanatum*	2,5-Dihydroxyacetophenone (DHAP)	Computational screening of DHAP was carried out using OAT1 as a model and a docking study was performed using in-house compound database of *G. applanatum*.	EC: 20, 40, and 80 mg kg^{-1}. Reduced the uric acid level in hyperuricemic mice by upregulating OAT1 and downregulating GLUT9, URAT1, OAT1, and activity of xanthine oxidase.	Liang et al., 2018

6.	*Ganoderma lucidum*	Crude extract of DMSO, acetone, and water.	Shade-dried powder of fruit bodies of *G. lucidum* was used as starting material. From this DMSO extract was prepared using soxhlet apparatus. Water extract was prepared following decoction technique. Acetone extract was filtered and then evaporated to one-fourth of original filtrate volume and finally powdered in oven.	IC_{50}: 40, 50.5, and 56 μg ml^{-1} for DMSO, water, and acetone extracts respectively. All three extracts showed anti-arthritic potential in the in vitro protein denaturation model.	Amin Mir et al., 2017
7.	*Ganoderma tsugae* (air-dried fruit body)	Tsugaric acid F (1) (lanostanoid), 3oxo-5α-lanosta-8-en-21-oic acid (4) (lanostanoid)	Chloroform extract of dried fruit bodies was prepared first then further purified using silica gel column chromatography to purify compound 1. 3-oxo-5α-lanosta-8,24-dien-21oic acid (3) previously isolated from same fungus, was hydrogenated, then subjected to silica gel chromatography after dialysis to purify compound 4. Structure elucidation of compounds was further carried out using a combination of HREIMS, ^{1}H, and ^{13}C NMR.	IC_{50}: 313.3 ± 80.0 μM for (1) and 43.9 ± 29.9 μM for (4). Showed and inhibitory activity against xanthine oxidase (XO), thus reducing the chances of hyperuricemia and ultimately GA.	Lin et al., 2016
8.	*Ganoderma tsugae*	Tsugaric acid D (1), (lanostanoid)	Dichloromethane extract from dried fruit bodies was subjected to silica gel chromatography to purify tsugaric acid D (1).	IC_{50}: 90.2 ± 24.2 μM. Inhibited activity of XO.	Lin et al., 2013

anti-oxidant properties. These factors are also crucial for increasing yield and storage of any active principal form of different species of *Ganoderma*.

9.4.2 Combined Application and Synergistic Activity of Photochemicals from Different Sources

Several reports indicate that both the powder of dried fruit bodies and crude extracts of *Ganoderma* spp., when applied in combination with other medicinally important traditional drugs, showed strong anti-inflammatory as well as anti-arthritic potential. For instance, when the aqueous extract of Malaysian *G. lucidum* (GLE) and ethanolic extract of Egyptian *Chlorella vulgaris* (CVE) were applied together, they synergistically attenuated different inflammatory mediators, NF-κB, inducible nitric oxide (iNOS), and COX-2, and also reduced the levels of NO and TNF-α in LPS-induced white blood cells (WBCs). Nonetheless it was also found that the combined application of GLE and CVE showed greater anti-inflammatory potential compared to the commercially available anti-inflammatory drug dexamethasone (Abu-Serie et al., 2018).

Likewise, a potent Chinese traditional drug called San-Miao-San (SMS) with remarkable anti-inflammatory and anti-arthritic properties was originally prepared by mixing equal portions of medicinally important Cangzhu (*Rhizoma atractylodis*), Huangbai (cortex of *Phellodendri chinensis*), and Niuxi (radix of *Achyranthis bidentatae*) (Li et al., 2007). The combined application of SMS and *G. lucidum* (lingzhi, LZ) showed good analgesic effect among patients with RA (Li et al., 2007).

Additionally, it is also observed that the combined application of *G. lucidum* and SMS reduced the inflammatory response and exerted a good analgesic effect among arthritic rats where arthritis was induced using Freund's complete adjuvant (FCA). The traditional Chinese medicine (TCM) was prepared in physiological saline and was administrated by intraperitoneal injection as well as oral doses at a concentration of 50 mg kg^{-1} day^{-1} for 7 days before FCA induction to 7 days after. It was observed that intraperitoneal injection prevented allodynia, edema, and hyperemia, whereas oral administration of TCM suppressed only edema and hyperemia in arthritic rats. Administration of TCM via both routes was found to have satisfactory results in the prevention of joint cartilage erosion and infiltration of immune cells (Lam et al., 2008). Another report indicated the combined application of lingzhi and SMS significantly reduced the plasma concentration of IL-21, IL-10, and IL-17A against systemic lupus erythematosus (SLE) (Cai et al., 2016).

9.4.3 Chemical Modification of Active Constituents

This method also enhanced the anti-inflammatory potential of the extracts and compounds. For example, GLPss58, a sulfated derivative of GLP20 polysaccharide from *G. lucidum*, is reported to have strong anti-inflammatory activity, predominantly by inhibiting the action of L-selectin. Even so, the inhibitory activity of GLPss58 on pro-inflammatory cytokines TNF-α and IFN-α is comparable to the anti-inflammatory drug heparin (Zhang et al., 2018). Similarly, 3-oxo-5α-lanosta-8,24-dien-21oic acid (3) (initially isolated from *G. tsugae*) was hydrogenated for 5 hours in an autoclave in presence of 5% Pd/C and 60 kg cm^{-3} initial pressure to prepare 3-oxo-8-lanostan-21-oic-acid (4). It is a potent lanostanoid derivative with remarkable inhibitory properties against XO and thus can be helpful to reduce the GA (Lin et al., 2016).

9.4.4 Recombinant DNA Technology

Several molecular biology–based techniques are also found to be equally important in mass production and deciphering the mode of action of different therapeutic proteins from *Ganoderma* spp. It has been reported that recombinant FIP-glu (expressed in the yeast system *Pichia pastoris*) showed strong anti-inflammatory activity. Western blotting and SDS-PAGE (Sodium dodecyl sulfate-polyacrylamide gel electrophoresis) examination of peptide-N-glycosidase F–treated wild-type FIP-glu confirmed that the glycan molecule is N-linked. Nonetheless, to unravel the importance of N-linked glycosylation of FIP-glu in its anti-inflammatory activity, several FIP-glu mutants (N31S, T36N, and N31S/T36N) were produced. Polymerase chain reaction (PCR) based on site-directed mutagenesis, cloning in pPIC9K vector, and final transformation into *P. pastoris* GS115 were successively used in this process. It has been observed that N-linked glycans are crucial for mediating anti-inflammatory response in wild-type recombinant FIP-glu (Li, Chen et al., 2021; Li, Chang et al., 2019).

9.4.5 Technologies for Genome Editing as a Complete Genome Toolkit

For the first time in higher fungi, our study employed the *Ganoderma* species as a model organism to develop CRISPR-Cas9–assisted gene disruption. It has been proposed that higher fungi might serve as cell factories producing bioactive secondary metabolites, yet without genetic manipulation tools like gene disruption, research on the production and its regulation of such useful compounds is limited. CRISPR-Cas9 induced a double-strand break (DSB), and the repair mechanism known as non-homologous end joining (NHEJ) had a role in amplifying the disruption of the gene shown in Figure 9.3. As a proof of concept, the ura3 gene in *G. lucidum* 260125 and *G. lingzhi* was successfully disrupted using codon-optimized Cas9 and gRNA generated in vitro. In *G. lucidum*, an anti-cancer ganoderic acid (GA) producing basidiomycete, two homologs of a Cys2-His2 (C2H2)-type zinc finger transcription factor (CRZ1), GlCRZ1 and GlCRZ2, were found. Calcium signaling of GA synthesis was investigated by disrupting the genes encoding the two CRZ1 orthologs (glcrz1 and glcrz2) using the cutting-edge genome-editing technique CRISPR/Cas9. The fundamental *G. lucidum* biological components (Paterson, 2006) studied in humans and animals have demonstrated that GAs, a family of highly oxygenated triterpenoids, have immunomodulatory, carcinoma-suppressing, and cholesterol-inhibiting activities (Li et al., 2013).

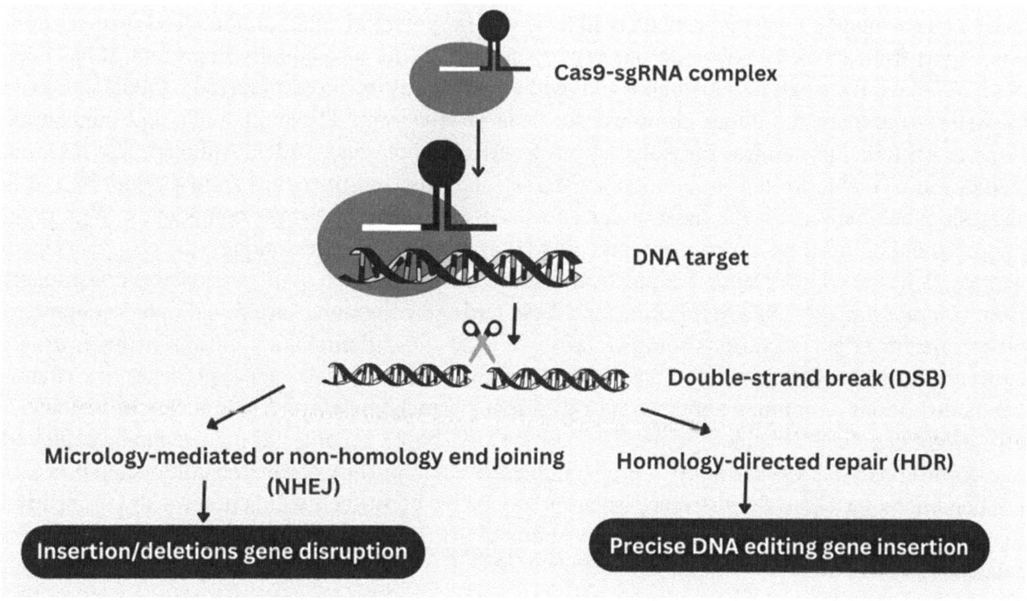

FIGURE 9.3 When CRISPR-Cas9 generated a double-strand break (DSB), it was followed by non-homo logous end joining (NHEJ), which increased the disruption of the gene.

9.4.6 Combined Application with Biomaterials

Biomaterials are the substances either of biological or synthetic origin which can be used for various medical purposes viz. as an implant, drug delivery system, support for regeneration of damaged tissues, etc. Reports suggest that biomaterials can be very useful at different stages of inflammation. Pathogen-associated molecular patterns (PAMPs), endogenous tissue damage-associated molecular patterns (DAMPs), reactive oxygen and nitrogen species (RONS), cell-free nucleic acids (cfNAs), and other inflammatory triggers all play a role in the initiation of the inflammatory response at its earliest stages. Biomaterials can function as RONS scavengers (e.g., artificial selenoenzymes, catalytic nanomaterials) or cfNA scavengers (e.g., nucleic acid binding polymers, fibers). Pro-inflammatory cytokines are produced in large quantities by immune cells during the intermediate stage of inflammation. In this case, various anionic biomaterials, such as liner polymers, dendritic polymers, etc., may prevent leukocytes from attaching to endothelial cells, limiting the migration of immune cells at the place of inflammation. Later, towards the end of the inflammatory process, biomaterials may be employed as drug delivery platforms to boost the effectiveness of anti-inflammatory medications by delivering them directly to the site of inflammation (Tu et al., 2022). Sulfated polysaccharides like GLPss58 (a sulfated derivative of GLP20 from *G. lucidum*) is a perfect example of engineered biomaterial. In mice, GPLss58 inhibits ILs and stops the lymphocytes from migrating to the spleen and lymph nodes during unchecked inflammation. This decreases the production of pro-inflammatory cytokines (Zhang et al., 2018).

9.5 DISCUSSION AND CONCLUDING REMARKS

The therapeutic potential of different species of *Ganoderma* is now well understood. This chapter summarizes certain facts which we strongly believe will be helpful in future research endeavors. It has been observed that among all the active constituents reported from different species of *Ganoderma*, the majority are terpenoids and polysaccharides, whereas only a very small fraction of such compounds are sterols, proteins, phenolics, and steroids. Concurrently, the majority of the work has been done on type species *G. lucidum* followed by *G. applanatum*. There are many species like *G. tsugae* and *G. cochlear* which could have similar potential and therefore need further attention. Apart from species of *Ganoderma* extraction solvents are also equally important. It has been observed that during extraction ethanol and methanol serve as the most preferred solvents, followed by water, to prepare the initial crude extract from *Ganoderma*. These ethanolic and methanolic extracts are found to be ideal for isolating terpenoids and polysaccharides. Although few research studies are available for anti-inflammatory sterols, steroids, and phenolics from *Ganoderma*, it is suggested that methanol is the most preferred solvent for isolation of these compounds. With rapid improvement of different chromatographic as well as spectroscopic techniques, the past years have witnessed increased involvement of purification and characterization of different active constituents from *Ganoderma* spp. On several occasions these purified compounds are found to have comparable therapeutic effects to various commercially available anti-inflammatory and anti-arthritic drugs. However, to further improve the efficacy of these compounds, several other approaches like chemical modifications, combined application of different phytochemicals and with different biomaterials, recombinant DNA technology, and CRISPER-Cas9–based genome editing need to be exploited to their full potential. Concurrently it is also imperative to have fully optimized culture and processing conditions for *Ganoderma* before isolating any active constituents. It is noteworthy to mention that we briefly discussed the immunopathogenesis of arthritic and inflammatory diseases and subsequently highlighted the main targets for development of suitable drugs against them. In discussing the targets here, we have observed that almost entirely different inflammatory cytokines like various interleukins (IL-6, IL-10, etc.), TNF-α, enzymes like COX2 and NO, etc., serve as prime targets for *Ganoderma*-derived anti-inflammatory phytochemicals. Similarly xanthine oxidase and hyperuricemia serve as primary targets for *Ganoderma*-derived anti-arthritic phytochemicals, whereas other inflammatory targets like FPRs and other arthritic targets like GLUT9, URAT1, and OAT1 remain much less explored. A similar trend is also observed for test organisms, where most of the anti-inflammatory studies are found to be centered on the RAW264.7 macrophage cell line from mice or on live mice. On the other hand, human cell lines like HaCaT and HL-60 cell lines, largely remain unexplored in terms of *Ganoderma*-derived anti-inflammatory and anti-arthritic compounds.

From our previous knowledge we know that arthritic and chronic inflammatory diseases cannot be cured completely. Factors like a stressful life, poor dietary habits, and obesity only worsen the conditions further. Application of NSAIDs is found to be the primary way to reduce the pain in these diseases, though it comes at a cost of associated side effects. As per *Ganoderma*-derived phytochemicals are concerned, almost all the active constituents showed no adverse effect on mammalian cell lines or on live mammals. Application of these compounds was not only found to be effective against inflammatory and arthritic diseases but also found to be beneficial in improving health conditions, boosting morale among patients, and on several occasions found to be comparable to commercially available drugs in terms of effectiveness. Considering this, it is evident that different species of *Ganoderma*, especially *G. lucidum*, *G. applanatum*, and *G. tsugae*, possess great anti-arthritis and anti-inflammatory potential and future prospects in this regard. These can also be the next wonder drug, provided all the facts and concerns raised here are thoroughly researched and analyzed before making them commercially available for public use.

ACKNOWLEDGEMENT

Kunal Kumar Saha is thankful to the Department of Science & Technology (DST), Ministry of Science and Technology, Govt. of India, for providing financial assistance through DST INSPIRE Fellowship under INSPIRE Program. Anik Barman acknowledges Council of Scientific and Industrial Research (CSIR), Govt. of India for providing financial assistance.

Abbreviations: ATPS – Alcohol/salt aqueous two phase system, **CNT2** – Gastrointestinal concentrative nucleoside transporter 2, **COX** – Cyclooxygenase, **DCs** – Dendritic cells, **DEPT** – Distortion less enhancement by polarization transfer, **EAC** – Ehrlich's ascites carcinoma, **ECD** – Electronic circular dichroism, **ESAC** – Ethanol-soluble acidic component, **FIP-glu** – *Ganoderma lucidum* fungal immunomodulatory protein, **FPRs** – Formyl peptide receptors, **GA** – Gouty arthritis, **GLS** – *Ganoderma lucidum* sterols, **GLUT9-** Glucose transporter 9, **HPLC-ELSD** – High-performance liquid chromatography–evaporative light scattering detector, **HPBLs** – Human peripheral blood lymphocytes, **HRESIMS** – High-resolution electrospray ionization mass spectroscopy, **IFN** – Interferon, **IMIDs** – Immune-mediated inflammatory diseases, **IL** – Interleukin, **LOX** – Lipoxygenase, **LPS** – Lipopolysaccharides, **MSLs** – Mouse spleen lymphocytes, **Ni – NTA** – Nickel-nitrilotriacetic acid agarose resin, **NK** – Natural killer cells, **NMR** – Nuclear magnetic resonance, **NO** – Nitric oxide, **NSAIDs** – Non-steroidal anti-inflammatory drugs, **OA** – Osteoarthritis, **RA** – Rheumatoid arthritis, **TNF** – Tumor necrosis factor, **UPLC/Q-TOF-MS** – Ultra-high-performance liquid chromatography–quadrupole time-of-flight mass spectrometry, **URAT** – Uric acid transporter 1, **XO** – Xanthine oxidase.

CONFLICT OF INTEREST

The authors declare that there is no conflict of interest.

REFERENCES

Abu-Serie, Marwa M., Noha H. Habashy, and Wafaa E. Attia. In vitro evaluation of the synergistic antioxidant and anti-inflammatory activities of the combined extracts from Malaysian *Ganoderma lucidum* and Egyptian Chlorella vulgaris. *BMC Complement Altern. Med.* 18, no. 1 (2018): 1–13.

Ahmad, Rizwan, Muhammad Riaz, Aslam Khan, et al. *Ganoderma lucidum* (Reishi) an edible mushroom: A comprehensive and critical review of its nutritional, cosmeceutical, mycochemical, pharmacological, clinical, and toxicological properties. *Phytother. Res.* 35, no. 11 (2021): 6030–6062.

Amin Mir, M., T. Sharma, K. Kiran Sharma, S. Saima Anjum, and M. Bilal Ahmad. Anti urolithiatic and anti-arthritis activity of various extracts of *Ganoderma lucidum. Stud. Nat. Prod. Chem.* 5, no. 297 (2017): 2.

Cai, Zhe, Chun Kwok Wong, Jie Dong, et al. Anti-inflammatory activities of *Ganoderma lucidum* (Lingzhi) and San-Miao-San supplements in MRL/LPR mice for the treatment of systemic lupus erythematosus. *Chin. Med.* 11, no. 1 (2016): 1–13.

Cao Tao, Chuanfeng Tang, Lezhen Xue, Mingzhu Cui, and Dan Wang. Protective effect of Ganoderic acid A on adjuvant-induced arthritis. *Immunol. Lett.* 226 (2020): 1–6.

Chen, Linlin, Huidan Deng, Hengmin Cui, et al. Inflammatory responses and inflammation-associated diseases in organs. *Oncotarget.* 9, no. 6 (2018): 7204.

Chen, Yu-Sheng, Quan-Zhan Chen, Zhen-Jiong Wang, and Chun Hua. Anti-inflammatory and hepatoprotective effects of *Ganoderma lucidum* polysaccharides against carbon tetrachloride-induced liver injury in Kunming mice. *Pharmacology.* 103, no. 3–4 (2019): 143–150.

Chertov, Oleg, De Yang, O. M. Howard, and Joost J. Oppenheim. Leukocyte granule proteins mobilize innate host defenses and adaptive immune responses. *Immunol. Rev.* 177 (2000): 68–78.

Chhem, Rethy K., Phoebe A. Kaplan, and Robert G. Dussault. Ultrasonography of the musculoskeletal system. *Radiol. Clin. North Am.* 32, no. 2 (1994): 275–289.

Choi, Solip, Nara Tae, Suhyun Lee, Sungwoo Ryoo, Byung-Sun Min, and Jeong-Hyung Lee. Anti-inflammatory and heme oxygenase–1 inducing activities of lanostane triterpenes isolated from mushroom *Ganoderma lucidum* in RAW264.7 cells. *Toxicol. Appl. Pharmacol.* 280, no. 3 (2014): 434–442.

Choy, Ernest H. S., and Gabriel S. Panayi. Cytokine pathways and joint inflammation in rheumatoid arthritis. *N. Engl. J. Med.* 344, no. 12 (2001): 907–916.

Cör, Darija, Željko Knez, and Maša Knez Hrnčič. Antitumour, antimicrobial, antioxidant and antiacetylcholinesterase effect of *Ganoderma lucidum* terpenoids and polysaccharides: A review. *Molecules.* 23, no. 3 (2018): 649.

Edwards, J. C. W., G. Cambridge, and V. M. Abrahams. Do self-perpetuating B lymphocytes drive human autoimmune disease? *Immunology* 97, no. 2 (1999): 188.

Feng, Xia, and Yan Wang. Anti-inflammatory, anti-nociceptive and sedative-hypnotic activities of lucidone D extracted from *Ganoderma lucidum. Cell. and Mol. Biol.* 65, no. 4 (2019): 37–42.

Ferrero-Miliani, Laura, O. H. Nielsen, P. S. Andersen, and S. E. Girardin. Chronic inflammation: Importance of NOD2 and NALP3 in interleukin–1β generation. *Clin. Exp. Immunol.* 147, no. 2 (2007): 227–235.

Fonseca, J. E., J. C. W. Edwards, S. Blades, and N. J. Goulding. Macrophage subpopulations in rheumatoid synovium: Reduced CD163 expression in CD4+ T lymphocyte-rich microenvironments. *Arthritis Rheum.* 46, no. 5 (2002): 1210–1216.

Garcia-Gordillo, Miguel A., Daniel Collado-Mateo, et al. Cost-utility analysis of a six-weeks *Ganoderma lucidum*-based treatment for women with fibromyalgia: A randomized double-blind, active placebo-controlled trial. *Myopain.* 23, no. 3–4 (2015): 188–194.

Giannini, Daiana, Matteo Antonucci, Fiorella Petrelli, et al. One year in review 2020: Pathogenesis of rheumatoid arthritis. *Clin. Exp. Rheumatol.* 38, no. 3 (2020): 387–397.

Gregersen, Peter K., Jack Silver, and Robert J. Winchester. The shared epitope hypothesis. An approach to understanding the molecular genetics of susceptibility to rheumatoid arthritis. *Arthritis Rheumatol.* 30, no. 11 (1987): 1205–1213.

Hu, Xiaoyi, Qiang Yu, Kunyou Hou, et al. Regulatory effects of *Ganoderma* atrum polysaccharides on LPS-induced inflammatory macrophages model and intestinal-like Caco–2/macrophages co-culture inflammation model. *Food Chem. Toxicol.* 140 (2020a): 111321.

Hu, Zhongpeng, Ruiping Du, Lei Xiu, et al. Protective effect of triterpenes of *Ganoderma lucidum* on lipopolysaccharide-induced inflammatory responses and acute liver injury. *Cytokine* 127 (2020b): 154917.

Huang, Chung-Hsiung, Tzu-Yu Chen, and Guo-Jane Tsai. Hypouricemic effect of submerged culture of *Ganoderma lucidum* in potassium Oxonate-induced Hyperuricemic rats. *Metabolites.* 12, no. 6 (2022): 553.

Jabbour, Henry N., Kurt J. Sales, Rob D. Catalano, and Jane E. Norman. Inflammatory pathways in female reproductive health and disease. *Reproduction.* 138, no. 6 (2009): 903.

Jessar, Ralph A., and Joseph Lee Hollander. Types of arthritis and their medical treatment. *Am. J. Nurs.* 55, no. 4 (1955): 426–428.

Joseph, Soniamol, Baby Sabulal, Varughese George, Kuttikkadan Rony Antony, and Kainoor Krishnankutty Janardhanan. Antitumor and anti-inflammatory activities of polysaccharides isolated from *Ganoderma lucidum*. *Acta. Pharm.* 61, no. 3 (2011): 335–342.

Komatsu, Noriko, and Hiroshi Takayanagi. Inflammation and bone destruction in arthritis: Synergistic activity of immune and mesenchymal cells in joints. *Front. Immunol.* 3 (2012): 77.

Koo, Man Hyung, Hae-Jung Chae, Jun Hyuck Lee, Sung-Suk Suh, and Ui Joung Youn. Antiinflammatory lanostane triterpenoids from *Ganoderma lucidum*. *Nat. Prod. Res.* 35, no. 22 (2021): 4295–4302.

Kuek, Annabel, Brian L. Hazleman, and Andrew J. K. Östör. Immune-mediated inflammatory diseases (IMIDs) and biologic therapy: A medical revolution. *Postgrad. Med. J.* 83, no. 978 (2007): 251–260.

Lam, Francis Fu Yuen, Iris Wai Man Ko, Ethel Sau Kuen Ng, et al. Analgesic and anti-arthritic effects of Lingzhi and San Miao San supplementation in a rat model of arthritis induced by Freund's complete adjuvant. *J Ethnopharmacol*, no. 1 (2008): 44–50.

Li, Edmund K., Lai-Shan Tam, Chun Kwok Wong, et al. Safety and efficacy of *Ganoderma lucidum* (Lingzhi) and San Miao San supplementation in patients with rheumatoid arthritis: A double-blind, randomized, placebo-controlled pilot trial. *Arthritis Care Res.* (Hoboken) 57, no. 7 (2007): 1143–1150.

Li, Qi-Zhang, Yu-Zhou Chang, Zhu-Mei He, Lei Chen, and Xuan-Wei Zhou. Immunomodulatory activity of *Ganoderma lucidum* immunomodulatory protein via PI3K/Akt and MAPK signaling pathways in RAW264.7 cells. *J. Cell. Physiol.* 234, no. 12 (2019): 23337–23348.

Li, Qi-Zhang, Xin Chen, Pei-Wen Mao et al. N-Glycosylated *Ganoderma lucidum* immunomodulatory protein improved anti-inflammatory activity via inhibition of the p38 MAPK pathway. *Food Funct.* 12, no. 8 (2021): 3393–3404.

Li, Ying-Bo, Ru-Ming Liu, and Jian-Jiang Zhong. A new ganoderic acid from *Ganoderma lucidum* mycelia and its stability. *Fitoterapia.* 84 (2013): 115–122.

Liang, Danling, Tianqiao Yong, Shaodan Chen, et al. Hypouricemic effect of 2, 5-dihydroxyacetophenone, a computational screened bioactive compound from *Ganoderma* applanatum, on hyperuricemic mice. *Int. J. Mol. Sci.* 19, no. 5 (2018): 1394.

Lin, Kai-Wei, Yen-Ting Chen, Shyh-Chyun Yang, et al. Xanthine oxidase inhibitory lanostanoids from *Ganoderma* tsugae. *Fitoterapia.* 89 (2013): 231–238.

Lin, Kai-Wei, Dravidum Maitraie, A. Mei Huang, Jih-Pyang Wang, and Chun-Nan Lin. Triterpenoids and an alkamide from *Ganoderma* tsugae. *Fitoterapia.* 108 (2016): 73–80.

Liu, Changda, David Dunkin, Joanne Lai, et al. Anti-inflammatory effects of *Ganoderma lucidum* triterpenoid in human crohn's disease associated with downregulation of NF-κB signaling. *Inflamm. Bowel Dis.* 21, no. 8 (2015): 1918–1925.

Liu, Jie-Qing, Chen-Lei Lian, Tian-Yong Hu, et al. Two new farnesyl phenolic compounds with anti-inflammatory activities from *Ganoderma* duripora. *Food Chem.* 263 (2018): 155–162.

Lu, Shuang-Yang, Xing-Rong Peng, Jin-Run Dong, et al. Aromatic constituents from *Ganoderma lucidum* and their neuroprotective and anti-inflammatory activities. *Fitoterapia.* 134 (2019): 58–64.

Lucey, Daniel R., Mario Clerici, and Gene M. Shearer. Type 1 and type 2 cytokine dysregulation in human infectious, neoplastic, and inflammatory diseases. *Clin. Microbiol. Rev.* 9, no. 4 (1996): 532–562.

Ma, Lingling, Ann Cranney, and Jayna M. Holroyd-Leduc. Acute monoarthritis: What is the cause of my patient's painful swollen joint? *Can. Med. Assoc. J.* 180, no. 1 (2009): 59–65.

McInnes, Iain B., and Ellen M. Gravallese. Immune-mediated inflammatory disease therapeutics: Past, present and future. *Nat. Rev. Immunol.* 21, no. 10 (2021): 680–686.

McInnes, Iain B., and Georg Schett. Cytokines in the pathogenesis of rheumatoid arthritis. *Nat. Rev. Immunol.* 7, no. 6 (2007): 429–442.

McInnes, Iain B., and Georg Schett. The pathogenesis of rheumatoid arthritis. *N. Engl. J. Med.* 365, no. 23 (2011): 2205–2219.

Medzhitov, Ruslan. Inflammation 2010: New adventures of an old flame. *Cell* 140, no. 6 (2010): 771–776.

Mei, Ren-Qiang, Feng-Jiao Zuo, Xiao-Yan Duan, Yi-Na Wang, Jun-Rong Li, Cheng-Zhen Qian, and Ji-Ping Xiao. Ergosterols from *Ganoderma* sinense and their anti-inflammatory activities by inhibiting NO production. *Phytochem. Lett.* 32 (2019): 177–180.

Nathan, Carl, and Aihao Ding. Nonresolving inflammation. *Cell.* 140, no. 6 (2010): 871–882.

Paterson, Russell R. M. *Ganoderma*: A therapeutic fungal biofactory. *Phytochemistry* 67, no. 18 (2006): 1985–2001.

Pazzi, Francesco, José Carmelo Adsuar, Francisco Javier Domínguez-Muñoz, et al. *Ganoderma lucidum* effects on mood and health-related quality of life in women with fibromyalgia. *Healthc (Amst).* 8, no. 4 (2020): 520. MDPI.

Peng, Xing-Rong, Qian Wang, Hui-Rong Wang, et al. FPR2-based anti-inflammatory and anti-lipogenesis activities of novel meroterpenoid dimers from *Ganoderma*. *Bioorg. Chem.* 116 (2021): 105338.

Qin, Fu-Ying, Hao-Xing Zhang, Qian-Qian Di, et al. *Ganoderma* cochlear metabolites as probes to identify a COX–2 active site and as in vitro and in vivo anti-inflammatory agents. *Org. Lett.* 22, no. 7 (2020): 2574–2578.

Rašeta, Milena, Mira Popović, Ivana Beara, et al. Anti-inflammatory, antioxidant and enzyme inhibition activities in correlation with mycochemical profile of selected indigenous *Ganoderma* spp. from Balkan Region (Serbia). *Chem. Biodivers.* 18, no. 2 (2021): e2000828.

Roberts, Ceri A., Abigail K. Dickinson, and Leonie S. Taams. The interplay between monocytes/macrophages and CD4+ T cell subsets in rheumatoid arthritis. *Front. Immunol.* 6 (2015): 571.

Ryu, Da Hye, Cho Jwa Yeong, Nooruddin Bin Sadiq, et al. Optimization of antioxidant, anti-diabetic, and anti-inflammatory activities and ganoderic acid content of differentially dried *Ganoderma lucidum* using response surface methodology. *Food Chem.* 335 (2021): 127645.

Sargowo, Djanggan, Indra Prasetya, Ria Ashriyah, et al. Anti inflammation and anti oxidant effect of active agent polysaccharide peptide (*Ganoderma lucidum*) in preventing atherosclerotic diseases. *Biomed. Pharmacol. J.* 8, no. 1 (2015): 27–33.

Shegarfi, Hamid, Fatemeh Naddafi, and Abbas Mirshafiey. Natural killer cells and their role in rheumatoid arthritis: Friend or foe? *Sci. World J.* 2012 (2012).

Shi, Qiangqiang, Yanjie Huang, Haiguo Su, et al. C28 steroids from the fruiting bodies of *Ganoderma* resinaceum with potential anti-inflammatory activity. *Phytochemistry*. 168 (2019): 112109.

Smolen, J. S., D. Aletaha, and I. B. McInnes. Rheumatoid arthritis. *Lancet*. 388, no. 10055 (2016): 2023–2038.

Su, Hai-Guo, Xing-Rong Peng, Qiang-Qiang Shi, et al. Lanostane triterpenoids with anti-inflammatory activities from *Ganoderma lucidum*. *Phytochemistry*. 173 (2020): 112256.

Sugiyama, E., A. Kuroda, H. Taki, et al. Interleukin 10 cooperates with interleukin 4 to suppress inflammatory cytokine production by freshly prepared adherent rheumatoid synovial cells. *J. Rheumatol.* 22, no. 11 (1995): 2020–2026.

Takemura, Seisuke, Piotr A. Klimiuk, Andrea Braun, Jörg J. Goronzy, and Cornelia M. Weyand. T cell activation in rheumatoid synovium is B cell dependent. *J. Immunol.* 167, no. 8 (2001): 4710–4718.

Takeuchi, O., and S. Akira. 2010. Pattern recognition receptors and inflammation. *Cell*, 140: 805–820.

Toh, Myew-Ling, and Pierre Miossec. The role of T cells in rheumatoid arthritis: New subsets and new targets. *Curr. Opin. Rheumatol.* 19, no. 3 (2007): 284–288.

Tran, Phuong Thao, Nguyen Tien Dat, Nguyen Hai Dang, et al. Ganomycin I from *Ganoderma lucidum* attenuates RANKL-mediated osteoclastogenesis by inhibiting MAPKs and NFATc1. *Phytomedicine*. 55 (2019): 1–8.

Tu, Jiajie, Wenming Hong, Pengying Zhang, Xinming Wang, Heinrich Körner, and Wei Wei. Ontology and function of fibroblast-like and macrophage-like synoviocytes: How do they talk to each other and can they be targeted for rheumatoid arthritis therapy? *Front. Immunol.* 9 (2018): 1467.

Tu, Zhaoxu, Yiling Zhong, Hanze Hu et al. Design of therapeutic biomaterials to control inflammation. *Nat. Rev. Mater.* (2022): 1–18.

Van Roon, J. A., J. L. Van Roy, Ashley Duits, F. P. Lafeber, and J. W. Bijlsma. Proinflammatory cytokine production and cartilage damage due to rheumatoid synovial T helper–1 activation is inhibited by interleukin–4. *Ann. Rheum. Dis.* 54, no. 10 (1995): 836–840.

Wang, Huirong, Xingrong Peng, Yunjun Ge, et al. Ye. A *Ganoderma*-derived compound exerts inhibitory effect through formyl peptide receptor 2. *Front. Pharmacol.* 11 (2020): 337.

Wehr, P., H. Purvis, S. C. Law, and R. Thomas. Dendritic cells, T cells and their interaction in rheumatoid arthritis. *Clin. Exp. Immunol.* 196, no. 1 (2019): 12–27.

Xu, Juan, Congmei Xiao, Haishun Xu, et al. Anti-inflammatory effects of *Ganoderma lucidum* sterols via attenuation of the p38 MAPK and NF-κB pathways in LPS-induced RAW 264.7 macrophages. *Food Chem. Toxicol.* 150 (2021): 112073.

Yao, Yu-Fei, Le-Feng Wang, Su-Mei Chen, et al. Antinociceptive and anti-inflammatory activities of ethanol-soluble acidic component from *Ganoderma* atrum by suppressing mannose receptor. *J. Funct. Foods*. 89 (2022): 104915.

Yong, Tianqiao, Shaodan Chen, Yizhen Xie, et al. Hypouricemic effects of *Ganoderma* applanatum in hyperuricemia mice through OAT1 and GLUT9. *Front. Pharmacol.* 8 (2018): 996.

Yoon, Hyun-Min, Kyung-Jun Jang, Min Seok Han, et al. *Ganoderma lucidum* ethanol extract inhibits the inflammatory response by suppressing the NF-κB and toll-like receptor pathways in lipopolysaccharide-stimulated BV2 microglial cells. *Exp. Ther. Med.* 5, no. 3 (2013): 957–963.

Zhang, Kai, Yanfang Liu, Xiangli Zhao, et al. Anti-inflammatory properties of GLPss58, a sulfated polysaccharide from *Ganoderma lucidum*. *Int. J. Biol. Macromol.* 107 (2018): 486–493.

Zhang, Wangxin, Quiling Zhang, Wen Deng, et al. Neuroprotective effect of pretreatment with *Ganoderma lucidum* in cerebral ischemia/reperfusion injury in rat hippocampus. *Neural. Regen. Res.* 9, no. 15 (2014): 1446–1452.

Zhou, Ying, Yan Hong, and Haihua Huang. Triptolide attenuates inflammatory response in membranous glomerulo-nephritis rat via downregulation of NF-κB signaling pathway. *Kidney Blood Press. Res.* 41, no. 6 (2016): 901–910.

10 Magical Mushroom *Ganoderma*—A Promising Treatment for Cancer

Sudeshna Nandi[1], *Annika Marial Paul*[2], *Anish Nag*[2], *and Krishnendu Acharya*[3]
1 University of Calcutta, Kolkata, India
2 Christ (Deemed to be University), Bangalore, Karnataka
3 University of Calcutta, Kolkata, India

10.1 INTRODUCTION

Around the world cancer has become an extending communal health issue. It is the principal cause of death in developed countries and the second most important reason for mortality in developing countries (Zhi et al., 2022). Contemporary medical sciences along with new synthetic remedies have moved forward the approaches in the fight against cancer, even though the increasing success of prevailing personalized cancer therapies, metastases and recurrence remain similar, based on the stage and type of cancer. Even significant changes in mortality rates were also not visible over the previous decade (Jeitler et al., 2020). Hence, the advancement of novel compounds to combat cancer is the need of the hour. As per the reports in 2022, 1,918,030 new cancer cases were noted with 609,360 cancer deaths projected to occur in the United States, which includes roughly 350 deaths per day from lung cancer, which is the leading cause of cancer mortality (Siegel et al., 2022). The treatment of this deadly disease has been a consistent battle with relatively less success. Recent possible options for cancer treatment include treating large accumulated biomass of cancer cells by radiation, surgically removing these biomasses, and systemic chemotherapy treatment for maintenance (Baskar et al., 2012). The widely used chemotherapeutic treatment possesses several disadvantages including drug resistance, toxic effects on non-targeted tissues which impair the use of anticancer drugs, recurrence of cancer, and limits on the quality of the patient's life (Choudhari et al., 2020). To overcome the problems of the present therapy, the applicable diagnosis and early detection might play a vital role in the management of cancer, along with the development or exploration of novel anticancer treatments with better efficacy and fewer side effects, which is a reassuring approach to aid clinical oncology in generating new anticancer drugs.

Phytochemicals and their derivatives existing in plants are considered a promising alternative to enhance treatment efficacy in cancer patients and reduce adverse side effects and implement safer substitutes (Dehelean et al., 2021). Around 50% of approved anticancer drugs from 1940 to 2014 originated or were derived from natural products. Scientific evidence also showed that phytochemicals have prominent antitumor potentiality (Newman and Cragg, 2016). The phytochemicals were tested at both in vitro and in vivo levels for anticancer efficacy. They exerted broad and complex types of actions on various molecular targets and signal transduction pathways (Choudhari et al., 2020). Traditionally, mushrooms were utilized for preventing and treating cancer and other chronic diseases (Figueiredo and Régis, 2017). There are approximately 270 species of mushrooms that were reported to be potentially beneficial for human wellness. Like other natural regimens, the treatment using mushrooms was observed to be quite safe, as could be expected from their long-established culinary and medicinal uses (Panda et al., 2022). A few of the mushroom species which

 DOI: 10.1201/9781003354789-10

exhibit promising anticancer activity and may contain potential anticancer compounds include *Agaricus, Calvatia, Albatrellus, Antrodia, Clitocybe, Flammulina, Cordyceps, Pleurotus, Fomes, Schizophyllum, Funlia, Trametes, Ganoderma, Inocybe, Phellinus, Inonotus, Russula, Lactarius, Suillus, Xerocomus*, etc. Research suggests that inclusion of more mushrooms in the diet will be a potential protective measure in treating cancer (Wasser, 2017).

Ganoderma, commonly called "lingzhi", is one of the most well-known medicinal species and well regarded as the "marvelous herb". The macrofungi is of great economic and medicinal value and has been used in Asia for about 4000 years. (Vaithanomsat et al., 2022). It is widely cultivated and used in many parts of the world, including America, Korea, China, Japan, and other countries (Wang et al., 2020). According to traditional Chinese medicine (TCM), for over 2000 years *Ganoderma* has been utilized as an herbal remedy possessing the ability to elevate body resistance (Yue et al., 2006). The production of *Ganoderma* occurs mostly through artificial cultivation, which has furnished a plethora of materials for the market; the yield has already surpassed that of wild *Ganoderma* (Chen et al., 2017). Species of *Ganoderma* have been considered to be a significant source of pharmacological active constituents. Due to the ancient remedial property of *Ganoderma* species, they have attracted the interest of scientists as well as researchers in terms of various medicinal applications against numerous diseases (Bishop et al., 2015). Mostly *Ganoderma* has been used for the clinical treatment of leukopenia, coronary heart disease, chronic bronchitis, acute infectious hepatitis, bronchial asthma, arrhythmia, and cancer. However, presently, it does not have the ability to be utilized solely as first-line therapy, but rather only as an addition to the conventional therapy in a clinical setting (Gao and Zhou, 2003; Unlu et al., 2016).

Research studies have established that different species of *Ganoderma*, including *G. hainanense G. sinense, G. atrum, G. neo-japonicum, G. tsugae*, and, the most widely used, *G. lucidum*, possess significant carcinopreventive properties. The biologically active substances like polysaccharides, sterols, triterpenoids, alkaloids, nucleosides, and proteins have been isolated, characterized, and identified from *Ganoderma* spp. Of these bioactive compounds, two main active elements, triterpenes and polysaccharides, have been observed to possess the promising anticancer effects both in vitro and in vivo (Suárez-Arroyo et al., 2017). Triterpenes are mostly isolated from the spores of *Ganoderma* spp., and studies have exhibited outstanding pharmacological and therapeutic properties against cancer (Yuen and Gohel, 2005; Shi et al., 2010). *Ganoderma* was observed to contain a prominent number of various metabolites like polysaccharides, alkaloids, terpenoids, steroids, etc., which were extracted from the fresh fruiting body, mycelia, and spores (Li et al., 2019; Sharma et al., 2019). The current chapter aims to evaluate all the available literature on the subject and to identify the species of *Ganoderma* which exhibit promising anticancer activity. Though numerous studies have been conducted on the anticancer activity of *G. lucidum*, no one has yet attempted to provide a critical discussion on the results of all the species of *Ganoderma* exhibiting anticancer property. Our study provides the first comprehensive meta-analysis of all the eligible results associated with the anticancer potential of *Ganoderma* species, along with its mode of action.

10.2 ANTICANCER EFFECTS OF *GANODERMA* AND ITS ACTIVE COMPONENTS ON VARIOUS CANCER TREATMENTS

Reports state that most chemotherapies were not effective in completely eradicating cancer cells. This is one of the prime reasons why researchers are concentrating on mycotherapy as an alternative means of curing cancer, and *Ganoderma* is known for its effectiveness in increasing immune responses and curing several chronic diseases (Hu et al., 2019; Cao et al., 2018). The review depicted that various species of *Ganoderma* have acted against several cancers to eradicate them in both in vitro and in vivo models (Figure 10.1). Many pharmacological and clinical reports demonstrated that *Ganoderma* is playing a significant anticancer role in various cancer models through the regulation of the wide range of signaling pathways (Table 10.1).

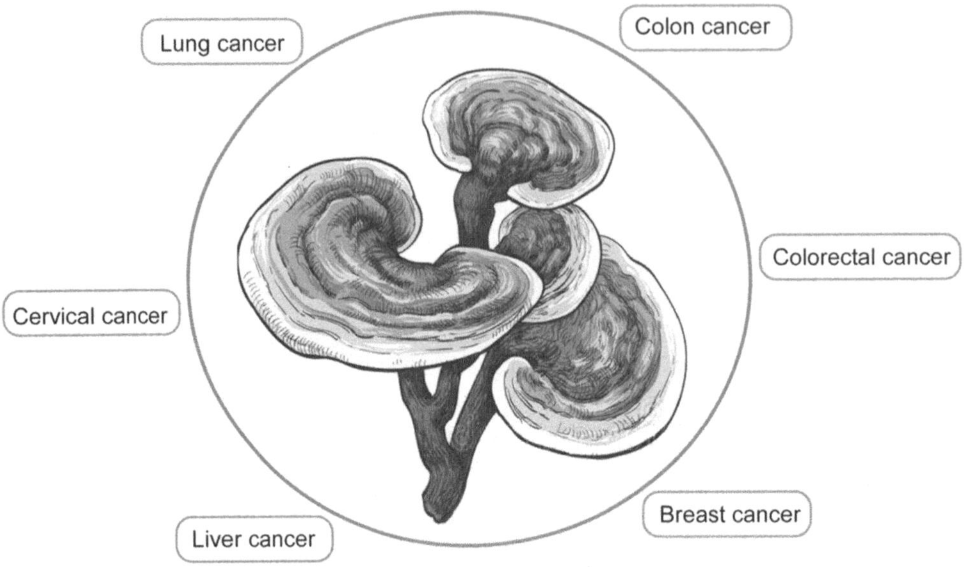

FIGURE 10.1 *Ganoderma* spp. exerting anticancer effects against multiple cell lines.

10.2.1 Breast Cancer

Breast cancer stem cells (BCSCs) are the subpopulation of triple-negative breast cancer (TNBC) cells. These cells are responsible for initiation, proliferation, metastasis, and drug resistance. The development and progression of the BCSCs depend on a signal transducer and activator of transcription 3 (STAT3) pathways. Furthermore, the calcium signaling cascade plays a vital role in the apoptosis of cancer cells (Park et al., 2019). Extraction using aqueous medium of G. *tsugae*, *G. lucidum*, and *G. sinense* were utilized in order to explore efficacy of these extracts against breast cancer cells, and the results exhibited a remarkable decrease in the proliferation of MCF-7 as well as in MDA-MB-231 cells based on varying concentrations. Amidst the different species evaluated, extracts of *G. tsugae* were the most potent against MCF-7 cells, and the capacity for inhibiting the MDA-MB-231 cell procreation was found to be homogenous among the various species of *Ganoderma* tested. Moreover, there was no toxic effect on human noncancerous mammary epithelial cells (HMECs) caused by the extract (Yue et al., 2006). The powdered extract (20:1) of *G. lucidum* which also contained the spores (ReishiMax GLp, GLE) inhibited the growth of MDA-MB-231 cells in a dose- as well as time-dependent manner. Treatment with GLE caused arrest in the G0/G1 phase of the cell cycle, causing the downregulation of cyclin D1 and CDK4 (Jiang et al., 2004). *G. lucidum* extract also showed a strong antitumor efficacy against noninvasive, estrogen-dependent MCF-7 cells (Jiang et al., 2006). GLE also demonstrated potential cytotoxic effects on the HER2 overexpressing MDA-MB-435 and TNBC cell lines, namely SUM-102 and MDA-MB-468. However, the noncancerous mammary epithelial cells, MCF-10A, were not affected by treatment (Martinez-Montemayor et al., 2011; Suárez-Arroyo et al., 2016). Treatment with GLE also caused a decrease in the expression of WEE and CCND1 (cyclin D1), which in turn was remarkably efficient in the reduction of CCNA2 (cyclin A2) and CCNB2 (cyclin B2) cell cycle as well as their gene expression that was abundant (Suarez-Arroyo et al., 2013). The aqueous extracts of *neo-japonicum* mycelia was utilized for the synthesis of silver nanoparticles which incited cellular death by the production of reactive oxygen species (ROS) and activation of caspase 3 fragmentation of DNA in MDA-MB-231 breast cancer cells (Gurunathan et al., 2013). Extraction was carried out for *Ganoderma* spp. using alcohol and was tested against breast cancer cells, and it depicted visible antiproliferative results. Furthermore, suppression based on varying times and dosages was observed with the ethanolic

TABLE 10.1
Bioactive Compounds from Different Species of *Ganoderma* and Their Anticancer Potential against Various Cancers

Ganoderma Species	Cancer Model	Mechanistic Pathway	Compound Isolated for Anticancer Activity	Reference
Ganoderma lucidum	EGFR-TKI–resistant human lung cancer A549	Inhibits the epidermal growth factor receptor and tyrosine kinase inhibitor (EGFR-TKI)–resistant human lung cancer A549	Gymnomitrane-3α,5α,9β,15-tetrol (1)	Binh et al., 2015
Ganoderma lucidum	LoVo human colon cancer cells	Exhibits cytotoxic effect by inhibiting migration of the cells, enhances the fragmentation of the DNA. Stimulated apoptosis by the activation of caspases-3, -8, and -9	*Ganoderma lucidum* polysaccharides (GLPs)	Liang et al., 2015
Ganoderma lucidum	Human colon cancer cells (HCT116)	Increased the production of caspase 3, PARP, caspase 7, and caspase 9 and reduced Bcl-2 protein levels. Elevates calcium levels in cell membrane and increased ROS,	Khz is a fusion of the mycelia of *Ganoderma lucidum* and *Polyporus umbellatus*	Kim et al., 2015
Ganoderma lucidum	HCT116 colorectal cancer cells	Inhibits tumor cells by arresting the cell cycle at the G1 phase by regulating the expression of cyclin D1 and p53. Induces apoptosis through the activation of caspase 9 and unfolded protein response.	Protein which was an RNA degrading enzyme (ribonuclease) (GLR)	Dan et al., 2016
Ganoderma lucidum	A549, H441, and H661	Inhibits Akt/mTOR signaling pathway	Ethanol/ethanol extract (E/E-SBGS) and ethanol/aqueous extract (E/A-SBGS)) of *G. lucidum*	Chen et al., 2016a
Ganoderma lucidum	Nonsmall-cell lung cancer (NSCLC) cells	Exhibits antimetastatic activity. Regulates cell mobility and epithelial-mesenchymal transition (EMT) by negatively modulating focal adhesion kinase (FAK)	Ling Zhi-8 (rLZ-8), a recombinant protein	Lin and Hsu, 2016
Ganoderma lucidum	TKI-resistant cancer cell lines (SUM-149)	Increases sensitivity of erlotinib by inactivation of AKT (serine/threonine-specific protein kinase) and extracellular signal-regulated kinase (ERK) pathways of cell signaling	Direct extract used	Suárez et al., 2016
Ganoderma lucidum	MCF-7 breast cancer cells.	Ca^{2+} signaling pathway elevates apoptosis of cancer cells	Khz, which is a fusion of two mycelia of *Ganoderma lucidum* and *Polyporus umbellatus*	Kim et al., 2016
Ganoderma lucidum	Breast cancer cell lines (MDA-MB-231) and mouse (4T1) cell lines	Decreases phosphorylation of LRP6 (LDL receptor–related protein 6) and suppresses Wnt3a-activated Wnt target auxin 2 expression, inhibits cell growth and proliferation.	Direct extract used	Wang et al., 2017

(Continued)

TABLE 10.1 *(Continued)*
Bioactive Compounds from Different Species of *Ganoderma* and Their Anticancer Potential against Various Cancers

Ganoderma Species	Cancer Model	Mechanistic Pathway	Compound Isolated for Anticancer Activity	Reference
Ganoderma lucidum	Human NSCLC adenocarcinoma cell lines A549, CL1–5[22], and LLC1 tumor-bearing mouse model	Arrests cell cycle and induced programmed cell death by downregulating the expression of wild and mutated EGFR	rLZ-8 is a recombinant protein that is isolated from *Ganoderma lucidum*	Lin et al., 2017
Ganoderma lucidum	In silico studies	Activates expression of mRNA of nuclear factor erythroid 2–related factor, increases antioxidant activity in cancer cells	Ganoderic acid A	Gill et al., 2017
Ganoderma lucidum	HCT116 cancer cells + mouse models	Disrupts cell cycle progression at the G2/M phase by downregulating the cyclins B1 and A2 and upregulating P21 at mRNA levels.	Polysaccharides from water extract	Na et al., 2017
Ganoderma lucidum	HCT116 cell lines	Arrests at G0/G1 phase and initiates apoptosis by suppressing the key genes and proteins like p21, p16, cyclin D1, Bcl-2, Bax, NAG-1, PARP, and caspase-3	Polysaccharides from ethanolic extract	Li et al., 2017
Ganoderma lucidum	CRC	Downregulates gene like Acaa1b, Fabp4, Mgll, and Scd1	*Ganoderma lucidum* polysaccharides (GLPs)	Luo et al., 2018
Ganoderma lucidum	A549 cells + mouse models	Improved specificity against cancer cells, represses growth of tumor cells.	Combination of *Ganoderma lucidum*–derived polysaccharides (GLP) with coix oil–based microemulsion	Guo et al., 2018
Ganoderma lucidum	Triple-negative breast cancer	Reduced cell viability by downregulating the STAT3 pathway.	GLE	Rios-Fuller et al., 2018
Ganoderma lucidum	H510A and A549 cells	Downregulation of ki6, PCNA, N-cadherin, vimentin, and Snail. Upregulation of Bcl-2, Bax, cleaved caspase 3, and cleaved PARP	Ganoderan A (GDNA), Ganoderan B (GDNB), and Ganoderan C (GDNC)	Wang et al., 2019
Ganoderma lucidum	Human adherent colorectal cancer cell lines HCT116, HT29, HCT116^{p53}	Increased ROS production, reduced proliferation of tumor cells	Co-treatment with 5-fluorouracil (5FU), the first-line chemotherapeutic treatment for colorectal cancer (CRC), and *Ganoderma lucidum* extract (GLE)	Opattova et al., 2019

Ganoderma lucidum	Human colorectal adenocarcinoma HCT-116 cells	Induces apoptosis by upregulating the BCL-2–associated X protein (Bax), phospho-extracellular regulated protein kinases (P-ERK), and cleaved caspase-3 along with downregulation of B-cell lymphoma-2 (Bcl-2), phospho-serine/threonine kinase 1 (p-Akt1), and cyclo-oxygenase (COX-2) expression	Hydrolyzed the *Ganoderma lucidum* polysaccharide (EGLP)	Bai et al., 2020
Ganoderma lucidum	Human cervical carcinoma cells C-33A	Arrests cell cycle and promotes apoptosis by inhibition of epithelial-mesenchymal and JAK/STAT5 pathways	*Ganoderma lucidum* polysaccharide (GLP)	Jin et al., 2020
Ganoderma lucidum	A549 cell lines and murine Lewis lung carcinoma (LLC1) and LLC1-bearing mice models	Reduces phosphorylation of ERK1/2 in cells by inducing the degradation of TGFβ and EGF.	Glucose-rich water-soluble polysaccharide (WSG)	Hsu et al., 2020
Ganoderma lucidum	BALB/c nude mice	Elevated expression of angiostatin, endostatin, and Bax protein responsible for tumor suppression	Combined treatment of *Ganoderma lucidum* triterpenoids with gefitinib (GEF) as an anticancer drug	Liu et al., 2020
Ganoderma lucidum	MDA-MB-231 cell lines	Inhibits cancer cells by upregulating the Bax and caspase-3	*G. lucidum* spore oil	Jiao et al., 2020
Ganoderma lucidum	Breast cancer patients	–	*Ganoderma* spore powder	Deng et al., 2021
Ganoderma lucidum	A549 cells, murine LLC1 cells and mice model studies	Cotreatment enhanced apoptosis by binding to N7 reactive center on purine residues of the cancer cell DNA	WSG along with cisplatin a chemotherapeutic drug	Qiu et al., 2021
Ganoderma lucidum	HT-29 colon cancer cells	Exerts cytotoxicity against cancer cells	Gold nanoparticles from the fruiting body of these mushrooms	Elumalai et al., 2021
Ganoderma lucidum	CRC cell line HT-29	Decreases AOM/DSS-induced colitis and tumorigenesis	Water soluble polysaccharides extracted from sporoderm-removed spores of *Ganoderma lucidum*	Guo et al., 2021
Ganoderma lucidum	A549 and HepG2 cell lines	Induced apoptosis by p53/caspase-3 pathway	Isolated new triterpene from 90% ethanol extract of the fruiting bodies of *G. lucidum*	Cao et al., 2022
Ganoderma lucidum	SUM-14, MDA-MB-231)	–	GLE	Suárez-Arroyo et al., 2022
Ganoderma applanatum	HT-29 colon adenocarcinoma cells	Inhibited proliferation	Lectins	Kumaran et al., 2017
Ganoderma applanatum	MCF-7 cell lines	Activates mitochondrial apoptosis signaling pathway, regulates MAPK signaling pathway	Homogenous polysaccharide called GAP-3S	Zhen et al., 2018

(*Continued*)

TABLE 10.1 *(Continued)*
Bioactive Compounds from Different Species of *Ganoderma* and Their Anticancer Potential against Various Cancers

Ganoderma Species	Cancer Model	Mechanistic Pathway	Compound Isolated for Anticancer Activity	Reference
Ganoderma applanatum	Colon cancer cell line (Caco-2)	Inhibit tumor proliferation through the initiation of apoptosis via p53-independent and p53-dependant pathway	80% ethanol extract	Elkhateeb et al., 2018
Ganoderma applanatum	4 T1 cells	–	Intratumoral polysaccharide	Tang et al., 2020
Ganoderma microsporum	A549 cells (ATCC, CCL-185) and CaLu-1 cells (ATCC, HTB-54)	Initiates apoptosis by autophagy through caspase-7–dependent and survivin- and ERCC1-independent pathways	GMI-TM, produced by Mycomagic Biotechnology Co., Ltd. (Taipei, Taiwan) was used along with anticancer drug cisplatin	Hsin et al., 2015
Ganoderma microsporum	MDR A549 lung cancer sublines	Upregulates autophagy by Akt/mTOR inhibition	GMI	Chiu et al., 2015
Ganoderma microsporum	NCl-H1355 cells	GMI downregulates GSK-3β, survivin, and cyclin-D1 by suppressing the expression of the β-catenin protein. Silencing of β-catenin increases PCD.	GMI is an important immunomodulatory protein	Hsin et al., 2018
Ganoderma microsporum	Pemetrexed-resistant lung cancer cells (A549/A400) and xenograft mice models	Inhibits cancer cells and initiates autophagy	*Ganoderma microsporum* immunomodulator-y (GMI) protein	Hsin et al., 2020
Ganoderma microsporum	A549 cells	Activates caspase 7	*Ganoderma microsporum*, along with saracatinib, an oral Src-kinase inhibitor	Chiu et al., 2021
Ganoderma resinaceum	Human breast cancer cell lines (MCF-7, MDA-MB-231)	Activates tumor suppressors like p53 and Bax with downregulation of cdk4/6 resulted in G0-G1 arrest	Phytosterol α-spinasterol	Sedky et al., 2018
Egyptian *Ganoderma resinaceum*	MCF-7 and MDA-MB-231 breast cancer cell lines	–	Ergosterol peroxide and ganoderic acid AMI	EL-Sherif et al., 2020
Ganoderma cochlear	Cell lines (H1975, PC9, A549)	–	Aqueous EtOH extract of the fruiting bodies	Cheng et al., 2018
Ganoderma cochlear	Triple-negative breast cancer cell lines.	Inhibits Cox-2 enzyme	Gancochlearols E-I (1, 3–6), ganomycin K (2)	Li et al., 2021
Ganoderma lingzhi	Human colorectal carcinoma (HCT-116, Caco-2),	–	Lucidumol C	Amen et al., 2016
Ganoderma lingzhi	Rat models were used for the study	Regulates microflora and secondary bile acids related to colon cancer	5% aqueous extract of either reishi mushroom (*Ganoderma lingzhi*) or the autodigested reishi *G. lingzhi* (AWGL)	Yang et al., 2017
Ganoderma atrum	MDA-MB-231 breast cancer cell lines	Arrests cell cycle at the G1/S transition phase and increases the apoptotic cell population	Fungal immunomodulatory proteins (FIP)	Xu et al., 2016

Ganoderma leucocontextum	MDA-MB-231	Arrests G1 phase by downregulating the cyclin D, CDK4, CDK6, cyclin E, and CDK2, and antiapoptotic c-Myc, Bcl-2, and Bcl-w	Ganoderiol F	Li et al., 2019
Ganoderma neo-japonicum	HCT 116 and HT 29 are human colonic carcinoma cell lines	Arrests cell cycle and exhibits antiproliferative effect	Hexane and chloroform extracts	Lau et al., 2022
Ganoderma sp.	MDA-MB-231 breast cancer cells	Inhibit JAK2 phosphorylation and downregulate STAT3 activation	Ganoderic acid A (GA-A)	Yang et al., 2018
Ganoderma sp.	Human triple-negative breast cancer (TNBC) cells (MDA-MB-231)	–	10 meroterpenoids, (±) dimercochlearlactones A–J (1–10, spirocochlealactone A (11)	Qin et al., 2022
Ganoderma colossum	Gefitinib-resistant H1650 cells and xenograft mice	Increased reactive oxygen species, DNA damage, and apoptosis, upregulates p53 protein. Colo H in combination with gefitinib exhibit anti-cancer effect (TKI) on the athymic mice.	Colossolactone H (colo H)	Chen et al., 2016b
Ganoderma sp.	CRC cell lines including HT29 and SW620	Affects energy metabolism through SIRT3 expression (sirtuin-3)	Ganoderic acid	Liu et al., 2018
Japanese *Ganoderma* sp.	HCT116 human colon carcinoma cell lines	–	80% methanolic extract of the fruiting bodies of Japanses *Ganoderma spp.*	Elkhateeb et al., 2019

extract of *G. lucidum* by the upregulation of p21/Waf1 and downregulating the cyclin D1. There was an elevation in the expression of Bax (proapoptotic protein) due to the extract, but no difference in the expression of Bcl-2 (antiapoptotic protein) was observed. The researchers also showed caspase-7 cleavage along with cleaved PARP expression (Hu et al., 2002).

In a study, the antiproliferative effect of *G. lucidum* was observed to be stronger than *G. sinense* extract. *G. lucidum* had the potential to decrease G1/S phase transition, while *G. sinense* had the capability to induce cell cycle arrest at the G2 phase (Liu et al., 2009). The effectiveness of the alcohol extract of *G. tsugae* (GTE) was tested against HER2-overexpressing cancer cells. The effect of the alcohol extract of *G. tsugae* (GTE) was assessed in HER2-overexpressing cancer cells. The analysis clearly depicted the inhibitory effects of the extract on BT-474 and SKBR-3 cell lines. Inhibitory effects in SKBR-3 cells caused an upregulation in G1 and downregulation in S and G2/M phases by regulating the cyclins (D1 and E) (Kuo et al., 2013). Recently, a study was conducted with five different solvents (ethanol-water, ethyl acetate, ethanol, ether and methanol) of *G. lucidum* extracts and the cytotoxic effects obtained against MCF-7 cells. *G. lucidum* ether extract (G.Ether) proved to have great antiproliferative activity against breast cancer cells than the others based on the results obtained from cytotoxicity studies (Atay et al., 2016; Gonul et al., 2015). A biologically active triterpene alcohol known as ganodermanontriol (GDNT) was isolated from *G. lucidum*. The cytotoxic effect of GDNT was noted in MDA-MB-231 cells, and it potently inhibited the growth of cells. Additionally, GDNT also inhibited the proliferation of MCF-7 cancer cells, and it also had a slight inhibitory effect on noncancerous MCF-10A mammary epithelial cells (Jiang et al., 2011). The ethanol-soluble and acidic component (ESAC) extracted from *G. lucidum* was evaluated for anticancer effect against breast cancer cells MCF-7 and MDA-MB-231 cells. The outcome obtained from the analysis depicted that ESAC notably reduced the viability of breast cancer cells mediating through G1 cell cycle arrest and apoptosis with increased expression of PARP cleavage (Wu et al., 2012b). Ganoderic acid DM (GADM), a *G. lucidum* triterpenoid, exhibited a reduction varying based on dose and time and was able to inhibit cancer cell viability even at a low concentration of the GADM extract, which efficaciously induced G1 cell cycle arrest in MCF-7 cells (Wu et al., 2012a). Ganoderic acid ME (GA-Me) isolated from *G. lucidum* restricted the proliferation of cells and induced the programmed cell death of breast cancer cells by decreasing the cell-cycle regulator cyclin D1 and prosurvival proteins BCL-2 and c-Myc, (Li et al., 2012). Recently 19 lanostane triterpenoids were isolated from a rare species of *Ganoderma*, *G. hainanense*. The group showed antiproliferative activities against MCF-7 cells with 16 compounds, including ganoderone A, lucidadiol, ganodermanontriol and 4,4,14a-trimethyl-3,7-dioxo-5a-chol-8-en-24-oic acid (Peng et al., 2015).

G. lucidum has the selenium (Se)-enriched mycelia (SeGLP-2B-1), which was utilized for purification of a polysaccharide from the mushroom (Shang et al., 2009). The inhibitory effect of SeGLP-2B-1 was analyzed in the breast cancer cell MCF-7, where cell viability was suppressed based on varying concentrations of the dose. The mode of action included the intrinsic and extrinsic programmed cell death pathways with an increase of caspases-8, -9, and -3 and cleavage of PARP. It also resulted in losing the action potential of mitochondria accompanied by the spillage of cytochrome *c* into the cytosol, representing that SeGLP-2B-1 resulted in cell death that was mediated by mitochondria (Shang et al., 2011). An active fraction that had a fucose-containing glycoprotein was extracted from Zhi (*G. lucidum*) extract (FFLZ), which had antiproliferative activity in cancer cells (Liao et al., 2013). The growth of mouse breast cancer cells 4T1 and MDA-MB-231 was decreased and the viability of both cell lines was controlled by the treatment of FLZZ in a dose-dependent fashion. FFLZ inhibits the colony formation of 4T1 .cells; this was evident as FFLZ-treated 4T1 cells formed fewer colonies than untreated cells (Tsao et al., 2016). Researchers isolated fungal immunomodulatory proteins (FIPs) from *G. atrum* and constructed a recombinant FIP-gat (rFIP-gat) in the *Escherichia coli* host for protein overexpression and studied the effect of its inhibition potential against the breast cancer cell line MDA-MB-231 (Xu et al., 2015). The results proved that FIPs could inhibit cancer cells by arresting the G1/S transition phase of the cell cycle and accelerating the apoptotic cell population (Xu et al., 2016). Lately a hydroethanolic extract (99%) of *G. lucidum* (reishi) (GLE) was also tested for its anticancer activity on the breast (MDA-MB-231) as well as mouse (4T1) cancer cell lines, and the study showed that cancer occurring due to the abnormalities in the activation and deactivation of the Wnt/β-catenin pathway could be suppressed with GLE. The extract has decreased the phosphorylation of LRP6 (LDL receptor–related protein 6) and suppressed Wnt3a-activated Wnt target Auxin 2 expression, inhibiting cellular growth and proliferation (Wang et al., 2017). Breast cancer can be treated effectively by targeting the epidermal growth factor receptor (EGFR) with the help of tyrosine kinase inhibitors (TKIs). Over a period of time, TKIs have become resistant to chemotherapeutics (De Melo Gagliato et al., 2016). The effect of GLE (ethanolic) on TKI-resistant cancer cell lines (SUM-149) was analyzed by Suárez-Arroyo et al., 2016. SUM-149 and erlotinib-resistant MDA-MB-231 cells were coupled with the chemotherapeutic agent erlotinib in combination with the GLE. The GLE efficiently increased the sensitivity of erlotinib by inactivation of AKT (serine/threonine-specific protein kinase) and ERK (extracellular signal-regulated kinase) cell signaling pathways. *Ganoderma* extract could effectively decrease the cell proliferation, migration, invasion, and viability of SUM-149 cells.

Khz is a fusion of *G. lucidum* and *Polyporus umbellatus* mycelia. The aqueous extract of Khz inhibited the proliferation of MCF-7 breast cancer cells in vitro (Kim et al., 2016). The effect of commercially available GLE on the BCSCs was studied in vitro by Rios-Fuller et al., 2018. An additional study was performed on the animal model carrying TNBC tumors in vivo. The GLE significantly reduced the TNBC cell viability by downregulating the STAT3 pathway (Rios-Fuller et al., 2018). Similarly, fungal triterpenoid ganoderic acid A showed an antineoplastic effect on MDA-MB-231 breast cancer cell lines. It inhibited JAK2 phosphorylation and downregulated STAT3 expression (Yang et al., 2018). Phytosterol α-spinasterol isolated from *G. resinaceum* was analyzed for its efficacy against human breast cancer cell lines (MCF-7, MDA-MB-231) in vitro. It induced the overexpression of some tumor suppressor proteins like p53 and Bax. Simultaneously, the compound downregulated cdk4/6, leading to G0/G1 cell cycle arrest in the cancer cells (Sedky et al., 2018). Ganoderiol F, isolated from *G. leucocontextum*, could arrest the cell cycle of breast cancer cell lines (MDA-MB-231) through downregulation of the cyclin D, CDK4, CDK6, cyclin E, and CDK2, and antiapoptotic c-Myc, Bcl-2, and Bcl-w genes. Further, it upregulated the tumor suppressor gene Foxo3 (Li et al., 2019). The antineoplastic effect of *G. lucidum* spore oil (GLSO) was extracted from the sporoderm of the fungi. GLSO was found to have an anticancer property through the upregulation of apoptotic proteins Bax and caspase-3 in the MDA-MB-231 cell line. It was believed to achieve this through the mitochondrial apoptotic pathway (Jiao et al., 2020). Two new metabolites, namely ergosterol peroxide and ganoderic acid, isolated from Egyptian *G. resinaceum* mushroom, showed an antitumor property when tested against MCF-7 and MDA-MB-231 breast cancer cell lines (El-Sherif et al., 2020). A commercially available *G. lucidum* extract (ReishiMax GLp) prepared from the combination of the fruiting body and cracked spores showed a promising anticancer property when tested against the arrest and most aggressive breast cancer type: inflammatory breast cancer (IBC). ReishiMax GLp combined with the anticancer drug carboplatin effectively regulated DNA damage response in the breast cancer cell lines *in vitro* (SUM-14, MDA-MB-231) and IBC xenograft model in vivo. This combination suppressed mammosphere formation and expression of cancer stemness proteins in the model systems (Suárez-Arroyo et al., 2022). Polysaccharides isolated from different species of *Ganoderma* are widely used as anticancer agents (Zhang et al., 2018; Wang et al., 2018). A homogenous polysaccharide called GAP-3S (*G. applanatum* polysaccharides) was tested for its anticancer activity on MCF-7 cell lines (Zhen et al., 2018). GAP-3S–treated breast cancer cells showed a significant collapse in the cellular mitochondrial potential. It also regulated mitogen activated protein kinase (MAPK) in the breast cancer cell line, thereby inhibiting the proliferation. The polysaccharide obtained from *G. applanatum*, combined with an anticancer drug paclitaxel, could inhibit the proliferation of 4T1 mammary carcinoma cells. Further, the combination of the polysaccharide and nanoparticle albumin-bound paclitaxel increased the therapeutic effect and reduced the chemotherapeutic side effects breast cancer model mice in vivo (Tang et al., 2020). Five new meroterpenoids (gancochlearols E–I [1, 3–6]) from *G. cochlear*, along with another compound called ganomycin K (2), were isolated. These compounds were analyzed for their antiproliferative activity against TNBC cell lines and were found to inhibit tumor cell proliferation by inhibiting the Cox-2 enzyme (Li et al., 2021). In a similar study, ten novel terpenoids, namely dimercochlearlactones A–J (1–10), along with spirocochlealactone A (11), were isolated from the mushroom. The compounds were found to be effective against human TNBC cells (MDA-MB-231) (Qin et al., 2022). All these data cumulatively suggested that *Ganoderma* spp. and their biologically active compounds are competent enough for inducing antiproliferative effects, cytotoxicity, and proapoptotic processes and arresting the cell cycle. A group of researchers treated a cohort of 120 breast cancer patients with *G. lucidum* spore powder (GLSP) along with the control. T-lymphocyte subsets as the biomarker and the ratio between the albumin-to-globulin ratio and neutrophil-to-lymphocyte ratio were analyzed statistically from the treated population. Results showed that GLSP was beneficial for immunological enhancement in post–breast cancer patients (Deng et al., 2021).

10.2.2 Lung Cancer

Lung cancer, categorized into non–small cell lung cancer (NSCLC) and small cell lung cancer (SCLC), is one of the most common malignancies worldwide in terms of both incidence and mortality (18% of total cancer deaths). Therapeutic options for treating this cancer include systemic treatments including chemotherapy, radiation therapy, hormonal therapy, surgery, targeted therapy, and immunotherapy (Yang et al., 2021). A widely used chemotherapeutic agent known by the name cisplatin is used worldwide for various cancers, including lung cancer. GMI (*G. microsporum* immunomodulatory protein) showed a promising role in enhancing the ability of cisplatin. GMI co-treated with cisplatin induced the cleavage of caspase-7 and poly (ADP-ribose) polymerase (PRAP) in the lung cancer cell line. The result showed that combining GMI and cisplatin could initiate programmed cell death by autophagy through caspase-7-dependent, survivin, and ERCC1-independent pathways (Hsin et al., 2015). Further, studies showed that treatment of GMI elevates intracellular calcium levels, thereby sensitizing the cellular death in the multidrug resistant (MDR) A549 lung cancer sublines via autophagy. These studies demonstrated that autophagy can play a vital role in the death of the MDR cells, and GMI could upregulate autophagy by Akt/mTOR inhibition to further aid in breaking the drug resistance of the cancer cells (Chiu et al., 2015). Sporoderm-broken spores of *G. lucidum* (SBGS) extracts (ethanol/ethanol extract [E/E-SBGS] and ethanol/aqueous extract [E/A-SBGS]) were examined for their antitumor activity against lung cancer. The results showed that E/E-SBGS could inhibit the migration of lung cancer cell lines (A549, H441, and H661) by arresting the cell cycle (G2/M). The extract also induced apoptosis by downregulating regulators of the cell cycle like cyclin B1 and cdc2 and antiapoptotic proteins (Bcl-2 and Bcl-xl) via the Akt/mTOR signaling pathway. E/E-SBGS and E/ASBGS both were administrated orally in the cancer model mice, effectively reducing the tumor volume (Chen et al., 2016a). The effect of GMI in inhibiting the β-catenin was explored, and it was found that GMI downregulates GSK-3β, survivin, and cyclin-D1 by suppressing the expression of the β-catenin protein. Silencing of β-catenin by the treatment of GMI enhanced the rate of apoptosis in H1355 cells. This study was able to give the first insight into the induction of apoptosis by the inhibition of β-catenin using GMI (Hsin et al., 2018). Ling Zhi-8 (rLZ-8), a recombinant protein from *G. lucidum*, was found to have cytotoxic effects against lung cancer in vivo. Treatment with rLZ-8 increased the rate of survival in the Lewis lung carcinoma mice. FAK plays a pivotal role in cancer metastasis through cell mobility regulation and EMT. The recombinant protein rLZ-8 downregulated FAK and induced ubiquitination, which led to degradation of metastasis marker Slug (SNAI2 gene) (Lin and Hsu, 2016).

Mutation of the EGFR leads to malignant lung cancer. The protein rLZ-8 downregulated EGFR along with AKT and ERK1/2 in the human NSCLC adenocarcinoma cell lines A549 and CL1–5[22] (Lin et al., 2017). The studies showed that GMI protein could inhibit pemetrexed-resistant lung cancer cells in both in vitro (MTT, clonogenic, tumor spheroid, and cancer stem cell sphere assays) and in vivo models. GMI was able to inhibit cancer cells through initiation of autophagy, but not programmed cell death, in the A549/A400 cells, leading to degradation of CD133 protein (Hsin et al., 2020). CD133 (prominin-1 a glycoprotein) is considered a possible target for cancer therapy, as it is believed to have EMT, stemness properties, and tumorigenicity by activating the Src signaling pathway. GMI also reduced the expression levels of CD44, NANOG, and OCT4, thereby reducing the growth of pemetrexed-resistant lung cancer. Saracatinib is an oral TKI. It also acts as a dual specific inhibitor of Src and Abl. A combination treatment of GMI and saracatinib activated caspase 7 and inhibited A549 cell proliferation (Chiu et al., 2021). The fruiting body of *G. lucidum* was the source for isolation of a gymnomitrane-type sesquiterpenoid [gymnomitrane-3α,5α,9β,15-tetrol (1)] and was found to be effective against EGFR-TKI–resistant human lung cancer A549 cell lines (Binh et al., 2015). The fruiting body of *G. colossum* served as a source for isolation of three pentacyclic triterpene dilactones. Among the isolated compounds Colossolactone H (Colo H) was found to have the most cytotoxic effect. Gene expression profiling indicated that the treatment with Colo H downregulated 398 genes that were mainly responsible for cell cycle progression in the gefitinib-resistant H1650 lung cancer cells. Colo H also upregulated around 252 genes, responsible for metabolic processes, cellular response to stimulus, and oxidation-reduction. It increased the ROS leading to DNA damage and apoptosis. Colo H could upregulate tumor suppressor p53 protein. When used in combination with the drug gefitinib in the tumor xenograft athymic mice, it was found to inhibit lung cancer (Chen et al., 2016b). A triterpene ganoderic acid, *G. lucidum* could modulate Kelch-like ECH-associated protein in 1-nuclear factor erythroid 2–related factor 2 signaling pathway through an in silico molecular docking experiment. Further experimentation also suggested that ganoderic acid A could inhibit lung cancer by activating the expression of messenger RNA of nuclear factor erythroid 2–related factor 2 in H460 cells (Gill et al., 2017). From the aqueous ethanolic extract of the fruiting bodies of *G. cochlear*, seven novel terpenoids, namely (+) and (−) gancochlearol C, (+) and (−) cochlearoid Q, and gancochlearol D, were isolated. These isolated compounds were tested for their anticancer effect against three lung cancer cell lines (H1975, PC9, and A549). The result suggests that among all seven isolated compounds, ganomycin F had a moderate inhibitory effect against the H1975 human lung cancer cell line, with an IC_{50} value of 19.47 uM (Cheng et al., 2018).

New triterpene [12β-acetoxy-3,7,11,15,23-pentaoxolanosta-8,20E(22)-dien-26-oic acid methyl ester] was obtained from 90% ethanolic extract fraction of the fruiting bodies of *G. lucidum*. The cytotoxic activity of this triterpenoid was tested in vitro on A549 and HepG2 cell lines using cisplatin as a positive control. The studies suggested that this new triterpene could induce apoptosis in the cancer cells through the p53/caspase-3 pathway (Cao et al., 2022). The combined treatment of GLT + GEF was found to be a very promising candidate against lung cancer. There was a high expression of angiostatin, endostatin, and Bax protein responsible for tumor suppression compared to GLT and GLE treatment alone in the BALB/c nude mice (Liu et al., 2020). A combination of GLP with coix oil-based micro-emulsion was used to determine its anticancer properties against lung cancer. The GLP combined coix oil-based micro-emulsion (MEs; PS-GLP) had similar physiochemical properties as the ME but with better stability and lowered zeta potential. The effect of MEs (PS-GLP) on A549 cells through in vivo studies was found to be effective in inhibiting the growth of the cancer cells in comparison with the control (Guo et al., 2018). A water-soluble polysaccharide (WSG), rich in glucose, was sequestrated from *G. lucidum*. The studies revealed that WSG was able to inhibit lung cancer cells by reducing phosphorylation of ERK1/2 in the A549 cell line. It induced the degradation of TGFβ and EGF. The in vivo study also suggested that WSG was effective in reducing the cancer cell mass growth in Lewis lung carcinoma (LLC1)–bearing mice after the treatment with WSG (Hsu et al., 2020). In another study, a combination of cisplatin and WSG was a potent extract that was effective against lung cancer cells; however, it reduced the toxicity of the cisplatin against both macrophages and normal lung fibroblasts. Also, the presence of WSG enhanced the apoptosis mechanism of cisplatin, thereby preventing lung carcinoma cell growth (Qiu et al., 2021). Isolated polysaccharides, namely ganoderan A, B, and C (GDNA, GDNB, and GDNC, respectively), inhibited the proliferation of cancer cells (H510A and A549) by suppression and downregulation of ki6, PCNA, N-cadherin, vimentin, and Snail. The in vivo study also confirmed the suppressed proliferation of NSCLC through modulating the ERK signaling pathway (Wang et al., 2019).

10.2.3 Colorectal Cancer

Malignant tumors in the inner wall of the large intestine lead to colorectal cancer; since they are very progressive and aggressive, they contribute to high mortality rates (Fang et al., 2015). Chemotherapeutic drugs that are used in the treatment, which include imatinib, doxorubicin, methotrexate, cisplatin, and vincristine, exhibited detrimental effects on the patients (Hildreth, 2008). The GLPs were isolated and experiments were carried out to evaluate their anticancer properties against LoVo human colon cancer cells. The results indicated that GLP exhibited a cytotoxic effect against LoVo cells by inhibiting migration of the cells, enhancing the fragmentation of the cancer cell DNA, etc. GLP stimulated apoptosis by the activation of caspases-3, 8, and 9. GLP also increased the production of Fas, a cell surface receptor that is responsible for activating apoptosis and caspase-3, while reducing the expression of the cleaved poly(ADP-ribose) polymerase, which is responsible for DNA repair. These studies indicated that GLP inhibited cancer cells through the Fas/caspasedependent apoptotic pathway (Liang et al., 2015). The impact of Khz on human colon cancer cells (HCT116) was studied. Khz increased the production of caspase 3, PARP, caspase 7, and caspase 9 and reduced Bcl-2 protein levels. Khz helped in increasing the calcium levels in the cell membrane and also increased the ROS, which was responsible for inhibiting the tumor progression. This was the first study that gave insights into the use of Khz as an anticancer agent for colon cancer (Kim et al., 2015). Six known compounds and a new triterpene of the oxygenated lanostane type of compound from *G. lingzhi*'s chloroform extract of the fruiting body were studied. These isolated compounds were analyzed for their cytotoxic activity against human colorectal carcinoma (HCT-116, Caco-2), human liver carcinoma (HepG2), and human cervical carcinoma (HeLa) cell lines. Among all the seven compounds, the one that exhibited the highest activity was lucidumol C, with an IC_{50} value of 7.86 ± 4.56 μg/mL for cytotoxicity against HCT-116 cells. Further studies on these compounds can be explored in order to study the underlying mechanism for their cytotoxic activity (Amen et al., 2016). *G. lucidum* is known for its therapeutic benefits, which is the reason why it has been extensively used in China for a very long time. A 17.4-kDa protein which was an RNA degrading enzyme (ribonuclease) (GLR), had the ability to act as an anticancer agent from *G. lucidum*. The ability of GLR on HCT116 colorectal tumor cells was tested in vitro, and its anticancer activity was analyzed. GLR was able to inhibit the cancer cells by arresting the cell cycle at the G1 phase by regulating the expression of cyclin D1 and P53. GLR was also able to initiate apoptosis through the activation of caspase 9 and unfolded protein response. GLR was efficient in suppressing autophagy, which is a stress-coping mechanism in cells in times of metabolic crisis (Dan et al., 2016).

The reishi mushroom is known for its ability to prevent colon cancer, but the underlying mechanism is not known. A high-fat dietary supplement from the 5% aqueous extract of either reishi mushroom (*G. lingzhi*) or the autodigested reishi *G. lingzhi* (AWGL) was fed to the rats for a period of 3 weeks. These extracts were efficient enough to reduce fecal secondary bile acids, such as lithocholic acid and deoxycholic acid (colon carcinogens). They were also efficient in the reduction of *Clostridium coccoides* and *Clostridium leptum* (secondary bile acids producing bacteria) per gram of cecal digesta. These results suggest that extracts from reishi mushroom have the ability to regulate microflora and secondary bile acids, etc., that is related to colon cancer (Yang et al., 2017). The growth of *G. applanatum* was optimized, and the optimum growth and production of lectin were obtained at pH 6.5 and temperature 26°C. Lectin expression was obtained after 5 days of culturing the mycelia in broth, and maximum growth was obtained after 15 days of incubation. Different carbon and nitrogen sources were also used to optimize the maximum production of lectin in which sucrose and yeast extract gave the highest yield. This lectin was tested for its cytotoxic activity against HT-29 colon adenocarcinoma cells and was found to be effective against them (Kumaran et al., 2017). Polysaccharides isolated from *Ganoderma* (GLP) species are widely used for anticancer activity. GLP isolated from the aqueous extract from sporoderm-broken spores of *G. lucidum* and its anticancer potential against colon cancer were tested both in vitro and in vivo. HCT116 cancer cells were successfully inhibited by the aqueous extract of sporoderm-broken spores, and the underlying mechanism responsible for inhibition is by disrupting the cell cycle progression at the G2/M phase of the cell cycle. The cell cycle was disrupted by downregulating the cyclins B1 and A2 and upregulating P21 at the mRNA levels. They also had the ability to induce apoptosis by downregulating Bcl-2 and survivin at mRNA levels and Bcl-2, PARP, pro-caspase-3, and pro-caspase-9 at protein levels. Both studies in vitro and in vivo suggested that the treatment with the aqueous extract of sporoderm-broken spore was able to activate the preapoptotic gene (NSAID-activated gene-1) (Na et al., 2017). All these studies suggested the effectiveness of using a water extract of sporoderm-broken spores as an antineoplastic agent. Similar work was carried out in ethanolic extracts of *G. lucidum* (sporoderm-broken spores), where the extract was tested on HCT116 cell lines. The extract was efficient in inhibiting the cancer lines by inhibiting the G0/G1 phase of the cell cycle and also in initiating apoptosis by suppressing the key genes and proteins (p21, p16, cyclin D1, Bcl-2, Bax, NAG-1, PARP, and caspase-3) (Li et al., 2017).

The possibility of GLP consumption can control the proliferation of colon cancer through the gut microflora. The study incorporated many parameters like the mortality, cecal microbiota, permeability, and gene expression profile of colonic epithelial cells of healthy and colorectal cancer mice consuming GLPs. Through the in vivo studies, it was found that GLPs had the ability to alleviate the effects of colorectal cancer by reducing certain bacteria and specific genes which are related to colorectal cancer. GLP treatment was not only efficient in reducing colon shortening and decreasing mortality by 30% in colorectal cancer mice but also reduced microflora, which was found to be high in mice that have colorectal cancer. It was also efficient in downregulating genes like Acaa1b, Fabp4, Mgll, and Scd1 that are related to cancer (Luo et al., 2018). The active secondary metabolites from *G. applanatum* was extracted using 80% ethanol, and cytotoxic activity was studied on a colon cancer cell line (Caco-2) using a total metabolite extract of *G. applanatum*. Through the in vitro and in vivo studies, it was understood that the extract of *G. applanatum* was successful in inhibiting tumor proliferation through the initiation of apoptosis through two pathways, the p53-independent pathway and p53-dependant, where p53 is a protein that functions as a tumor suppressor (Elkhateeb et al., 2018). Ganoderic acid is used as an antitumor agent; studies have explained that GAD has the ability to affect the energy metabolism in colon cancer through the SIRT3 expression (sirtuin-3), which is a critical modulator of tumorigenesis through upregulation via acetylated CypD inhibition. GAD inhibits energy programming of the colon cancer cells by inhibiting glucose uptake, pyruvate, lactate, and acetyl co-enzyme production, which inhibited the growth of the cancer cells (Liu et al., 2018). Earlier studies have suggested the treatment of anticancer agents with metabolites and polysaccharides from *Ganoderma* helped in improving the efficiency of the chemotherapeutic. The effect of co-treatment with GLE with 5-fluorouracil (5FU) in the first-line chemotherapeutic of colorectal cancer was studied. Through the in vitro (human adherent colorectal cancer cell lines HCT116, HT29, HCT116^{p53}) and in vivo studies, it was understood that the combined therapy improved the specificity by increasing the selective cancer cell death more than nonmalignant cells and was also efficient in decreasing the adversities of the chemotherapeutic treatment. The GLE treatment increased the accumulation of ROS in colorectal cancer cell lines, which caused oxidative DNA damage, which killed the malignant cells but protected the nonmalignant cells from oxidative DNA damage. It was also effective in reducing the proliferation of tumor cells (Opattova et al., 2019).

The neoplastic activity of 80% methanolic extract of the fruiting bodies of Japanses *Ganoderma* sp was explored where the *in vitro* studies suggested that the methanolic extract was effective against the HCT116 human colon carcinoma cell lines (Elkhateeb et al., 2019). Polysaccharides from *Ganoderma* used as anticancer drugs are very common, but lately researchers enzymatically hydrolyzed the *G. lucidum* polysaccharide (EGLP), which was used to check its cytotoxic ability against colorectal cancer. The in vitro studies for cytotoxicity were carried out in human colorectal adenocarcinoma HCT-116 cells. The EGLP had the ability to induce apoptosis by upregulating the BCL-2 associated X protein (Bax), phospho-extracellular regulated protein kinases (P-ERKs), and cleaved caspase-3 expression and downregulating B-cell lymphoma-2 (Bcl-2), phospho-serine/threonine kinase 1 (p-Akt1), and cyclooxygenase 2 (COX-2) expression (Bai et al., 2020). The polysaccharide of GLP had the ability to alleviate colon cancer cell proliferation by blocking the cell cycle and further promote apoptosis by the inhibition of the epithelial-mesenchymal and JAK/STAT5 pathways. Through Western blotting, it was understood that after the treatment with GLP there was an increase in pro-apoptotic proteins like Bax and cleaved caspases 3 and 9 while anti-apoptotic genes like Bcl-2 decreased after the treatment (Jin et al., 2020). Elumalai et al. (2021) successfully synthesized gold nanoparticles from the fruiting bodies of *Ganoderma*. The nanoparticles had varying shapes from oval, to spherical, to irregular and sizes from 1 to 100 nm were targeted for cytotoxicity against HT-29 colon cancer cells. The results suggested that these nanoparticles exhibited cytotoxic activity against HT-29 colon cancer cells with an IC_{50} value of 84.58 lg mL^{-1}; it also had the ability to increase the apoptosis in a dose-dependent manner. Water-soluble polysaccharides extracted from sporoderm-removed spores of *G. lucidum* against azoxymethane (AOM)/dextran sodium sulfate (DSS) induced inflammation, tumorigenesis, and gut microbiota modification. The studies revealed that the usage of these polysaccharides (200 and 300 mg/kg) was efficient in decreasing the AOM/DSS-induced colitis and tumorigenesis; it was also successful in decreasing the tumor size. There was a significant increase in the numbers of goblet cells, MUC2 secretion, and tight junction protein expressions, which suggests that there was an improvement in the gut microflora after the treatment with GLP (Guo et al., 2021). These studies suggested that GLP is a very good prebiotic which can be further explored, studied, and used in the prevention of colorectal cancer. The patients undergoing chemotherapy for colon cancer often face high blood glucose levels as a result of attenuating the cytotoxicity of the chemotherapeutic agent. The possibility to use *G. neo-japonicum* as a natural antiproliferative agent against colon cancer was studied with both hexane and chloroform extracts, and different parameters such as oxidative stress, cell cycle, and apoptosis were analyzed between the cancer cells and normal cells. Both hexane and chloroform fractions were able to arrest the cell cycle and had an antiproliferative effect against cancer cells. These fractions were able to induce apoptosis, and they were also able to manage hyperglycemia-associated with colorectal cancer (Lau et al., 2022).

10.3 CONCLUSION

Cancer has become a soaring global public health issue, and preventing cancer with natural alternatives like *Ganoderma* is the need for the entire globe. *Ganoderma* is turning into one of the most globally used antitumor agents due to its efficiency for used as an immunotherapy agent and its ability to have very low cytotoxicity in combination therapy. The current review furnishes the most up-to-date analysis of *Ganoderma* research over 20 years. The pharmacologically active compounds of *Ganoderma* may be a probable alternative to the battle against cancer growth with conventional and advanced therapies as natural substitutes. This review recapitulated the area of research carried out on *Ganoderma* species (Figure 10.2). There are numerous facts in favor of bioactive compounds extracted from different species of *Ganoderma* to prove anticancer properties through different mechanisms. However, there are various mechanistic pathways which lack specificity and have not properly selected their specific targets; additionally, very few results are derived from in vivo studies; almost all results are collected from in vitro studies. In many cases the species were not identified properly, which requires further attention. Future upcoming studies should target the combinational therapies of *Ganoderma* and clinical chemotherapy medicine to alleviate the side effects of these drugs. The major bioactive compounds should be inspected further and corresponding *in vivo* pharmacokinetic studies should be carried out. Still there are various triterpenes

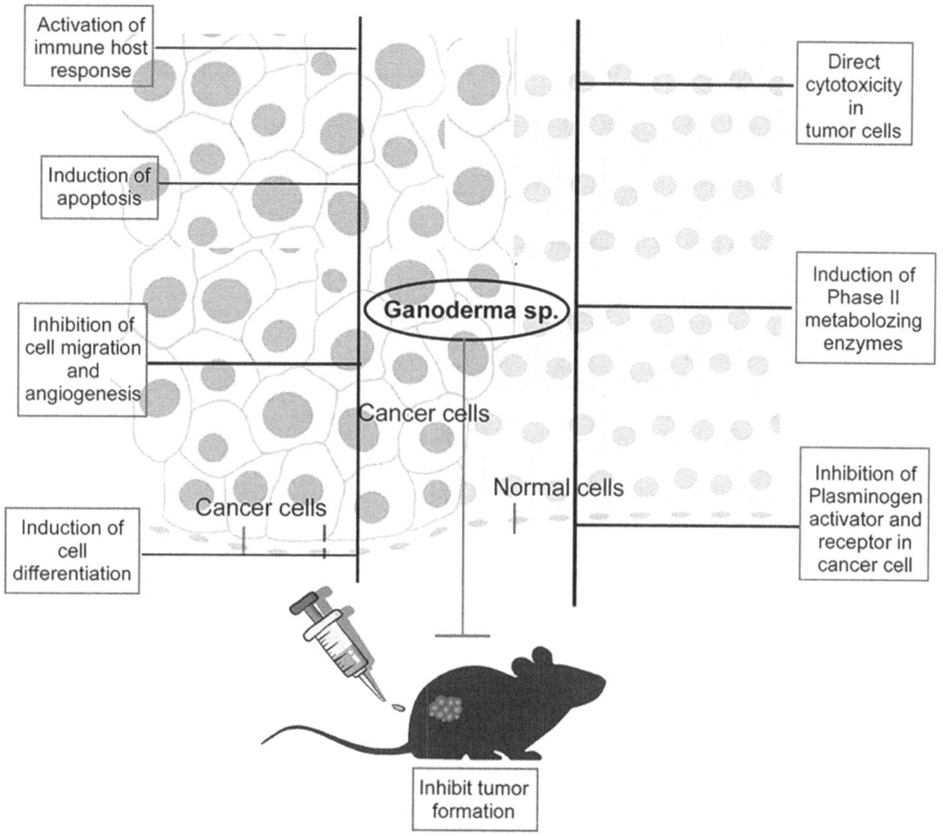

FIGURE 10.2 Mode of action through which *Ganoderma* species exerts anticancer properties.

and polysaccharides of species of *Ganoderma* whose anticancer mechanisms are unexplored. The mode of action of biologically active constituents from *Ganoderma* against cancer should be further explored in detail for the prevention and treatment of various other types of cancers.

REFERENCES

Amen, Y.M., Zhu, Q., Tran, H.B., Afifi, M.S., Halim, A.F., Ashour, A., Mira, A., Shimizu, K. (2016). Lucidumol C, a new cytotoxic lanostanoid triterpene from *Ganoderma lingzhi* against human cancer cells. *J. Nat. Med.* 70(3), 661–666.

Atay, S., Ak, H., Kalmis, E., Kayalar, H., Aydin, H.H. (2016). Diverse effects of the lingzhi or Reishi medicinal mushroom, *Ganoderma lucidum* (Agaricomycetes), in combination with tamoxifen citrate and doxorubicin in MCF-7 breast cancer cells. *Int. J. Med. Mushrooms.* 18, 489–499. doi: 10.1615/IntJMedMushrooms.v18.i6.30.

Bai, J.H., Xu, J., Zhao, J., Zhang, R. (2020). *Ganoderma lucidum* polysaccharide enzymatic hydrolysate suppresses the growth of human colon cancer cells via inducing apoptosis. *Cell Transplant.* 29, 0963689720931435.

Baskar, R., Lee, K.A., Yeo, R., Yeoh, K.W. (2012). Cancer and radiation therapy: Current advances and future directions. *Int. J. Med. Sci.* 9(3), 193–199. doi: 10.7150/ijms.3635.

Binh, P.T., Descoutures, D., Dang, N.H., Dai Nguyen, N.P., Dat, N.T. (2015). A new cytotoxic gymnomitrane sesquiterpene from *Ganoderma lucidum* fruiting bodies. *Nat. Prod. Commun.* 10(11), 1934578X1501001125.

Bishop, K.S., Kao, C.H., Xu, Y., Glucina, M.P., Paterson, R.R.M., Ferguson, L.R. (2015). From 2000 years of *Ganoderma lucidum* to recent developments in nutraceuticals. *Phytochemistry.* 114, 56–65.

Cao, L., Jin, H., Liang, Q., Yang, H., Li, S., Liu, Z., Yuan, Z. (2022). A new anti-tumor cytotoxic triterpene from *Ganoderma lucidum*. *Na. Pro. Res.* 36(16), 4125–4131.

Cao, Y., Xu, X., Liu, S., Huang, L., Gu, J. (2018). *Ganoderma*: A cancer immunotherapy review. *Front Pharmacol.* 9, 1217.

Chen, B., Ke, B., Ye, L., Jin, S., Jie, F., Zhao, L., Wu, X. (2017). Isolation and varietal characterization of *Ganoderma resinaceum* from areas of *Ganoderma lucidum* production in China. *Sci. Hortic.* 224, 109–114. doi: 10.1016/j.scienta.2017.06.002.

Chen, S.Y., Chang, C.L., Chen, T.H., Chang, Y.W., Lin, S.B. (2016b). Colossolactone H, a new *Ganoderma* triterpenoid exhibits cytotoxicity and potentiates drug efficacy of gefitinib in lung cancer. *Fitoterapia*, 114, 81–91.

Chen, Y., Lv, J., Li, K., Xu, J., Li, M., Zhang, W., Pang, X. (2016a). Sporoderm-broken spores of *Ganoderma lucidum* inhibit the growth of lung cancer: Involvement of the Akt/mTOR signaling pathway. *Nutr. Cancer.* 68(7), 1151–1160.

Cheng, L.Z., Qin, F.Y., Ma, X.C., Wang, S.M., Yan, Y.M., Cheng, Y.X. (2018). Cytotoxic and n-acetyltransferase inhibitory meroterpenoids from *Ganoderma* cochlear. *Molecules.* 23(7), 1797.

Chiu, L.Y., Hsin, I.L., Tsai, J.N., Chen, C.J., Ou, C.C., Wu, W.J., Sheu, G.T., Ko, J.L. (2021). Combination treatment of Src inhibitor Saracatinib with GMI, a *Ganoderma microsporum* immunomodulatory protein, induce synthetic lethality via autophagy and apoptosis in lung cancer cells. *J. Cell. Physiol.* 236(2), 1148–1157.

Chiu, L.Y., Hu, M.E., Yang, T.Y., Hsin, I.L., Ko, J.L., Tsai, K.J., Sheu, G.T. (2015). Immunomodulatory protein from *Ganoderma* microsporum induces pro-death autophagy through Akt-mTOR-p70S6K pathway inhibition in multidrug resistant lung cancer cells. *PLoS One.* 10(5), e0125774.

Choudhari, A.S., Mandave, P.C., Deshpande, M., Ranjekar, P., Prakash, O. (2020). Phytochemicals in cancer treatment: From preclinical studies to clinical practice. *Front. Pharmacol.* 10, 1614. doi: 10.3389/fphar.2019.01614.

Dan, X., Liu, W., Wong, J.H., Ng, T.B. (2016). A ribonuclease isolated from wild *Ganoderma lucidum* suppressed autophagy and triggered apoptosis in colorectal cancer cells. *Front Pharmacol.* 7, 217.

Dehelean, C.A., Marcovici, I., Soica, C., Mioc, M., Coricovac, D., Iurciuc, S., Cretu. O.M., Pinzaru, I. (2021). Plant-derived anticancer compounds as new perspectives in drug discovery and alternative therapy. *Molecules.* 26(4), 1109. doi: 10.3390/molecules26041109.

De Melo Gagliato, D., Jardim, D.L.F., Marchesi, M.S.P., Hortobagyi, G.N. (2016). Mechanisms of resistance and sensitivity to anti-HER2 therapies in HER2+ breast cancer. *Oncotarget.* 7(39), 64431.

Deng, Y., Ma, J., Tang, D., Zhang, Q. (2021). Dynamic biomarkers indicate the immunological benefits provided by *Ganoderma* spore powder in post-operative breast and lung cancer patients. *Clin. Transl. Oncol.* 23(7), 1481–1490.

Elkhateeb, W.A., Daba, G.M., Sheir, D., El-Dein, A.N., Fayad, W., Elmahdy, E.M., Shaheen M.N.F., Thomas, P.W., Wen, T.C. (2019). GC-MS analysis and in-vitro hypocholesterolemic, anti-rotavirus, anti-human colon carcinoma activities of the crude extract of a Japanese *Ganoderma* spp. *Egypt Pharm. J.*, 18(2), 102–110.

Elkhateeb, W.A., Zaghlol, G.M., El-Garawani, I.M., Ahmed, E.F., Rateb, M.E., Moneim, A. E.A. (2018). *Ganoderma applanatum* secondary metabolites induced apoptosis through different pathways: In vivo and *in vitro* anticancer studies. *Biomed Pharmacother.* 101, 264–277.

El-Sherif, N.F., Ahmed, S.A., Ibrahim, A.K., Habib, E.S., El-Fallal, A.A., El-Sayed, A.K., Wahba, A.E. (2020). Ergosterol peroxide from the Egyptian red Lingzhi or Reishi mushroom, *Ganoderma resinaceum* (Agaricomycetes), showed preferred inhibition of MCF-7 over MDA-MB-231 breast cancer cell lines. *Int. J. Med. Mushrooms.* 22(4).

Elumalai, D., Suman, T.Y., Hemavathi, M., Swetha, C., Kavitha, R., Arulvasu, C., Kaleena, P.K. (2021). Biofabrication of gold nanoparticles using *Ganoderma lucidum* and their cytotoxicity against human colon cancer cell line (HT-29). *Bull. Mater. Sci.* 44(2), 1–6.

Fang, J.Y., Dong, H.L., Sang, X.J., Xie, B., Wu, K. S., Du, P.L., Xu, Z.X., Jia, X.Y., Lin, K. (2015). Colorectal cancer mortality characteristics and predictions in China, 1991–2011. *Asian Pac. J. Cancer Prev.* 16(17), 7991–7995.

Figueiredo, L., Régis, W.C.B. (2017). Medicinal mushrooms in adjuvant cancer therapies: An approach to anticancer effects and presumed mechanisms of action. *Nutrire.* 42, 28.

Gao, Y.H., Zhou, S.F. (2003). Cancer prevention and treatment by *Ganoderma*, a mushroom with medicinal properties. *Food. Rev. Int.* 19, 275–325. doi: 10.1081/FRI-120023480.

Gill, B.S., Kumar, S., Navgeet, S. (2017). Ganoderic acid targeting nuclear factor erythroid 2-related factor 2 in lung cancer. *Tumor. Biol.* 39(3), 1010428317695530.

Gonul, O., Aydin, H.H., Kalmis, E., Kayalar, H., Ozkaya, A.B., Atay, S., Ak, H. (2015). Effects of *Ganoderma lucidum* (higher basidiomycetes) extracts on the mirna profile and telomerase activity of the MCF-7 breast cancer cell line. *Int. J. Med. Mushrooms.* 17, 231–239. doi: 10.1615/IntJMedMushrooms.v17.i3.30.

Guo, C., Guo, D., Fang, L., Sang, T., Wu, J., Guo, C., Wang, Y., Wang, Y., Chen, C., Chen, J., Chen, R., Wang, X (2021). *Ganoderma lucidum* polysaccharide modulates gut microbiota and immune cell function to inhibit inflammation and tumorigenesis in colon. *Carbohydr. Polym.* 267, 118231.

Guo, J., Yuan, C., Huang, M., Liu, Y., Chen, Y., Liu, C., Chen, Y. (2018). *Ganoderma lucidum*-derived polysaccharide enhances coix oil-based microemulsion on stability and lung cancer-targeted therapy. *Drug. Deliv.* 25(1), 1802–1810.

Gurunathan, S., Raman, J., Abd Malek, S.N., John, P.A., Vikineswary, S. (2013). Green synthesis of silver nanoparticles using *Ganoderma neo-japonicum* imazeki: A potential cytotoxic agent against breast cancer cells. *Int. J. Nanomed.* 8, 4399–4413.

Hildreth, C.J. (2008). Vitamin C and chemotherapy. *JAMA*. 300(21), 2476–2476.

Hsin, I.L., Chiu, L.Y., Ou, C.C., Wu, W.J., Sheu, G.T., Ko, J.L. (2020). CD133 inhibition via autophagic degradation in pemetrexed-resistant lung cancer cells by GMI, a fungal immunomodulatory protein from *Ganoderma microsporum. Br. J. Cancer.* 23(3), 449–458.

Hsin, I.L., Hsu, J.C., Wu, W.J., Lu, H.J., Wu, M.F., Ko, J.L. (2018). GMI, a fungal immunomodulatory protein from *Ganoderma microsporum*, induce apoptosis via β-catenin suppression in lung cancer cells. *Environ. Toxicol.* 33(9), 955–961.

Hsin, I.L., Ou, C.C., Wu, M.F., Jan, M.S., Hsiao, Y.M., Lin, C.H., Ko, J.L. (2015). GMI, an immunomodulatory protein from *Ganoderma microsporum*, potentiates cisplatin-induced apoptosis via autophagy in lung cancer cells. *Mol. Pharm.* 12(5), 1534–1543.

Hsu, W.H., Qiu, W.L., Tsao, S.M., Tseng, A.J., Lu, M.K., Hua, W.J., Cheng, H.C., Hsu, H.Y., Lin, T.Y. (2020). Effects of WSG, a polysaccharide from *Ganoderma lucidum*, on suppressing cell growth and mobility of lung cancer. *Int. J. Biol. Macromol.* 165, 1604–1613.

Hu, H., Ahn, N.S., Yang, X., Lee, Y.S., Kang, K.S. (2002). *Ganoderma lucidum* extract induces cell cycle arrest and apoptosis in MCF-7 human breast cancer cell. *Int. J. Cancer.* 102, 250–253. doi: 10.1002/ijc.10707.

Hu, W., Wang, G., Huang, D., Sui, M., & Xu, Y. (2019). Cancer immunotherapy based on natural killer cells: Current progress and new opportunities. *Front. Immunol.* 10, 1205.

Jeitler, M., Michalsen, A., Frings, D., Hübner, M., Fischer, M., Koppold-Liebscher, D.A., Murthy, V., Kessler, C.S. (2020). Significance of medicinal mushrooms in integrative oncology: A narrative review. *Front. Pharmacol.* 11, 580656.

Jiang, J., Jedinak, A., Sliva, D. (2011). Ganodermanontriol (GDNT) exerts its effect on growth and invasiveness of breast cancer cells through the down-regulation of CDC20 and uPA. *Biochem Biophys Res Commun.* 415, 325–329. doi: 10.1016/j.bbrc.2011.10.055.

Jiang, J., Slivova, V., Harvey, K., Valachovicova, T., Sliva, D. (2004). *Ganoderma lucidum* suppresses growth of breast cancer cells through the inhibition of Akt/Nf-Kappab signaling. *Nutr. Cancer.* 49, 209–216. doi: 10.1207/s15327914nc4902_13.

Jiang, J., Slivova, V., Sliva, D. (2006). *Ganoderma lucidum* inhibits proliferation of human breast cancer cells by down-regulation of estrogen receptor and nf-kappab signaling. *Int. J. Oncol.* 29, 695–703.

Jiao, C., Chen, W., Tan, X., Liang, H., Li, J., Yun, H., He, C., Chen, J., Ma, X., Xie, Y., Yang, B.B. (2020). *Ganoderma lucidum* spore oil induces apoptosis of breast cancer cells *in vitro* and in vivo by activating caspase-3 and caspase-9. *J. Ethnopharmacol.* 247, 112256.

Jin, H., Song, C., Zhao, Z., & Zhou, G. (2020). *Ganoderma lucidum* polysaccharide, an extract from *Ganoderma lucidum*, exerts suppressive effect on cervical cancer cell malignancy through mitigating epithelial-mesenchymal and JAK/STAT5 signaling pathway. *Pharmacology.* 105(7–8), 461–470.

Kim, T.H., Kim, J.S., Kim, Z.H., Huang, R.B., Chae, Y.L., Wang, R.S. (2015). Khz (fusion product of *Ganoderma lucidum* and Polyporus umbellatus mycelia) induces apoptosis in human colon carcinoma HCT116 cells, accompanied by an increase in reactive oxygen species, activation of caspase 3, and increased intracellular Ca2+. *J. Med. Food.* 18(3), 332–336.

Kim, T.H., Kim, J.S., Kim, Z.H., Huang, R.B., Chae, Y. L., Wang, R.S. (2016). Induction of apoptosis in MCF-7 human breast cancer cells by Khz (fusion of *Ganoderma lucidum* and Polyporus umbellatus mycelium). *Mol. Med. Rep.* 13(2), 1243–1249.

Kumaran, S., Pandurangan, A.K., Shenbhagaraman, R., Esa, N.M. (2017). Isolation and characterization of lectin from the artist's conk medicinal mushroom, *Ganoderma applanatum* (Agaricomycetes), and evaluation of its antiproliferative activity in HT-29 colon cancer cells. *Int. J. Med. Mushrooms.* 19(8).

Kuo, H.P., Hsu, S.C., Ou, C.C., Li, J.W., Tseng, H.H., Chuang, T.C., Liu, J.Y., Chen, S.J., Su, M.H., Cheng, Y.C., Chou, W.Y., Kao, M.C. (2013). *Ganoderma tsugae* extract inhibits growth of her2-overexpressing cancer cells via modulation of HER2/PI3K/Akt signaling pathway. *Evid. Based Complement Alternat. Med.* 2013, 219472. doi: 10.1155/2013/219472.

Lau, M.F., Chua, K.H., Sabaratnam, V., Kuppusamy, U. R. (2022). *In vitro* Anti-colorectal cancer potential of the medicinal mushroom *Ganoderma neo-japonicum* Imazeki in hyperglycemic condition: Impact on oxidative stress, cell cycle and apoptosis. *Nutr. Cancer.* 74(3), 978–995.

Li, F., Wang, Y., Wang, X., Li, J., Cui, H., Niu, M. (2012). Ganoderic acids suppress growth and angiogenesis by modulating the Nf-kappab signaling pathway in breast cancer cells. *Int. J. Clin. Pharmacol. Ther.* 50, 712–721. doi: 10.5414/CP201663.

Li, K., Na, K., Sang, T., Wu, K., Wang, Y., Wang, X. (2017). The ethanol extracts of sporoderm-broken spores of *Ganoderma lucidum* inhibit colorectal cancer *in vitro* and in vivo. *Oncol Rep.* 38(5), 2803–2813.

Li, X., Xie, Y., Peng, J., Hu, H., Wu, Q., Yang, B.B. (2019). Ganoderiol F purified from *Ganoderma* leucocontextum retards cell cycle progression by inhibiting CDK4/CDK6. *Cell Cycle*, 18(21), 3030–3043.

Li, Y.P., Jiang, X.T., Qin, F.Y., Zhang, H.X., Cheng, Y.X. (2021). Gancochlearols E– I, meroterpenoids from *Ganoderma* cochlear against COX-2 and triple negative breast cancer cells and the absolute configuration assignment of ganomycin K. *Bioorg. Chem.* 109, 104706.

Liang, Z.E., Yi, Y.J., Guo, Y.T., Wang, R.C., Hu, Q.L., Xiong, X. Y. (2015). Inhibition of migration and induction of apoptosis in LoVo human colon cancer cells by polysaccharides from *Ganoderma lucidum. Mol. Med. Rep.* 12(5), 7629–7636.

Liao, S.F., Liang, C.H., Ho, M.Y., Hsu, T.L., Tsai, T.I., Hsieh, Y.S., Tsai, C.M., Li, S.T., Cheng, Y.Y., Tsao, S.M., Lin, T.Y., Lin, Z.Y., Yang, W.B., Ren, C.T., Lin, K.I., Khoo, K.H., Lin, C.H., Hsu, H.Y., Wu, C.Y., Wong, C.H. (2013). Immunization of fucose-containing polysaccharides from Reishi mushroom induces antibodies to tumor-associated globo H-series epitopes. *Proc. Natl. Acad. Sci. USA.* 110, 13809–13814. doi: 10.1073/pnas.1312457110.

Lin, T.Y., Hsu, H.Y. (2016). Ling Zhi-8 reduces lung cancer mobility and metastasis through disruption of focal adhesion and induction of MDM2-mediated Slug degradation. *Cancer Lett.* 375(2), 340–348.

Lin, T.Y., Hsu, H.Y., Sun, W.H., Wu, T.H., Tsao, S.M. (2017). Induction of CBL-dependent epidermal growth factor receptor degradation in Ling Zhi-8 suppressed lung cancer. *Int. J. Cancer.* 140(11), 2596–2607.

Liu, W., Yuan, R., Hou, A., Tan, S., Liu, X., Tan, P., Huang, X., Wang, J. (2020). *Ganoderma* triterpenoids attenuate tumor angiogenesis in lung cancer tumor-bearing nude mice. *Pharm. Biol.* 58(1), 1070–1077.

Liu, Y.W., Gao, J.L., Guan, J., Qian, Z.M., Feng, K., Li, S.P. (2009). Evaluation of antiproliferative activities and action mechanisms of extracts from two species of *Ganoderma* on tumor cell lines. *J. Agric. Food Chem.* 57, 3087–3093. doi: 10.1021/jf900011f.

Liu, Z., Li, L., Xue, B. (2018). Effect of ganoderic acid D on colon cancer Warburg effect: Role of SIRT3/cyclophilin D. *Eur. J. Pharmacol.* 824, 72–77.

Luo, J., Zhang, C., Liu, R., Gao, L., Ou, S., Liu, L., Peng, X. (2018). *Ganoderma lucidum* polysaccharide alleviating colorectal cancer by alteration of special gut bacteria and regulation of gene expression of colonic epithelial cells. *J. Func. Foods.* 47, 127–113.

Martinez-Montemayor, M.M., Acevedo, R.R., Otero-Franqui, E., Cubano, L.A., Dharmawardhane, S.F. (2011). *Ganoderma lucidum* (Reishi) inhibits cancer cell growth and expression of key molecules in inflammatory breast cancer. *Nutr. Cancer.* 63, 1085–1094. doi: 10.1080/01635581.2011.601845.

Na, K., Li, K., Sang, T., Wu, K., Wang, Y., Wang, X. (2017). Anticarcinogenic effects of water extract of sporoderm-broken spores of *Ganoderma lucidum* on colorectal cancer *in vitro* and in vivo. *Int. J. Oncolo.* 50(5), 1541–1554.

Newman, D.J., Cragg, G.M. (2016). Natural products as sources of new drugs from 1981 to 2014. *J. Nat. Prod.* 79(3), 629–661. doi: 10.1021/acs.jnatprod.5b01055.

Opattova, A., Horak, J., Vodenkova, S., Kostovcikova, K., Cumova, A., Macinga, P., . . . Vodicka, P. (2019). *Ganoderma lucidum* induces oxidative DNA damage and enhances the effect of 5-Fluorouracil in colorectal cancer *in vitro* and in vivo. *Mutat. Res. Genet. Toxicol. Environ. Mutagen.* 845, 403065.

Panda, S.K., Sahoo, G., Swain, S.S., Luyten, W. (2022). Anticancer activities of mushrooms: A neglected source for drug discovery. *Pharmaceuticals (Basel).* 15(2), 176. doi: 10.3390/ph15020176.

Park, S.Y., Choi, J.H., Nam, J.S. (2019). Targeting cancer stem cells in triple-negative breast cancer. *Cancers (Basel).* 11(7), 965. doi: 10.3390/cancers11070965.

Peng, X., Liu, J., Xia, J., Wang, C., Li, X., Deng, Y., Bao, N., Zhang, Z., Qiu, M. (2015). Lanostane triterpenoids from *Ganoderma hainanense* J. D. Zhao. *Phytochemistry.* 114, 137–145. doi: 10.1016/j.phytochem.2014.10.009.

Qin, F.Y., Chen, Y.Y., Zhang, J.J., Cheng, Y.X. (2022). Meroterpenoid dimers from *Ganoderma* mushrooms and their biological activities against triple negative breast cancer cells. *Front. Chem.* 10.

Qiu, W.L., Hsu, W.H., Tsao, S.M., Tseng, A.J., Lin, Z.H., Hua, W.J., Yeh, H., Lin, T.E., Chen, C.C., Chen, L.S., Lin, T.Y. (2021). WSG, a glucose-rich polysaccharide from *Ganoderma lucidum*, combined with cisplatin potentiates inhibition of lung cancer *in vitro* and in vivo. *Polymers.* 13(24), 4353.

Rios-Fuller, T.J., Ortiz-Soto, G., Lacourt-Ventura, M., Maldonado-Martinez, G., Cubano, L.A., Schneider, R.J., Martinez-Montemayor, M.M. (2018). *Ganoderma lucidum* extract (GLE) impairs breast cancer stem cells by targeting the STAT3 pathway. *Oncotarget.* 9(89), 35907.

Sedky, N.K., El Gammal, Z.H., Wahba, A.E., Mosad, E., Waly, Z.Y., El-Fallal, A.A., Arafa, R.K., El-Badri, N. (2018). The molecular basis of cytotoxicity of α-spinasterol from *Ganoderma resinaceum*: Induction of apoptosis and overexpression of p53 in breast and ovarian cancer cell lines. *J. Cell. Biochem,* 119(5), 3892–3902.

Shang, D., Li, Y., Wang, C., Wang, X., Yu, Z., Fu, X. (2011). A novel polysaccharide from Se-enriched *Ganoderma lucidum* induces apoptosis of human breast cancer cells. *Oncol. Rep.* 25, 267–272. doi: 10.3892/or_00001070.

Shang, D., Zhang, J., Wen, L., Li, Y., Cui, Q. (2009). Preparation, characterization, and antiproliferative activities of the Se-containing polysaccharide SeGLP-2B-1 from Se-enriched *Ganoderma lucidum. J. Agric. Food Chem.* 57, 7737–7742. doi: 10.1021/jf9019344.

Sharma, C., Bhardwaj, N., Sharma, A., Tuli, H.S., Batra, P., Beniwal, V., Gupta, G.K., Sharma, A.K. (2019). Bioactive metabolites of *Ganoderma lucidum*: Factors, mechanism and broad-spectrum therapeutic potential. *J. Herb. Med.* 17, 100268.

Shi, L., Ren, A., Mu, D., Zhao, M. (2010). Current progress in the study on biosynthesis and regulation of ganoderic acids. *Appl. Microbiol Biotechnol.* 88, 1243–1251. doi: 10.1007/s00253-010-2871-1.

Siegel, R.L., Miller, K.D., Fuchs, H.E., Jemal, A. (2022). Cancer statistics, 2022. *CA Cancer J. Clin.* 72(1), 7–33. doi: 10.3322/caac.21708.

Suárez-Arroyo, I.J., Acevedo-Díaz, A., Ríos-Fuller, T.J., Ortiz-Soto, G., Vallejo-Calzada, R., Reyes-Chea, J., Maldonado-Martínez, G., Schneider, R.J., Martínez-Montemayor, M.M. (2022). *Ganoderma lucidum* enhances carboplatin chemotherapy effect by inhibiting the DNA damage response pathway and stemness. *Am. J. Cancer Res.* 12(3), 1282.

Suárez-Arroyo, I.J., Loperena-Alvarez, Y., Rosario-Acevedo, R., Martínez-Montemayor, M.M. (2017). *Ganoderma* spp.: A promising adjuvant treatment for breast cancer. *Medicines (Basel).* 4(1) (March), 15. doi: 10.3390/medicines4010015.

Suárez-Arroyo, I.J., Rios-Fuller, T.J., Feliz-Mosquea, Y.R., Lacourt-Ventura, M., Leal-Alviarez, D.J., Maldonado-Martinez, G., Cubano, L.A., Martínez-Montemayor, M.M. (2016). *Ganoderma lucidum* combined with the EGFR tyrosine kinase inhibitor, erlotinib synergize to reduce inflammatory breast cancer progression. *J. Cancer.* 7, 500–511. doi: 10.7150/jca.13599.

Suarez-Arroyo, I.J., Rosario-Acevedo, R., Aguilar-Perez, A., Clemente, P.L., Cubano, L.A., Serrano, J., Schneider, R.J., Martínez-Montemayor, M.M. (2013). Anti-tumor effects of *Ganoderma lucidum* (Reishi) in inflammatory breast cancer in in vivo and *in vitro* models. *PLoS One.* 8, e57431. doi: 10.1371/journal.pone.0057431.

Tang, L., Zhu, Z.F., Cao, L.P., Shen, M., Gao, Y., Tu, C.J., Zhang, Z.H., Shan, W.G. (2020). Thermosensitive gel of polysaccharide from *Ganoderma* applanatum combined with paclitaxel for mice with 4T1 breast cancer. *Zhongguo Zhong Yao Za Zhi.* 45(11), 2533–2539.

Tsao, S.M., Hsu, H.Y. (2016). Fucose-containing fraction of Lingzhi enhances lipid rafts-dependent ubiquitination of TGFβ receptor degradation and attenuates breast cancer tumorigenesis. *Sci. Rep.* 6. doi: 10.1038/srep36563.

Unlu, A., Nayir, E., Kirca, O., and Ozdogan, M. (2016). *Ganoderma lucidum* (Reishi mushroom) and cancer. *J. Buon.* 21, 792–798.

Vaithanomsat, P., Boonlum, N., Chaiyana, W., Tima, S., Uchapreeda, S., Trakunjae, C., Apiwatanapiwat, W., Janchai, P., Boondaeng, A., Nimitkeatkai, H., Jarerat, A. (2022). Mushroom Anβ-Glucan recovered from antler-type fruiting body of *Ganoderma lucidum* by enzymatic process and its potential biological activities for cosmeceutical applications. *Polymers.* 14(19), 4202.

Wang, C., Shi, S., Chen, Q., Lin, S., Wang, R., Wang, S., Chen, C. (2018). Antitumor and immunomodulatory activities of *Ganoderma lucidum* polysaccharides in glioma-bearing rats. *Integr. Cancer Ther.* 17(3), 674–683.

Wang, J., Cao, B., Zhao, H., Feng, J. (2017). Emerging roles of *Ganoderma lucidum* in anti-aging. *Aging Dis.* 8(6), 691.

Wang, L., Li, J. Q., Zhang, J., Li, Z. M., Liu, H. G., Wang, Y. Z. (2020). Traditional uses, chemical components and pharmacological activities of the genus *Ganoderma* P. Karst.: A review. *RSC Adv.* 10(69), 42084–42097.

Wang, W., Gou, X., Xue, H., Liu, K. (2019). Ganoderan (GDN) regulates the growth, motility and apoptosis of non-small cell lung cancer cells through ERK signaling pathway *in vitro* and in vivo. *OncoTargets Ther.* 12, 8821–8832. doi: 10.2147/OTT.S221161.

Wasser, S.P. (2017). Medicinal mushrooms in human clinical studies. Part I. Anticancer, oncoimmunological, and immunomodulatory activities: A review. *Int. J. Med. Mushrooms.* 19, 279–317.

Wu, G.S., Lu, J.J., Guo, J.J., Li, Y.B., Tan, W., Dang, Y.Y., Zhong, Z.F., Xu, Z.T., Chen, X.P., Wang, Y.T. (2012a). Ganoderic acid DM, a natural triterpenoid, induces DNA damage, G1 cell cycle arrest and apoptosis in human breast cancer cells. *Fitoterapia.* 83, 408–414. doi: 10.1016/j.fitote.2011.12.004.

Wu, G.S., Qian, Z., Guo, J., Hu, D., Bao, J., Xie, J., Xu, W., Lu, J., Chen, X., Wang, Y. (2012b). *Ganoderma lucidum* extract induces G1 cell cycle arrest, and apoptosis in human breast cancer cells. *Am. J. Chin. Med.* 40, 631–642. doi: 10.1142/S0192415X12500474.

Xu, H., Kong, Y.Y., Chen, X., Guo, M.Y., Bai, X.H., Lu, Y.J., Li, W., Zhou, X.W. (2016). Recombinant FIP-gat, a fungal immunomodulatory protein from *Ganoderma* atrum, induces growth inhibition and cell death in breast cancer cells. *J. Agri. Food Chem.* 64(13), 2690–2698.

Yang, Y., Li, N., Wang, T.M., Di, L. (2021). Natural products with activity against lung cancer: A review focusing on the tumor microenvironment. *Int. J. Mol. Sci.* 22(19), 10827. doi: 10.3390/ijms221910827.

Yang, Y., Nirmagustina, D.E., Kumrungsee, T., Okazaki, Y., Tomotake, H., Kato, N. (2017). Feeding of the water extract from *Ganoderma lingzhi* to rats modulates secondary bile acids, intestinal microflora, mucins, and propionate important to colon cancer. *Biosci Biotechnol Biochem.* 81(9), 1796–1804.

Yang, Y., Zhou, H., Liu, W., Wu, J., Yue, X., Wang, J., Lian, X., Zhang, Q. (2018). Ganoderic acid A exerts antitumor activity against MDA-MB-231 human breast cancer cells by inhibiting the Janus kinase 2/signal transducer and activator of transcription 3 signaling pathway. *Oncol Lett.* 16(5), 6515–6521.

Yue, G.G., Fung, K.P., Tse, G.M., Leung, P.C., Lau, C.B. (2006). Comparative studies of various *Ganoderma* species and their different parts with regard to their antitumor and immunomodulating activities *in vitro*. *J. Altern. Complement. Med.* 12, 777–789. doi: 10.1089/acm.2006.12.777.

Yuen, J.W., Gohel, M.D. (2005). Anticancer effects of *Ganoderma lucidum*: A review of scientific evidence. *Nutr. Cancer.* 53, 11–17. doi: 10.1207/s15327914nc5301_2.

Zhang, K., Liu, Y., Zhao, X., Tang, Q., Dernedde, J., Zhang, J., Fan, H. (2018). Anti-inflammatory properties of GLPss58, a sulfated polysaccharide from *Ganoderma lucidum*. *Int. J. Biol. Macromol.* 107, 486–493.

Zhen, D., Su, L., Miao, Y., Zhao, F., Ren, G., Mahfuz, S., Song, H. (2018). Purification, partial characterization and inducing tumor cell apoptosis activity of a polysaccharide from *Ganoderma applanatum*. *Int. J. Biol. Macromol.* 115, 10–17.

Zhi, X., Kuang, X.H., Liu, K., Li, J. (2022). The global burden and temporal trend of cancer attributable to high body mass index: Estimates from the global burden of disease study 2019. *Front. Nutr.* 9, 918330. doi: 10.3389/fnut.2022.918330.

11 Hypolipidemic and Cholesterol-Lowering Effects of *Ganoderma*

Aloke Saha[1] *and Somanjana Khatua*[2]
1 University of Kalyani, Nadia, India
2 University of Allahabad, Prayagraj, India

11.1 INTRODUCTION

Increased lipid levels and excessive amounts of cholesterol pose significant global health issues, greatly impacting cardiovascular well-being. As per the World Health Organization (WHO), cardiovascular ailments are the primary cause of mortality worldwide, resulting in 17.9 million fatalities in 2019 (Lopez-Jimenez et al. 2022). Elevated total cholesterol concentrations in the bloodstream constitute a substantial risk factor for cardiovascular diseases like coronary heart disease and stroke (Peters et al. 2016). It is estimated that high cholesterol levels contribute to roughly 4.4 million deaths each year across the globe (World Heart Federation 2023). In India, the burden of cardiovascular ailments is also substantial, with an estimated 2.8 million deaths in 2016. The Indian population exhibits a considerable prevalence of high cholesterol levels, affecting around 40% of adults aged 30–69 years (Gupta et al. 2017). Various factors contribute to the widespread occurrence of hyperlipidemia and elevated cholesterol levels in India, including the consumption of unhealthy diets, sedentary lifestyles, and genetic predisposition (Sarat Chandra et al. 2014).

Despite the presence of effective medications, adjustments in lifestyle, and dietary interventions, the control of hyperlipidemia and elevated cholesterol levels remains a notable challenge worldwide, including in India. Furthermore, the adverse effects linked to pharmacological interventions have prompted an amplified demand for alternative treatments (Grundy et al. 2019). The potential impact of *Ganoderma*, a conventional medicinal fungus, in reducing lipid levels and cholesterol has aroused substantial interest among researchers and the general public (Berger et al. 2004). The exploration of alternative treatments such as *Ganoderma* to manage hyperlipidemia and high cholesterol levels is crucial to address the mounting burden of cardiovascular diseases globally (Klupp et al. 2015). *Ganoderma*, also known as lingzhi in China and reishi in Japan, is a medicinal mushroom that has been employed in traditional Chinese medicine for centuries. It comprises diverse bioactive compounds, including polysaccharides, triterpenes, and sterols, which have exhibited numerous pharmacological activities, such as immunomodulatory, antioxidant, and anticancer effects (Oke et al. 2022). A more profound comprehension of the effectiveness and safety of *Ganoderma* in handling hyperlipidemia and high cholesterol levels could pave the way for development of novel therapeutic alternatives for individuals affected by these conditions (El Sheikha 2022).

Regarding cultivation, *Ganoderma* can be grown both indoors and outdoors (Zhou et al. 2012). It is feasible to cultivate the mushroom on a small scale by individual farmer or on a larger scale by commercial farms (Li and Hu 2014; Grimm and Wösten 2018). The cultivation process involves growing *Ganoderma* on a substrate like sawdust or grain in a controlled environment. One advantage of *Ganoderma* cultivation is its compatibility with low-cost materials and equipment, making it a viable option for small-scale farmers lacking access to expensive facilities and resources. Furthermore, the cultivation process can be easily adapted to suit local conditions, rendering it a sustainable and environmentally friendly choice for farmers (Zhou 2017; Azizi et al. 2012). In terms

DOI: 10.1201/9781003354789-11

of industrial production, *Ganoderma* can be cultivated in large-scale facilities utilizing advanced technologies like bioreactors. This approach enhances production efficiency and delivers a consistent, high-quality product. Industrial production of *Ganoderma* can meet the increasing demand for functional foods and pharmaceuticals while contributing to the advancement of the bioeconomy (Seethapathy et al. 2023; Hu et al. 2018; Ye et al. 2018).

The main aim of this review is to assess the existing proof concerning the hypolipidemic and cholesterol-reducing impact of *Ganoderma*. The review will delve into the mechanisms through which *Ganoderma* operates, its possible therapeutic advantages, and any unfavorable consequences linked to its usage. Additionally, the review will address the constraints of the current evidence and pinpoint areas that necessitate further research.

11.2 LIPID METABOLISM PROCESS AND REASONS BEHIND HYPERCHOLESTEROLEMIA

Lipid metabolism encompasses the intricate progression of lipids, encompassing fats, oils, waxes, sterols, and phospholipids, involving their synthesis, degradation, and conveyance. Lipids fulfill crucial functions within the body, serving as an energy reservoir, a structural element of cell membranes, and a precursor for the production of hormones and other signal-carrying molecules (De Carvalho and Caramujo 2018). The course of lipid metabolism entails various stages, encompassing digestion, assimilation, transportation, and utilization (Olsen et al. 2021). Dietary lipids are initially disassembled by digestive enzymes in the small intestine, leading to the formation of constituent fatty acids and glycerol, which are subsequently absorbed into the bloodstream and conveyed to the liver (Iqbal and Hussain 2009).

In the liver, fatty acids undergo either oxidation to produce energy or undergo the process of triglyceride synthesis, after which they are packaged into particles called very low-density lipoproteins (VLDLs) and discharged into the bloodstream. Lipoprotein lipase (LPL) metabolizes VLDL particles in peripheral tissues, including muscle and adipose tissue, where they are utilized for energy or stored as triglycerides (Alves-Bezerra and Cohen 2017). Once stored in adipose tissue, triglycerides are broken down by lipases to liberate free fatty acids, which are then transported to other tissues for energy generation or utilized in the production of other lipids, like phospholipids and cholesterol (Ahmadian et al. 2007).

Cholesterol, a vital type of lipid molecule, plays a crucial role in various biological processes within the body. It acts as a key constituent of cell membranes, preserving their stability and fluidity (Yang et al. 2016). Additionally, cholesterol serves as a precursor for the synthesis of significant molecules like steroid hormones, bile acids, and vitamin D (Morzycki 2014). While the liver is the primary site for cholesterol production, other tissues, such as the adrenal glands and intestines, can also synthesize cholesterol to some degree (Arnold and Kwiterovich 2003). The process of cholesterol synthesis involves a series of enzymatic reactions that transform mevalonic acid, a precursor molecule, into cholesterol. The initial step is the conversion of acetyl-CoA to 3-hydroxy-3-methylglutaryl-CoA (HMG-CoA) by the enzyme HMG-CoA synthase (Hu et al. 2010). HMG-CoA is then converted to mevalonic acid by HMG-CoA reductase, the rate-limiting enzyme in cholesterol synthesis and the target of cholesterol-lowering medications called statins. Mevalonic acid further undergoes enzymatic reactions along the mevalonate pathway to generate isopentenyl pyrophosphate (IPP) and dimethylallyl pyrophosphate (DMAPP) (Friesen and Rodwell 2004). These compounds serve as precursors for the synthesis of various molecules, including cholesterol. Subsequently, six molecules of IPP and one molecule of DMAPP condense to form squalene, which undergoes a series of enzymatic reactions to transform into lanosterol (Karlic and Varga 2017). Lanosterol is then modified through the removal of three methyl groups and the introduction of a double bond to yield cholesterol (Nes 2011). The regulation of cholesterol synthesis is influenced by factors such as dietary intake, hormonal signals, and feedback inhibition. High dietary cholesterol levels can decrease cholesterol synthesis in the liver, whereas low dietary cholesterol levels can increase synthesis (Trapani 2012). Hormonal signals, including insulin and thyroid hormones, also impact cholesterol synthesis. Furthermore, cholesterol acts as a feedback inhibitor of HMG-CoA reductase, maintaining cholesterol levels within a narrow range (Duan et al. 2022).

Hypercholesterolemia denotes a condition marked by excessively elevated cholesterol levels in the bloodstream. Heightened cholesterol levels pose a substantial hazard for the development of cardiovascular ailments, which stand as the primary cause of mortality on a global scale (Soran et al. 2018). Cholesterol exists in two variants: low-density lipoprotein (LDL) and high-density lipoprotein (HDL). LDL, often labeled "bad" cholesterol, has the propensity to accumulate within artery walls, forming plaques that may lead to the narrowing and hardening of arteries, a condition known as atherosclerosis (Rafieian-Kopaei et al. 2014). Conversely, HDL, often recognized as "good" cholesterol, plays a crucial role in eliminating surplus cholesterol from the bloodstream, facilitating its transport back to the liver for disposal (Daniels et al. 2009).

Multiple factors can contribute to the emergence of hypercholesterolemia, including genetic predisposition, dietary habits, and lifestyle choices. Familial hypercholesterolemia represents a hereditary condition distinguished by elevated levels of LDL cholesterol and an augmented susceptibility to premature cardiovascular ailments (Ison et al. 1993). Genetic mutations passed down through generations, particularly in genes governing cholesterol metabolism, such as the LDL receptor gene, can result in the accumulation of LDL particles within the circulatory system (Lui et al. 2021).

Dietary elements that can contribute to hypercholesterolemia encompass the intake of foods rich in saturated and trans-fats, like red meat, butter, and fried cuisine. These types of fats escalate LDL cholesterol levels and diminish HDL cholesterol levels (Soliman 2018). In contrast, adopting a heart-healthy eating regimen, such as the Mediterranean diet, which is low in saturated and trans fats and high in fiber, has demonstrated substantial efficacy in reducing LDL cholesterol levels and mitigating the risk of cardiovascular disease (Wickman et al. 2021). Lifestyle elements that can contribute to hypercholesterolemia consist of physical inactivity, smoking, and obesity. Engaging in regular physical exercise can elevate HDL cholesterol levels and lower LDL cholesterol levels, while smoking and obesity have been found to increase LDL cholesterol levels and reduce HDL cholesterol levels (Mannu et al. 2013).

11.3 MEDICATIONS USED FOR TREATMENT OF HYPERCHOLESTEROLEMIA AND THEIR ADVERSE EFFECTS

Hypercholesterolemia refers to a state marked by elevated cholesterol levels within the bloodstream. It stands as a significant determinant for cardiovascular ailments, including heart attacks and strokes (Nelson 2013). Numerous medications exist for managing hypercholesterolemia, each exhibiting distinct rates of effectiveness and potential undesirable consequences (Table 11.1).

Statins represent the most frequently prescribed medications for managing hypercholesterolemia. Their mechanism of action involves impeding the activity of HMG-CoA reductase, an enzyme engaged in cholesterol synthesis within the liver. Through curtailing cholesterol production, statins have the capability to diminish LDL cholesterol levels in the bloodstream (Sizar et al. 2023). Research has demonstrated that statins can mitigate the risk of heart attack, stroke, and other cardiovascular incidents. The effectiveness of statins may vary, depending on the specific medication, dosage, and characteristics of the patient population, yet they have the potential to reduce LDL cholesterol levels by 20–60% (Rossini et al. 2022). Common undesirable outcomes associated with statin use include muscular discomfort, liver impairment, and an augmented likelihood of developing diabetes (Jose 2016).

Ezetimibe functions by impeding the assimilation of cholesterol in the small intestine. It can be utilized either independently or in conjunction with statins to further diminish LDL cholesterol levels (Toth et al. 2012). Although the effectiveness of ezetimibe is comparatively lower than that of statins, it can still reduce LDL cholesterol levels by approximately 15–20%. Diarrhea and

TABLE 11.1
Adverse Effects and Mode of Action of Some Anti-Hypercholesterolemic Drugs

Class	Representative Drugs	Chemical Structure	Mode of Action	Adverse Effects	Refs.
Statins	Atorvastatin	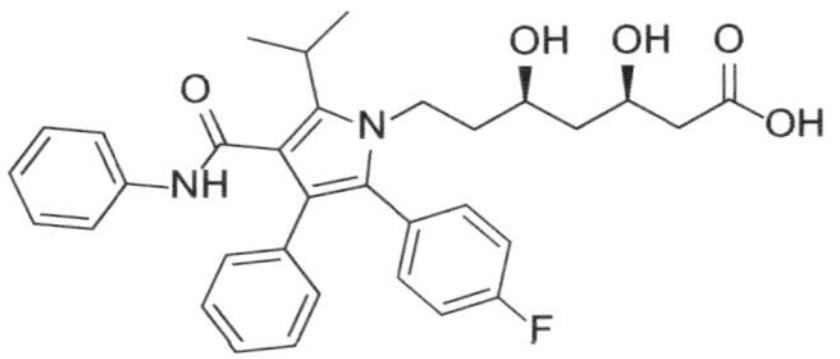	Inhibit HMG-CoA reductase enzyme	Muscle pain, liver damage, gastrointestinal issues	(Ramkumar et al. (2016)
	Simvastatin	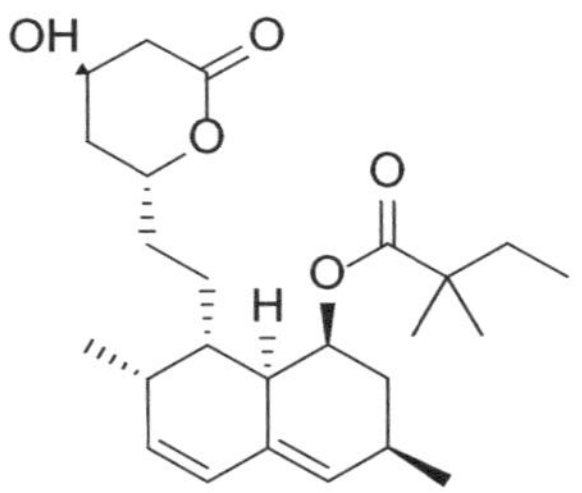			
Bile acid sequestrants	Cholestyramine	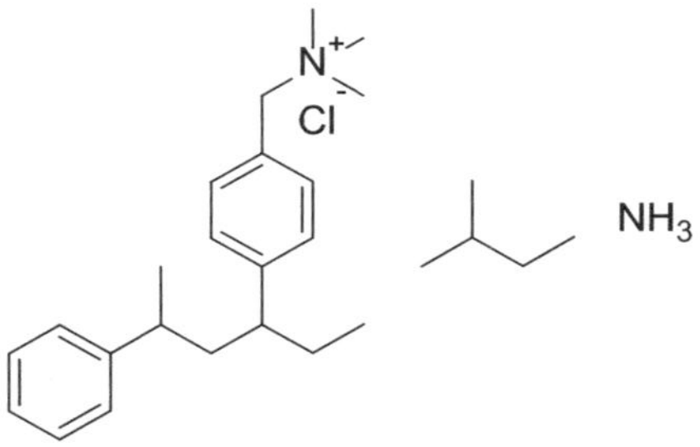	Bind to bile acids and prevent reabsorption	Constipation, bloating, vitamin deficiencies	(Maíz et al. (1990)

Ezetimibe	Ezetimibe	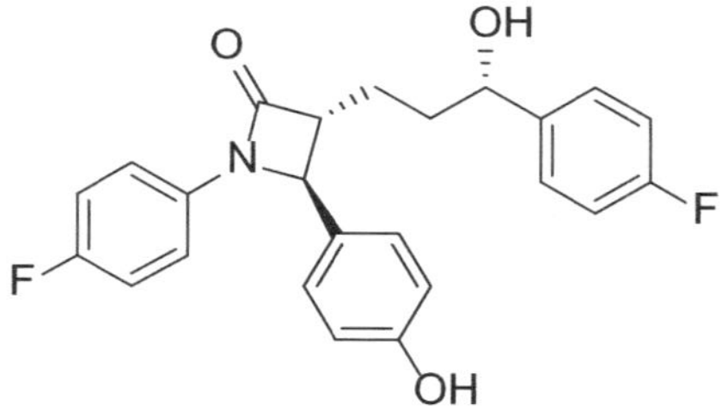	Inhibits intestinal cholesterol absorption	Diarrhea, abdominal pain, allergic reactions	(Florentin et al. (2007)
PCSK9 inhibitors	Evolocumab, alirocumab	–	Inhibits PCSK9 and increases LDL receptor	Injection site reactions, flu-like symptoms	(Gürgöze et al. (2019)
Fibrates	Fenofibrate	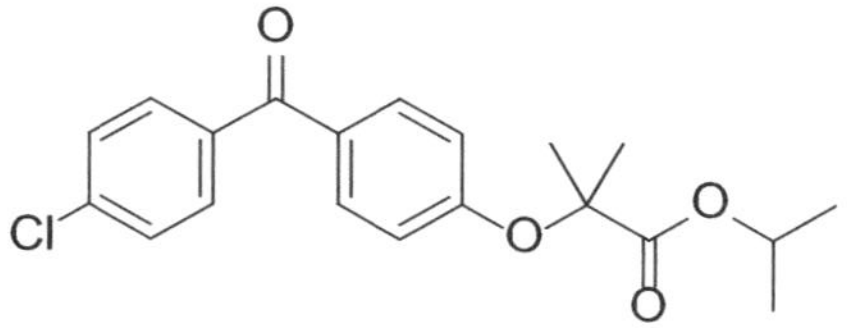	Activates PPAR-alpha and decreases triglycerides	Upset stomach, gallstones, liver problems	(Staels et al. (2010)
	Gemfibrozil	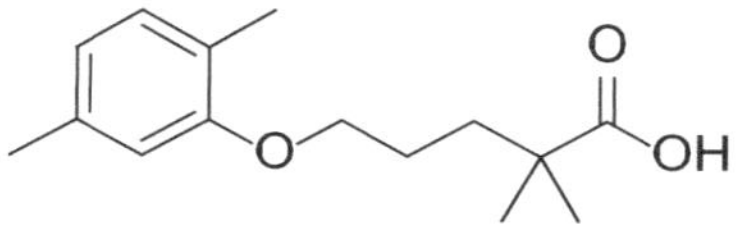			
Niacin	Niacin	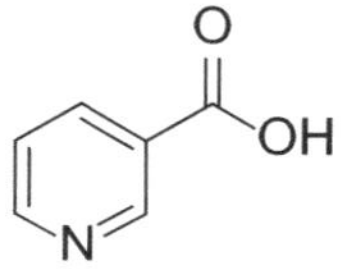	Decreases VLDL production and increases HDL	Flushing, itching, liver damage	(Žák (2015)
Omega-3 fatty acid derivatives	Icosapent ethyl	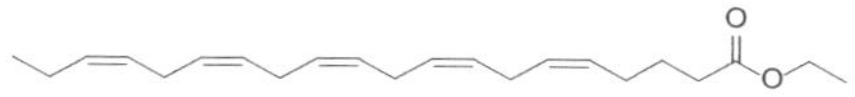	Reduce triglyceride synthesis	Fishy aftertaste, indigestion, bleeding disorders	(Sherratt et al. (2023)

Note: All chemical structures have been drawn using ChemDraw software.

abdominal pain are common undesired consequences associated with ezetimibe usage (Mikhailidis et al. 2011).

Proprotein convertase subtilisin/kexin type 9 (PCSK9) inhibitors represent a novel category of medications that operate by inhibiting a protein implicated in the degradation of LDL receptors in the liver. By augmenting the quantity of LDL receptors, PCSK9 inhibitors can curtail LDL cholesterol levels in the bloodstream (Chaudhary et al. 2017). These inhibitors have demonstrated efficacy in diminishing the risk of cardiovascular events, although their cost is higher than other cholesterol-lowering medications (Lepor and Kereiakes 2015). The effectiveness of PCSK9 inhibitors is notable, with certain medications capable of reducing LDL cholesterol levels by up to 60%. Injection site reactions and flu-like symptoms are frequently encountered as adverse effects of PCSK9 inhibitors (Stoekenbroek et al. 2018).

Bile acid sequestrants operate by attaching to bile acids within the small intestine, hindering their reabsorption. As a result, cholesterol metabolism in the liver intensifies, leading to diminished LDL cholesterol levels in the bloodstream (Staels et al. 2010). The effectiveness of bile acid sequestrants is relatively lower than that of statins; nevertheless, they can still reduce LDL cholesterol levels by approximately 10–20%. Common unfavorable effects associated with bile acid sequestrants include constipation, bloating, and excessive gas production (Insull 2006).

Fibrates, on the other hand, function by stimulating a receptor in the liver responsible for breaking down triglycerides and generating HDL cholesterol. They can also have a modest impact on lowering LDL cholesterol levels (Staels et al. 1998). Fibrates have exhibited a capacity to reduce the likelihood of cardiovascular incidents among individuals with elevated triglyceride levels, although their effectiveness in reducing LDL cholesterol levels is comparatively less potent than other medications (Jacobson and Zimmerman 2006). Common adverse effects of fibrates encompass muscular discomfort, liver impairment, and an increased susceptibility to gallstones (Elisaf et al. 2023).

In cases where severe hypercholesterolemia persists despite lifestyle adjustments and medication usage, LDL apheresis may be suggested as a potential course of action. LDL apheresis entails the extraction of LDL cholesterol from the bloodstream using a specialized machine, akin to the principles of dialysis (Bambauer et al. 2012).

11.4 NATUROPATHIC TREATMENT AS AN ALTERNATIVE OPTION FOR TREATMENT OF HYPERCHOLESTEROLEMIA

In addition to their positive effects, synthetic medications also carry a range of potential unfavorable impacts. Muscle pain or weakness, liver damage, digestive issues, neurological effects, an increased risk of diabetes, and interactions with other medications are among the possible adverse effects associated with these cholesterol-lowering medications (Ward et al. 2019). Statins, for instance, may induce muscle pain, cramps, or weakness in certain individuals, impairing their ability to engage in physical activity and perform daily tasks (Di Stasi et al. 2010). In rare instances, statins have been linked to memory loss, confusion, or other neurological symptoms. Studies have also suggested a potential elevated risk of developing diabetes with long-term statin use (Mach et al. 2018). Consequently, naturopathic treatment serves as an alternative option to assist in managing hypercholesterolemia. Phytochemicals, which are plant-derived compounds with therapeutic effects on the body, have been discovered to possess cholesterol-lowering properties (Santini and Novellino 2017).

Plant sterols and stanols, for instance, are nature-derived compounds that share a structural resemblance to cholesterol. They can diminish the absorption of dietary cholesterol in the intestine by competing with cholesterol for uptake by intestinal cells (Plat et al. 2019). Through this mechanism, plant sterols and stanols aid in lowering LDL cholesterol levels, which tend to be elevated in individuals with hypercholesterolemia (Trautwein et al. 2018). Numerous plant-based foods, including fruits, vegetables, nuts, and seeds, contain plant sterols and stanols. Achieving the recommended daily intake of approximately 2 grams can be accomplished by consuming fortified foods or supplements (Trautwein and McKay 2020).

Soluble fiber, present in oats, barley, legumes, and fruits, can aid in the reduction of cholesterol by limiting its absorption in the gastrointestinal tract (Lattimer and Haub 2010). When soluble fiber interacts with bile acids derived from cholesterol, it forms a gel-like substance that impedes their reabsorption into the bloodstream, leading to their elimination from the body (Jesch and Carr 2017). Consequently, the liver increases cholesterol utilization to produce more bile acids, leading to reduction in the overall levels of cholesterol. It is recommended to consume approximately 25–30 grams of fiber per day, which can be achieved by incorporating a variety of fruits, vegetables, whole grains, and legumes into the diet (Soliman 2019).

Polyphenols, a group of plant-based compounds known for their antioxidant properties, have the potential to alleviate inflammation within the body (Zhang et al. 2022). Some specific types of polyphenols have been found to possess cholesterol-lowering effects. For instance, flavonoids, which are present in tea, cocoa, and berries, have been shown to enhance lipid profiles by reducing total cholesterol, LDL cholesterol, and triglycerides (Sun et al. 2021). Polyphenols may operate by hindering cholesterol absorption in the gastrointestinal tract and by promoting the excretion of bile acids. However, the precise mechanisms through which polyphenols lower cholesterol levels necessitate further investigation and research (Feldman et al. 2021).

Omega-3 fatty acids, classified as beneficial fats, can be obtained from fatty fish, flaxseeds, and walnuts. These fatty acids have demonstrated their ability to improve lipid profiles by decreasing triglyceride levels and elevating HDL cholesterol (Bradberry and Hilleman 2013). Furthermore, omega-3 fatty acids contribute to the reduction of inflammation within the body, a condition often associated with elevated cholesterol levels. The recommended daily intake of omega-3 fatty acids is approximately 250–500 mg (Skulas-Ray et al. 2019; Troesch et al. 2020).

Garlic, a bulbous plant with a long history of medicinal use, has been found to possess cholesterol-lowering properties in various studies. It is known to lower total cholesterol, LDL cholesterol, and triglyceride levels (Yeh and Liu 2001). Although the exact active compounds responsible for these effects in garlic are not yet fully understood, they may function by inhibiting cholesterol synthesis in the liver or reducing cholesterol absorption in the gastrointestinal tract (Ansary et al. 2020).

Curcumin, an active component found in the spice turmeric frequently utilized in Indian cooking, exhibits antioxidant and anti-inflammatory characteristics (Abd El-Hack et al. 2021). It has been discovered to possess cholesterol-reducing effects by diminishing total cholesterol, LDL cholesterol, and triglyceride levels (Sharifi-Rad et al. 2020). Although the precise mechanisms by which curcumin decreases cholesterol levels are not fully comprehended, potential actions may involve impeding cholesterol synthesis in the liver, promoting the elimination of bile acids, and enhancing endothelial function, which pertains to the inner lining of blood vessels (Hong et al. 2022; Kim and Kim 2010).

Red yeast rice, a variety of rice subjected to fermentation using *Monascus purpureus* yeast species, results in the production of monacolin K (Klimek et al. 2009). This compound possesses an identical chemical structure to lovastatin, a medication employed for lowering cholesterol (Xiong et al. 2019). Numerous studies have demonstrated that red yeast rice can effectively reduce total cholesterol, LDL cholesterol, and triglyceride levels (Cicero et al. 2019).

Certain varieties of fungi encompass compounds that could potentially serve as a natural remedy for hypercholesterolemia by exerting cholesterol-lowering effects. Among these compounds is beta-glucan, a form of soluble fiber present in the cell walls of macrofungi (Chugh et al. 2022). Research indicates that beta-glucan aids in diminishing cholesterol absorption in the gastrointestinal tract, thereby potentially reducing levels of LDL ("bad") cholesterol in the bloodstream (Jefferson 2009). Furthermore, mushrooms harbor additional bioactive constituents such as ergosterol, sterols, and triterpenes, which may also contribute to cholesterol-lowering effects. Studies propose that integrating mushrooms into a well-balanced diet may assist in reducing cholesterol levels and improving other cardiovascular risk factors, including blood pressure and inflammation (Alam et al. 2009).

Various other plant-based treatments have undergone scrutiny for their potential to reduce cholesterol, including guggul, artichoke leaf extract, fenugreek, and berberine (Wider et al. 2013). These remedies may function by inhibiting cholesterol synthesis in the liver, curbing cholesterol absorption in the gut, or enhancing endothelial function (Frigerio et al. 2021).

11.5 HYPOLIPIDEMIC AND CHOLESTEROL-LOWERING PROPERTIES OF *GANODERMA*

Ganoderma, an ancient medicinal fungus with roots in traditional Chinese medicine, has garnered attention in recent scientific research for its potential in combating heart disease through its hypolipidemic and cholesterol-reducing properties (Lee et al. 2012). Numerous studies have presented evidence of *Ganoderma*'s ability to diminish overall cholesterol levels, including LDL cholesterol and triglycerides, while simultaneously elevating levels of HDL cholesterol (Ekiz et al. 2023). The active constituents within *Ganoderma* believed to be responsible for these effects are beta-glucans, triterpenoids, and polysaccharides (Mirończuk-Chodakowska et al. 2021). Additionally, *Ganoderma* exhibits anti-inflammatory and antioxidant qualities, which may further contribute to its cholesterol-lowering abilities. Given that chronic inflammation is a significant risk factor for heart disease, antioxidants can impede the oxidation of LDL cholesterol, thereby preventing arterial plaque formation (Chan et al. 2021; Ahmad 2018).

11.5.1 Potential Mechanisms of Action

Numerous investigations have provided evidence of the hypolipidemic and cholesterol-reducing characteristics of *Ganoderma* spp. and its bioactive constituents, as presented in Table 11.2, although the precise mechanisms by which they operate remain incompletely comprehended. The following are a few of the plausible mechanisms of action that have been attributed to *Ganoderma* spp. and its bioactive compounds.

11.5.1.1 Inhibition of HMG-CoA Reductase Activity

The primary manner in which statins reduce cholesterol is by impeding the activity of HMG-CoA reductase, an enzyme crucial for cholesterol synthesis (Zhou and Liao 2023). *Ganoderma lucidum* contains triterpenoids like ganoderic acid and its variations that have demonstrated the ability to inhibit HMG-CoA reductase activity (Chan et al. 2021). A study involving hypercholesterolemic rats revealed that ganoderic acid A and B obtained from *G. lucidum* restrained the activity of HMG-CoA reductase and led to a decrease in serum cholesterol levels (Li et al. 2006). Additionally, a controlled trial employing a randomized, double-blind, placebo approach demonstrated that the administration of *G. lucidum* powder over a 16-week period significantly lowered serum total cholesterol, LDL cholesterol, and triglyceride levels in individuals with hypercholesterolemia (Klupp et al. 2016). These findings suggest that the suppression of HMG-CoA reductase activity by ganoderic acid and its derivatives could be one of the mechanisms accountable for the cholesterol-lowering and hypolipidemic effects of *G. lucidum* (Chen et al. 2017).

11.5.1.2 Activation of AMP-Activated Protein Kinase

AMP-activated protein kinase (AMPK), a critical regulator of lipid and glucose metabolism, plays a significant role in reducing lipid levels in the bloodstream when activated (Srivastava et al. 2012). *G. lucidum* contains polysaccharides such as β-D-glucans and heteropolysaccharides, which have been found to activate AMPK (Seweryn et al. 2021). In a study involving hyperlipidemic mice, a polysaccharide extract derived from *G. lucidum* was shown to enhance AMPK phosphorylation, resulting in decreased levels of total cholesterol, triglycerides, and LDL cholesterol in the serum (Jing et al. 2022). Another study conducted on hyperlipidemic rats demonstrated that oral administration of *G. lucidum* polysaccharides reduced serum lipid levels and boosted the expression of

TABLE 11.2
***Ganoderma* spp.–Derived Phytochemicals and Their Mechanisms of Action in Lowering Cholesterol**

Bioactive Constituent	Species of *Ganoderma*	Chemical Structure	Extract Type	Mechanism of Action	Refs.
Ganoderic acid A	*Ganoderma lucidum*	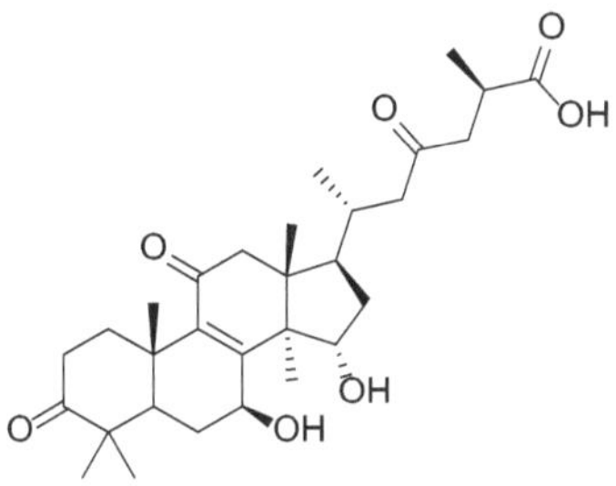	Ethanol extract	Inhibition of cholesterol synthesis, promotion of cholesterol efflux through gut microbiota modulation	Guo et al. 2020
Ganoleucoin K	*Ganoderma leucocontextum*	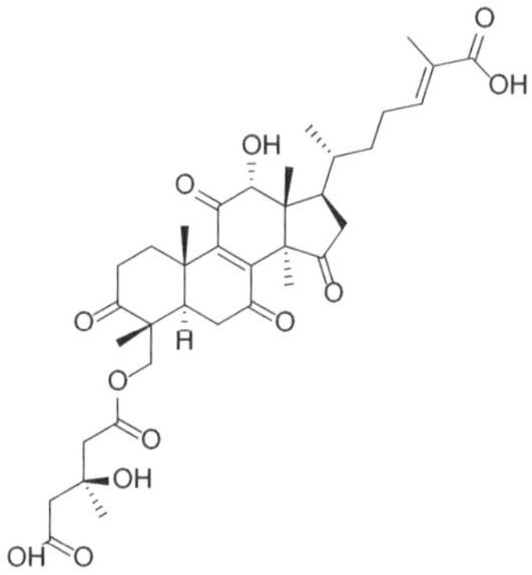	Ethanol extract	Inhibition of HMG-CoA reductase	Wang et al. 2015
Ganoleucoin T	*Ganoderma leucocontextum*	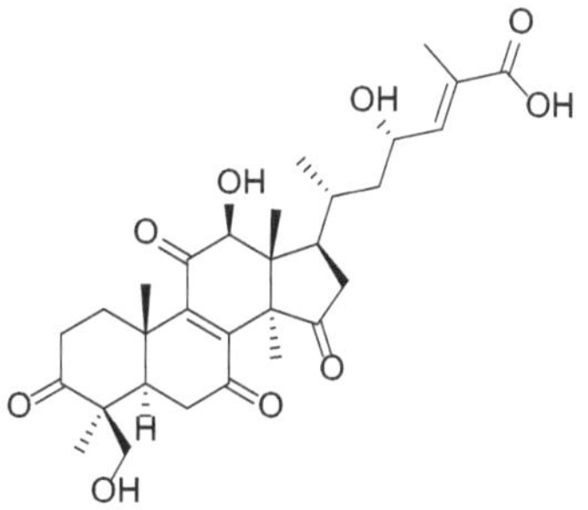	Ethanol extract	Inhibition of HMG-CoA reductase	Zhang et al. 2018
Ganoleucoin Y		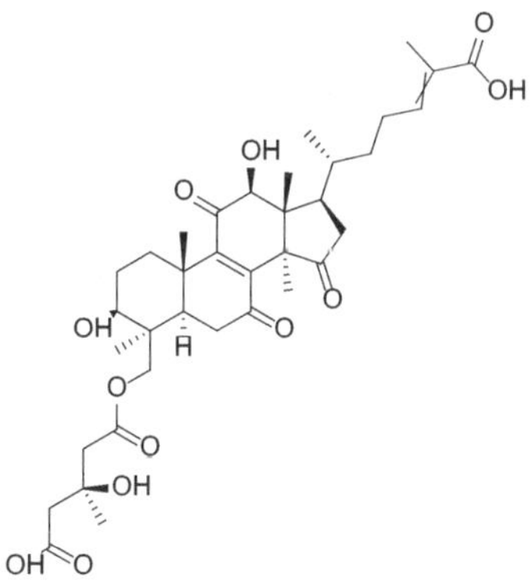			

(Continued)

TABLE 11.2 *(Continued)*
***Ganoderma* spp.–Derived Phytochemicals and Their Mechanisms of Action in Lowering Cholesterol**

Bioactive Constituent	Species of *Ganoderma*	Chemical Structure	Extract Type	Mechanism of Action	Refs.
Ganoleucoin Z		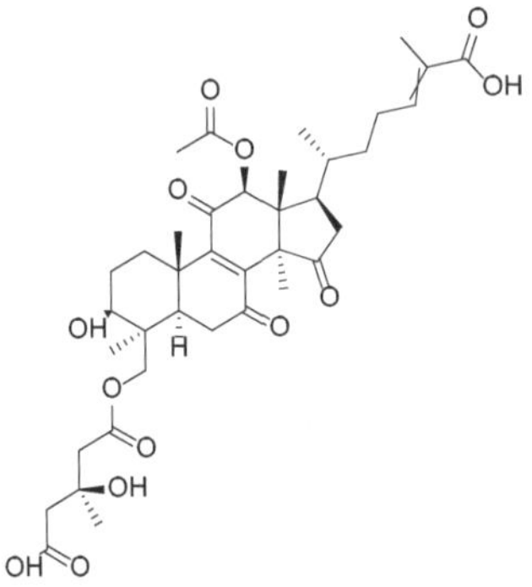			
Adenosine	*Ganoderma lucidum*	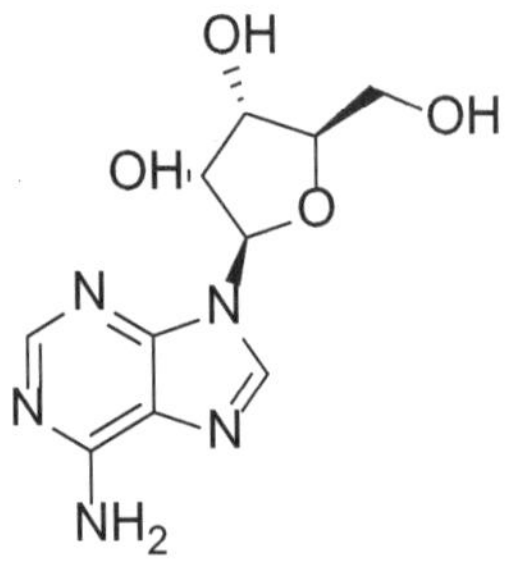	Ethanol extract	Inhibition of cholesterol absorption, regulation of lipoprotein metabolism	Li et al. 2022
Ganoderal B	*Ganoderma lucidum*	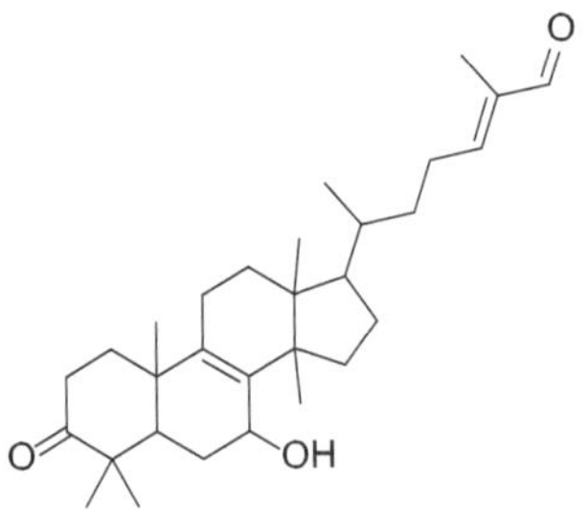	Aqueous extract	Inhibit conversion of 24,25-dihydro-lanosterol to cholesterol at the lanosterol 14 α-demethylase step and also indirectly to inhibit HMG-CoA reductase activity.	Berger et al. 2004
5α,8α-epidiox-yergosta-6,22-dien-3β-ol	*Ganoderma lucidum*	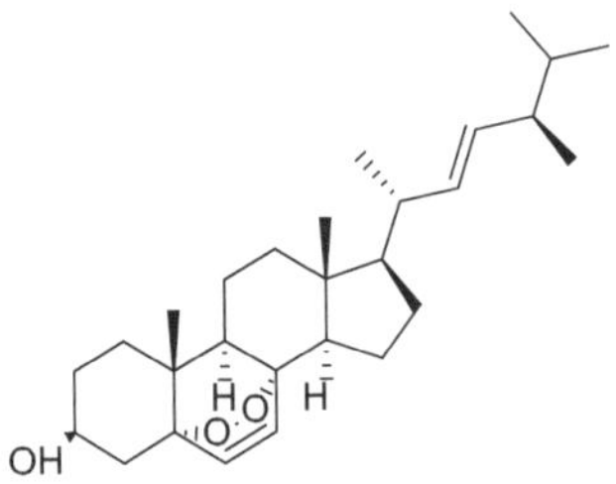	Methanol extract	Inhibit cholesterol esterase activity with an IC50 value of 42 μM	Kim 2010
Ganoderic acid C2	*Ganoderma lucidum*	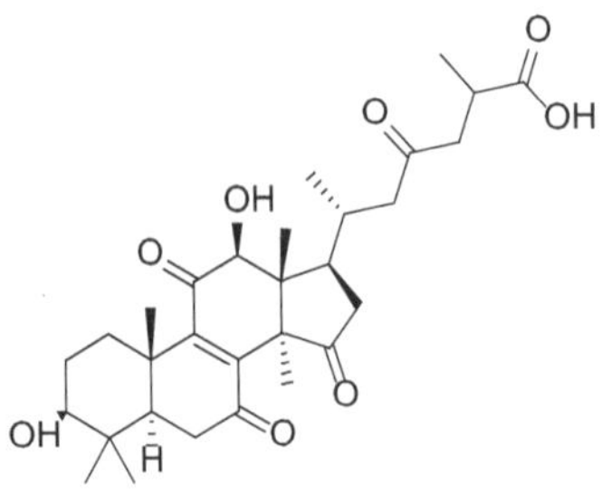	Ethanol extract	Decrease the mRNA levels of FAS, ACAT2, SREBP-1C, and HMGCR and increase the mRNA levels of CYP7A1, PPARα, ApoB, and Acox1; decreased synthesis of cholesteryl ester and a subsequent increase in cholesterol excretion.	Guo et al. 2018

TABLE 11.2 *(Continued)*

***Ganoderma* spp.–Derived Phytochemicals and Their Mechanisms of Action in Lowering Cholesterol**

Bioactive Constituent	Species of *Ganoderma*	Chemical Structure	Extract Type	Mechanism of Action	Refs.
Ganoderic acid η		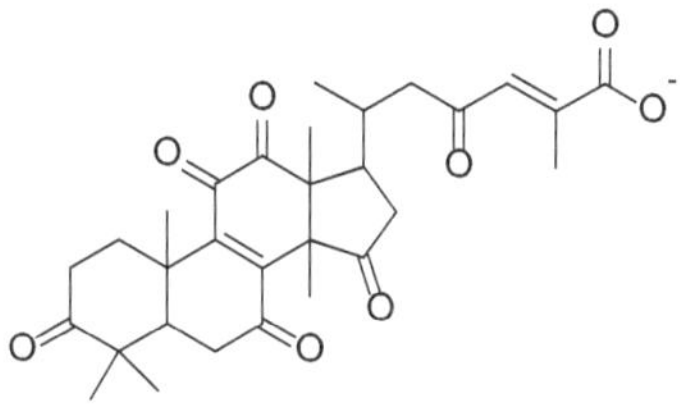			
Ganoderic acid Me		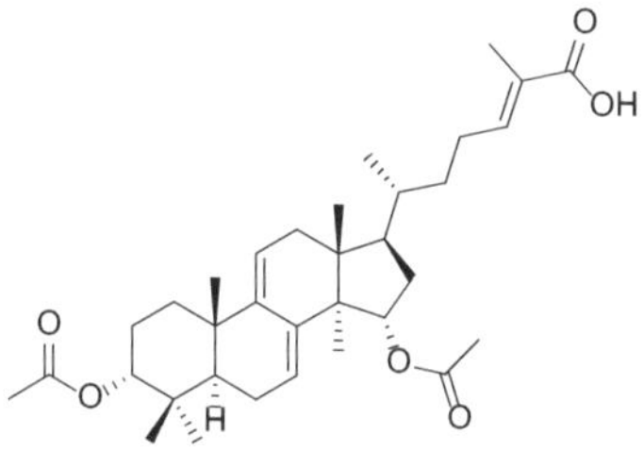			
Poricoic acid HM		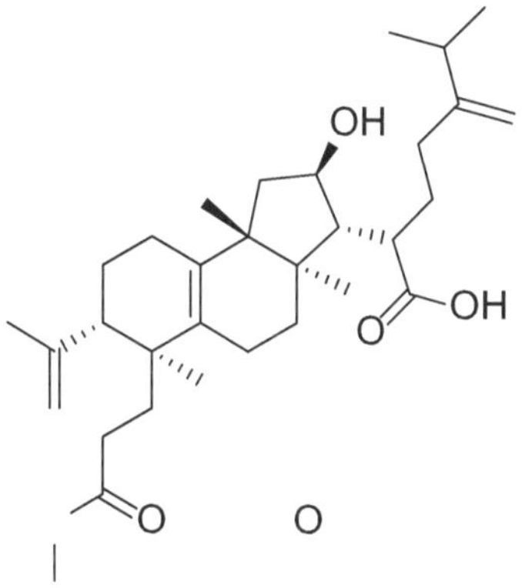			
Ganoderic acid G		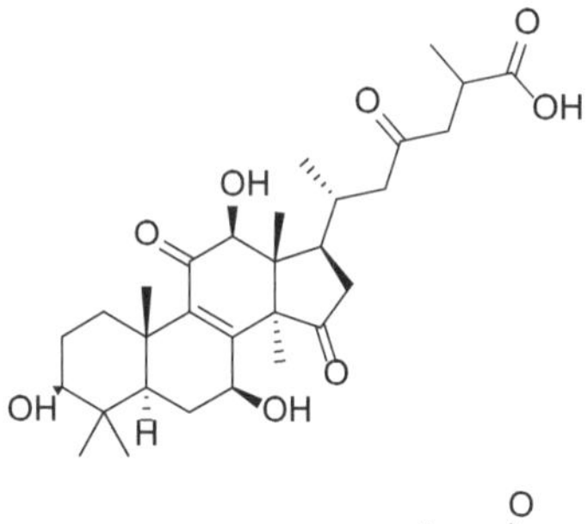			
Ganoderic acid F		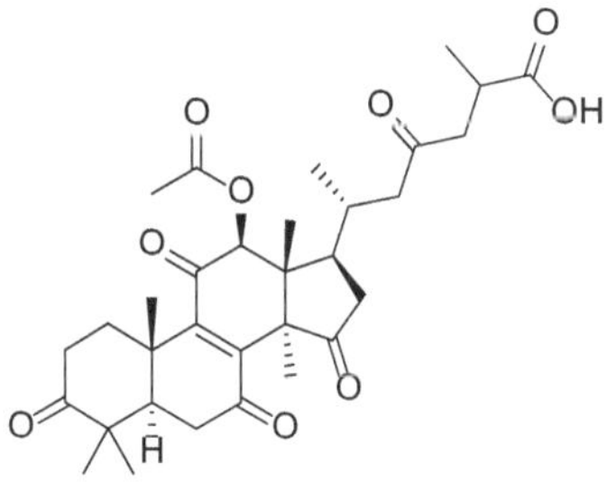			

(Continued)

TABLE 11.2 *(Continued)*
Ganoderma spp.–Derived Phytochemicals and Their Mechanisms of Action in Lowering Cholesterol

Bioactive Constituent	Species of *Ganoderma*	Chemical Structure	Extract Type	Mechanism of Action	Refs.
1,2-Benzene-dicarboxylic acid, 3-nitro	*Ganoderma* spp.	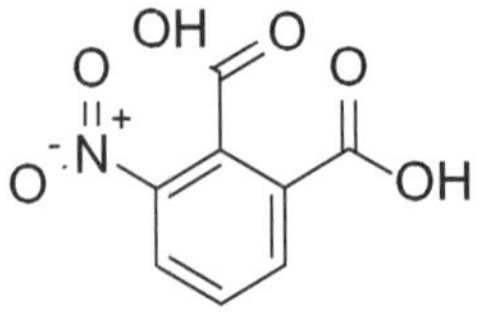	Methanol extract	Inhibition of cholesterol absorption, regulation of lipoprotein metabolism	Elkhateeb et al. 2019
Ganoder-manontriol	*Ganoderma* spp.	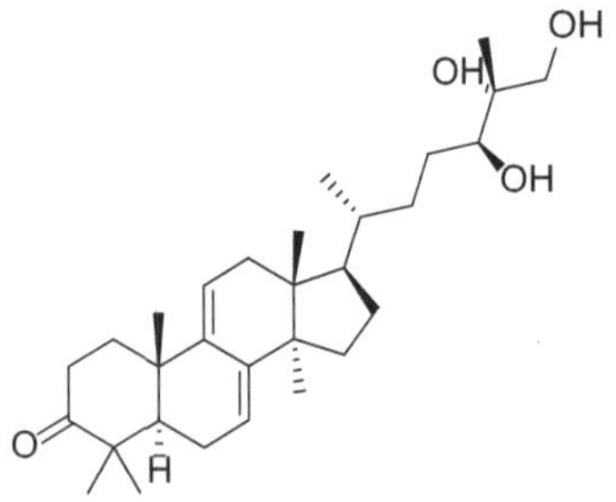	Ethanol extract	Modulation of SREBP-2 and HMG-CoA reductase, regulation of LDL receptor expression	Keong 2015
Ganomycin I	*Ganoderma leucocontextum*	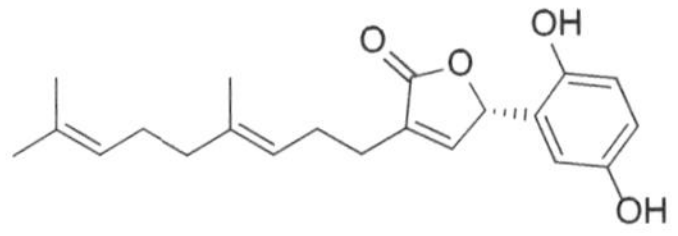	Ethanol	Inhibition of HMG-CoA reductase	Wang et al. 2017
Ganoderic acid DM	*Ganoderma leucocontextum*	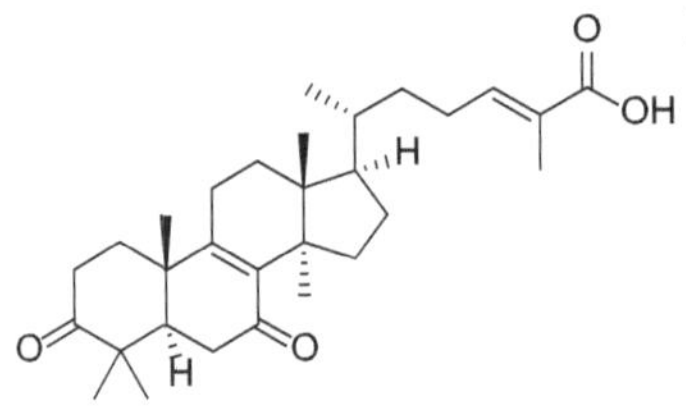	Ethanol extract	Inhibition of HMG-CoA reductase and 5α-reductase	Wang et al. 2015
Ganoderol B	*Ganoderma lucidum*	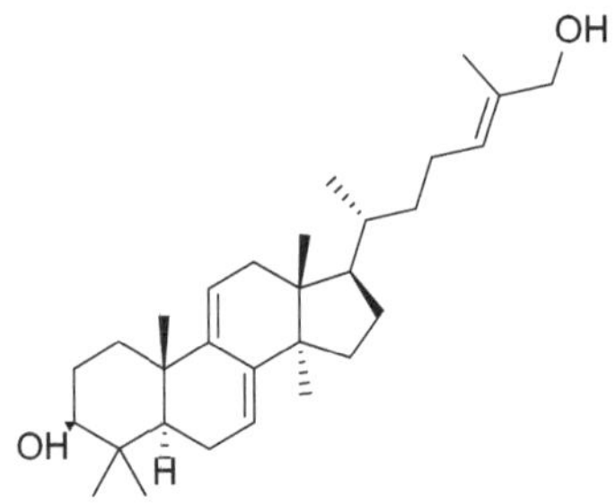	MeOH extract	Inhibition of cholesterol absorption, inhibition of cholesterol biosynthesis via conversion of acetate or mevalonate as a precursor of cholesterol, inhibition of HMG-CoA reductase, inhibition of the lanosterol 14α-demethylase	Hajjaj et al. 2005
Ganoderol A		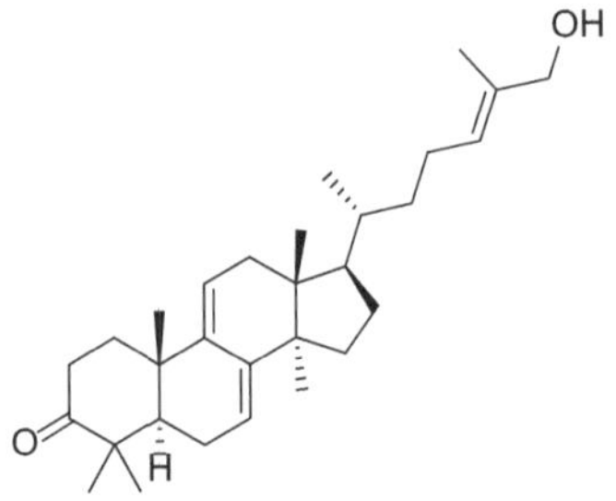			

TABLE 11.2 ***(Continued)***
***Ganoderma* spp.–Derived Phytochemicals and Their Mechanisms of Action in Lowering Cholesterol**

Bioactive Constituent	Species of *Ganoderma*	Chemical Structure	Extract Type	Mechanism of Action	Refs.
Ganoderal A		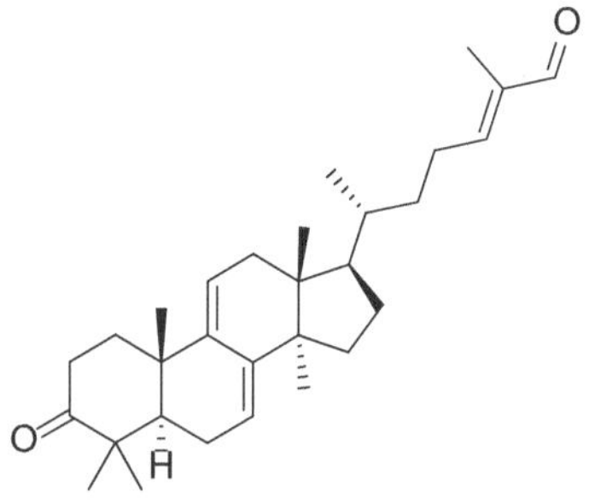			
Ganoderic acid Y		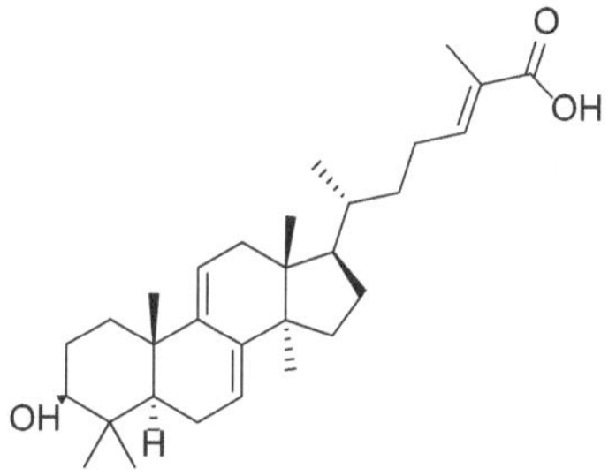			
Ganoderic Acid B	*Ganoderma lucidum*	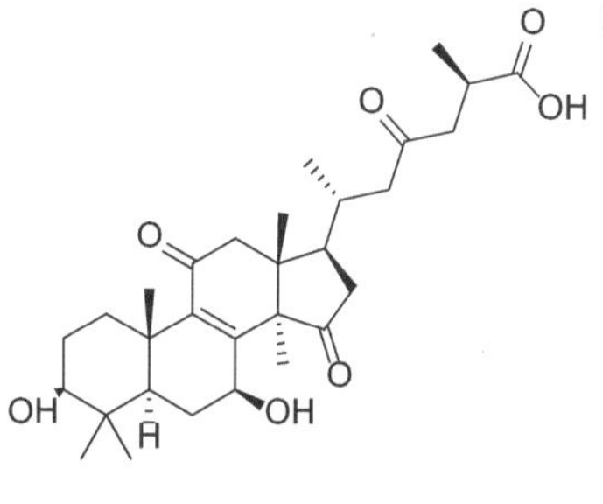	Ethanol extract	Inhibition of cholesterol synthesis through inhibition of HMG-CoA reductase	Komoda et al. 1989

Note: All chemical structures have been drawn using ChemDraw software.

AMPK (Arunachalam et al. 2022). These findings suggest that the activation of AMPK by polysaccharides found in *G. lucidum* may contribute to its cholesterol-lowering and hypolipidemic effects.

11.5.1.3 Inhibition of Intestinal Cholesterol Absorption

The regulation of cholesterol balance heavily relies on the absorption of cholesterol in the intestines. Niemann-Pick C1-like 1 (NPC1L1), a cholesterol transporter situated in the brush border membrane of enterocytes, plays a crucial role in transporting dietary and biliary cholesterol from the intestinal lumen into enterocytes (Betters and Yu 2010). Inhibiting the activity of NPC1L1 is a target for addressing hypercholesterolemia (Altmann et al. 2004). *G. lucidum* contains triterpenoids such as ganoderic acid and its derivatives, which have been found to impede the activity of NPC1L1 (Hsu and Yen 2014). Several studies revealed that ganoderic acid A derived from *G. lucidum* reduced the expression of NPC1L1 and hindered cholesterol uptake in Caco-2 cells (Wong et al. 2017; Caz et al. 2015). Additionally, another study demonstrated that administering *G. lucidum* extract orally reduced serum levels of total cholesterol and LDL cholesterol, and also decreased NPC1L1 expression in the intestines of hypercholesterolemic rats (Lai et al. 2020). These findings suggest that the inhibition of cholesterol absorption in the intestines by ganoderic acid and its derivatives may serve as another mechanism contributing to the cholesterol-lowering and hypolipidemic effects of *G. lucidum* (Guo et al. 2022).

11.5.1.4 Modulation of Gut Microbiota

The role of the gut microbiota in host metabolism and well-being has gained significant recognition. Growing evidence indicates that changes in the composition and functioning of the gut microbiota are linked to hyperlipidemia and dyslipidemia (Vourakis et al. 2021). *G. lucidum* contains polysaccharides such as β-D-glucans and heteropolysaccharides, which have been found to influence the gut microbiota (Jayachandran et al. 2017). A study revealed that orally administering polysaccharides from *G. lucidum* increased the presence of *Bifidobacterium* and *Lactobacillus* in the intestines of hyperlipidemic rats, which was associated with reduced levels of total cholesterol and LDL cholesterol in the bloodstream (Romero-Córdoba et al. 2020). Another study demonstrated that a polysaccharide extract derived from *G. lucidum* enhanced the population of beneficial bacteria while diminishing the presence of harmful bacteria in the intestines of hyperlipidemic mice, leading to decreased levels of total cholesterol and triglycerides in the bloodstream (Meneses et al. 2023). These findings indicate that the modulation of the gut microbiota by polysaccharides in *G. lucidum* may contribute to its hypolipidemic and cholesterol-lowering effects.

11.5.1.5 Anti-Inflammatory and Antioxidant Effects

Inflammation and oxidative stress play a role in the development of hyperlipidemia and dyslipidemia (Boarescu et al. 2022). *G. lucidum* contains a variety of active components, such as polysaccharides, triterpenoids, and polyphenols, which possess anti-inflammatory and antioxidant properties (El Sheikha 2022). A study demonstrated that an extract of polysaccharides from *G. lucidum* reduced levels of total cholesterol and LDL cholesterol in the bloodstream and also suppressed the expression of inflammatory cytokines in rats with hyperlipidemia (Peng et al. 2023). Another study revealed that an extract rich in polyphenols from *G. lucidum* decreased levels of total cholesterol and triglycerides in the bloodstream, while enhancing the activity of antioxidant enzymes in hyperlipidemic rats (Sang et al. 2021). These findings indicate that the anti-inflammatory and antioxidant effects of the active components in *G. lucidum* may contribute to its ability to lower lipid levels and cholesterol.

11.5.1.6 Regulation of Gene Expression

G. lucidum contains a variety of active components that can influence the expression of genes involved in lipid metabolism. A study revealed that an extract of triterpenoids from *G. lucidum* increased the expression of peroxisome proliferator-activated receptor alpha (PPARα) and its associated genes responsible for fatty acid oxidation. This increase in expression was linked to reduced levels of triglycerides in the bloodstream of hyperlipidemic rats (Huang et al. 2011). Another study found that an extract of polysaccharides from *G. lucidum* boosted the expression of cholesterol 7α-hydroxylase (cytochrome P450 7A1 or CYP7A1), an enzyme crucial in converting cholesterol into bile acids (Jing et al. 2022). This increase in expression was associated with decreased levels of total cholesterol and LDL cholesterol in the bloodstream of hyperlipidemic rats (Li et al. 2011). These findings suggest that the regulation of gene expression by the active components in *G. lucidum* may contribute to its ability to lower lipid levels and cholesterol.

In summary, *Ganoderma* spp. and its active compounds have demonstrated their potential to reduce lipid levels and cholesterol in both animal and human studies. These effects are likely achieved through various mechanisms, including the inhibition of cholesterol synthesis, the promotion of cholesterol excretion, the suppression of cholesterol absorption, the modulation of gut microbiota, the anti-inflammatory and antioxidant properties, and the regulation of gene expression. The active compounds responsible for these effects encompass triterpenoids, polysaccharides, polyphenols, and other natural chemicals. Further research is necessary to uncover the precise mechanisms underlying the hypolipidemic and cholesterol-lowering effects of *Ganoderma* spp. and to determine its potential therapeutic applications in preventing and treating hyperlipidemia and dyslipidemia.

11.6 BIOACTIVE COMPONENTS OF *GANODERMA* SPP. RESPONSIBLE FOR HYPOLIPIDEMIC AND CHOLESTEROL-LOWERING EFFECTS

Ganoderma spp. have been recognized for harboring an assortment of bioactive constituents that exhibit promising medicinal properties. Within these constituents, polysaccharides stand out as prominent elements accountable for the hypolipidemic and cholesterol-reducing properties attributed to *Ganoderma* spp. (Seweryn et al. 2021).

11.6.1 Polysaccharides

Polysaccharides are intricate carbohydrates that consist of repetitive units of simple sugars like glucose, galactose, and mannose (Seweryn et al. 2021). As an example, a polysaccharide derived from *G. lucidum* was discovered to have a ratio of glucose to mannose at 2.2:1, with a molecular weight of around 10 kDa (Xu et al. 2011). Another study identified a polysaccharide from *Ganoderma atrum* comprised of glucose, galactose, and arabinose with a ratio of 8.2:1.7:1 and a molecular weight of roughly 104 kDa (Ferreira et al. 2015). The primary polysaccharide in *G. lucidum* is β-D-glucan, characterized by a linear chain of β-D-glucose units connected by β-1,3 and β-1,6 glycosidic bonds (Hung et al. 2008). Conversely, polysaccharides from *Ganoderma sinense* contain a complex blend of glucose, galactose, and mannose units linked by various types of glycosidic bonds (Jiang et al. 2017). Polysaccharides have displayed hypolipidemic effects by heightening bile acid excretion and diminishing cholesterol absorption within the intestine (Meneses et al. 2016). Furthermore, polysaccharides have been shown to lower the expression of genes associated with cholesterol synthesis and uptake in the liver, thus resulting in reduced serum cholesterol levels (Romero-Córdoba et al. 2020). For instance, Wang et al. (2023) conducted a study illustrating that a polysaccharide extract from *G. lucidum* can lower the levels of total cholesterol, triglycerides, and LDL cholesterol in rats on a high-fat diet. The polysaccharide was observed to inhibit the expression of genes linked to cholesterol synthesis and uptake, such as HMG-CoA reductase and LDL receptors (Wang et al. 2023). Multiple studies have demonstrated the hypolipidemic and cholesterol-lowering effects of polysaccharides obtained from various *Ganoderma* spp. Lai et al. (2020) conducted a study revealing that polysaccharides extracted from *G. lucidum* can reduce the levels of total cholesterol, triglycerides, and LDL cholesterol in hyperlipidemic rats. The polysaccharides were shown to regulate the expression of genes involved in lipid metabolism, such as HMG-CoA reductase and PPARα (Lai et al. 2020). Similarly, Huang et al. (2021) investigated the hypolipidemic effects of polysaccharides from *G. sinense* in rats fed a high-fat diet. The results exhibited that the polysaccharides can decrease the levels of total cholesterol, triglycerides, and LDL cholesterol, while elevating the levels of HDL cholesterol (Huang et al. 2021). Furthermore, the polysaccharides were found to suppress the expression of genes implicated in lipid synthesis and uptake, such as sterol regulatory element binding protein 1c (SREBP-1c) and CD36 (Eilam et al. 2022). In conclusion, polysaccharides serve as significant bioactive constituents in *Ganoderma* spp. responsible for their hypolipidemic and cholesterol-lowering effects. These polysaccharides manifest a diverse array of biological activities and possess potential health advantages beyond their hypolipidemic properties (Pan et al. 2021). For instance, β-D-glucans from *G. lucidum* have been reported to exhibit immunomodulatory effects by stimulating the production of cytokines like interleukin-2 and interferon-γ (Ahmad 2018). Heteropolysaccharides from *G. atrum* have been identified to demonstrate antioxidant activity by scavenging free radicals and inhibiting lipid peroxidation (Chen et al. 2008).

11.6.2 Triterpenes

Another group of biologically active compounds discovered in *Ganoderma* spp. that have been documented to demonstrate hypolipidemic and cholesterol-lowering properties are triterpenoids (Martínez-Montemayor et al. 2019). Triterpenoids are a class of compounds derived from isoprene units and are widely distributed in plants and fungi (Zhao et al. 2023). *Ganoderma* spp. contain various types of triterpenoids, including ganoderic acids, lucidenic acids, and ergosterols (Wang et al. 2020). Ganoderic acid A, obtained from *G. lucidum*, is a triterpene featuring a lanostane framework and a carboxyl group (Meng et al. 2020). Another triterpene, ganoderic acid DM, feature a lanostane skeleton and two hydroxyl groups (M. Johnson 2010). Ganoderic acids are among the most abundant and extensively studied triterpenoids in *Ganoderma* spp. They are tetracyclic triterpenoids comprising a carboxylic acid group and several hydroxyl groups. Structurally similar to steroids, ganoderic acids have demonstrated diverse biological activities, including antitumor, anti-inflammatory, and hypolipidemic effects (Galappaththi et al. 2022; Wang et al. 2020). Lucidenic acids, on the other hand, are pentacyclic triterpenoids that incorporate a lactone ring and multiple hydroxyl groups (Galappaththi et al. 2022). Several studies have explored the hypolipidemic and cholesterol-lowering effects of ganoderic acids derived from different *Ganoderma* spp. For instance, Zhu et al. (2018) conducted a study illustrating that ganoderic acid A from *G. lucidum* can decrease the levels of total cholesterol, triglycerides, and LDL cholesterol in hyperlipidemic rats. The compound was observed to inhibit the expression of genes implicated in cholesterol synthesis and uptake, such as HMG-CoA reductase and LDL receptors (Zhu et al. 2018). Li et al. 2006 investigated the inhibitory potential of *G. lucidum*–derived bioactive triterpenoids ganoderiol F and ganodermic acid Q against HMG Co-A reductase (Li et al. 2006). Triterpenes obtained from *Ganoderma* spp. commonly possess lanostane or ergostane frameworks and have been shown to exhibit hypolipidemic and cholesterol-lowering effects (Xia et al. 2014). Lee et al. (2010) conducted a study demonstrating that a triterpene extract from *G. lucidum* can decrease the levels of total cholesterol, triglycerides, and LDL cholesterol in rats fed a high-fat diet. The triterpene was found to inhibit the expression of genes involved in cholesterol synthesis and uptake, such as HMG-CoA reductase and LDL receptors (Lee et al. 2010). Similarly, the hypolipidemic effects of lucidenic acids from *G. sinense* in rats fed a high-fat diet were also investigated in various studies (Zheng et al. 2023a). The results indicated that lucidenic acids can lower the levels of total cholesterol, triglycerides, and LDL cholesterol, while elevating the levels of HDL cholesterol (Baby et al. 2015). Moreover, the compound was found to regulate the expression of genes related to lipid metabolism and inflammation, such as PPARα and NF-κB. Aside from their hypolipidemic and cholesterol-lowering effects, triterpenoids derived from *Ganoderma* spp. have been associated with other health benefits (Zheng et al. 2023a). For instance, ganoderic acids have been reported to exhibit antitumor effects by inducing apoptosis and inhibiting angiogenesis (Radwan 2012). Ergosterol, another triterpenoid found in *Ganoderma* spp., has shown anti-inflammatory effects by inhibiting the production of pro-inflammatory cytokines (Xu et al. 2021).

11.6.3 Sterols

The sterols present in *Ganoderma* spp. share a structural resemblance to cholesterol, featuring a hydroxyl group at the C-3 position of the steroid ring (Grienke et al. 2015). Sterols derived from *Ganoderma* spp. have demonstrated hypolipidemic and cholesterol-lowering effects. Ergosterol, an isolated sterol from *G. lucidum*, possesses a steroid ring and a side chain incorporating a double bond (Jeong and Park 2020). Similarly, ganoderol B comprises a steroid ring and a side chain with a hydroxyl group (Feng et al. 2018). A study conducted by Lee et al. (2020) showed that sterol extracts from *G. lucidum* can diminish the levels of total cholesterol, triglycerides, and LDL cholesterol in rats fed a high-fat diet (Lee et al. 2020). The sterol was identified to impede the expression of genes associated with cholesterol synthesis and uptake, such as HMG-CoA reductase and LDL receptors (Berger et al. 2004).

11.6.4 Alkaloids

The alkaloids present in *Ganoderma* spp. showcase a wide range of structures, featuring a nitrogenous base with heterocyclic properties and a variety of functional groups, including hydroxyl, methyl, and acetyl groups (Liu et al. 2011). Ganoderic acid DM, an alkaloid derived from *G. lucidum*, possesses a lanostane skeleton, a carboxyl group, and two hydroxyl groups. This alkaloid has demonstrated the ability to reduce serum cholesterol levels in hyperlipidemic rats (Liu et al. 2012). Ganoderic acid DM was found to impede cholesterol biosynthesis in the liver by suppressing the expression of key enzymes involved in cholesterol synthesis (Shiao 2003). Alkaloids derived from *Ganoderma* spp. have exhibited hypolipidemic and cholesterol-lowering effects. For example, a study conducted by Aboraya et al. (2022) revealed that an alkaloid extract obtained from *G. lucidum* can decrease the levels of total cholesterol, triglycerides, and LDL cholesterol in rats subjected to a high-fat diet (Aboraya et al. 2022). The alkaloid was observed to hinder the expression of genes associated with cholesterol synthesis and uptake, such as HMG-CoA reductase and LDL receptors (Eilam et al. 2022).

11.6.5 Flavonoids

The flavonoids present in *Ganoderma* spp. consist of a backbone of either flavone or flavonol, accompanied by various hydroxyl groups (Atiq and Parhar 2020). Quercetin, a flavonoid obtained from *G. lucidum*, encompasses a flavone backbone and three hydroxyl groups (Huh et al. 2019). Another flavonoid, apigenin, features a flavone backbone and four hydroxyl groups (Kolniak-Ostek et al. 2022). Flavonoids derived from *Ganoderma* spp. have demonstrated hypolipidemic and cholesterol-lowering effects. For example, a study conducted by Romero-Córdoba et al. (2020) revealed that a flavonoid extract derived from *G. lucidum* can diminish the levels of total cholesterol, triglycerides, and LDL cholesterol in rats subjected to a high-fat diet. The flavonoid was observed to impede the expression of genes associated with cholesterol synthesis and uptake, such as HMG-CoA reductase and LDL receptors (Romero-Córdoba et al. 2020).

11.6.6 Other Bioactive Compounds

In addition to polysaccharides and triterpenoids, *Ganoderma* spp. harbor various other biologically active substances that possess potential health advantages. These encompass proteins, peptides, nucleotides, and additional small molecules like alkaloids and flavonoids (Galappaththi et al. 2022). Proteins and peptides derived from *Ganoderma* spp. have been documented to exhibit a diverse array of biological activities, encompassing immunomodulatory, antitumor, and hypolipidemic effects (Ahmad 2018). For instance, a study conducted by Meneses et al. (2023) demonstrated that extract from *G. lucidum* can lower the levels of total cholesterol, triglycerides, and LDL cholesterol in rats subjected to a high-fat diet. The protein was observed to hinder the expression of genes associated with cholesterol synthesis and uptake, such as HMG-CoA reductase and LDL receptors (Meneses et al. 2023). Nucleotides originating from *Ganoderma* spp. have also been reported to display hypolipidemic and cholesterol-lowering effects (Gao et al. 2007). Furthermore, Liang et al. (2019) unveiled the potent cholesterol-lowering effects of ganoderic acid C2, an isolated triterpenoid compound from *G. lucidum*, in hyperlipidemic rats. Ganoderic acid C2 was found to diminish serum cholesterol levels by inhibiting cholesterol absorption in the intestine and enhancing the excretion of bile acids (Liang et al. 2019). Another compound existing in *Ganoderma* spp., known as ganodermanontriol, has also demonstrated hypolipidemic properties (El Sheikha 2022). Ganodermanontriol, a triterpene alcohol extracted from *G. lucidum* and other *Ganoderma* spp., was observed in a study by Zheng et al. (2023b) to reduce serum triglyceride and cholesterol levels in mice with diet-induced obesity. The authors suggested that the hypolipidemic effects of ganodermanontriol were attributed to its capacity to boost fatty acid oxidation and diminish lipogenesis in the liver (Zheng et al. 2023b). On the whole, the bioactive compounds inherent in *Ganoderma* spp. have exhibited promising outcomes in reducing serum cholesterol levels and enhancing lipid metabolism. Nevertheless, further investigations are required to comprehensively understand the precise mechanisms of action employed by these compounds and to ascertain their effectiveness and safety in human subjects.

11.7 CONCLUSION

To summarize, *Ganoderma* is a type of mushroom that exhibits potential in reducing lipid levels and lowering cholesterol. Its bioactive constituents, such as triterpenoids and polysaccharides, have demonstrated the ability to decrease serum lipid levels, impede cholesterol absorption, and enhance bile acid excretion. The observed impacts suggest that *Ganoderma* has the potential to be considered as a viable option for the prevention and treatment of hyperlipidemia, a significant risk factor for cardiovascular disease. Numerous studies have been conducted to explore *Ganoderma*'s hypolipidemic and cholesterol-lowering properties, yielding promising results. However, further research is necessary to comprehend its mechanisms of action and establish the optimal dosage and treatment duration. One possible application of *Ganoderma* lies in the development of functional foods or nutraceuticals. As a natural product, it has the advantage of generally being safe and well-tolerated. Given the increasing prevalence of hyperlipidemia and cardiovascular disease, there is a rising demand for alternative and complementary therapies that can aid in managing these conditions. Another potential avenue for *Ganoderma* is in pharmaceutical development. Its bioactive compounds have exhibited diverse pharmacological properties, including anti-inflammatory, antioxidant, and immunomodulatory effects. These attributes position it as an encouraging candidate for drug development targeting various diseases such as cancer, diabetes, and autoimmune disorders. *Ganoderma* cultivation can be done indoors or outdoors, making it accessible for both individual farmers and commercial farms. It can be grown using low-cost materials, making it suitable for small-scale farmers, and it can be adapted to local conditions, making it environmentally friendly. Additionally, large-scale industrial production using advanced technologies like bioreactors ensures efficient production and high-quality output, meeting the demand for functional foods and pharmaceuticals in the bioeconomy.

In conclusion, *Ganoderma* exhibits promising hypolipidemic and cholesterol-lowering effects. Its bioactive components hold potential for use in functional foods and pharmaceuticals. Nevertheless, caution should be exercised when using *Ganoderma*, and consultation with a healthcare professional is advised. By following recommended guidelines for safe usage, individuals may be able to incorporate *Ganoderma* into their overall health regimen safely and effectively. With further research and development, *Ganoderma* can emerge as a valuable resource for enhancing human health and well-being, being globally accessible and cost-effective.

REFERENCES

Abd El-Hack, M.E., M.T. El-Saadony, A.A. Swelum, M. Arif, M.M. Abo Ghanima, M. Shukry, A. Noreldin, A.E. Taha, and K.A. El-Tarabily. 2021. Curcumin, the active substance of turmeric: Its effects on health and ways to improve its bioavailability. *Journal of the Science of Food and Agriculture* 101(14): 5747–5762.

Aboraya, A.O., Y.A.E.E. Elhassaneen, and O.M. Nassar. 2022. Reishi mushroom (*Ganoderma lucidum*) intervention improves lipids profile and paraoxonase/arylesterase activities in serum as well as enhances haemostatic effects in streptozotocin-induced diabetic rats. *Alexandria Science Exchange Journal* 43(4): 593–608.

Ahmad, M.F. 2018. *Ganoderma lucidum*: Persuasive biologically active constituents and their health endorsement. *Biomedicine & Pharmacotherapy* 107: 507–519.

Ahmadian, M., R.E. Duncan, K. Jaworski, E. Sarkadi-Nagy, and H. Sook Sul. 2007. Triacylglycerol metabolism in adipose tissue. *Future Lipidology* 2(2): 229–237.

Alam, N., R. Amin, A. Khan, I. Ara, M.-J. Shim, M.-W. Lee, U.-Y. Lee, and T.-S. Lee. 2009. Comparative effects of oyster mushrooms on lipid profile, liver and kidney function in hypercholesterolemic rats. *Mycobiology* 37(1) (March): 37–42.

Altmann, S.W., H.R. Davis, L. Zhu, X. Yao, L.M. Hoos, G. Tetzloff, S.P.N. Iyer, et al. 2004. Niemann-Pick C1 like 1 protein is critical for intestinal cholesterol absorption. *Science* 303(5661): 1201–1204.

Alves-Bezerra, M., and D.E. Cohen. 2017. Triglyceride metabolism in the liver. In *Comprehensive Physiology*, ed. R. Terjung, 1–22. 1st ed. Wiley.

Ansary, J., T.Y. Forbes-Hernández, E. Gil, D. Cianciosi, J. Zhang, M. Elexpuru-Zabaleta, J. Simal-Gandara, F. Giampieri, and M. Battino. 2020. Potential health benefit of garlic based on human intervention studies: A brief overview. *Antioxidants* 9(7): 619.

Arnold, D.R., and P.O. Kwiterovich. 2003. Cholesterol|Absorption, function, and metabolism. In *Encyclopedia of Food Sciences and Nutrition*, 1226–1237. Elsevier.

Arunachalam, K., P.S. Sreeja, and X. Yang. 2022. The antioxidant properties of mushroom polysaccharides can potentially mitigate oxidative stress, beta-cell dysfunction and insulin resistance. *Frontiers in Pharmacology* 13: 874474.

Atiq, A., and I. Parhar. 2020. Anti-neoplastic potential of flavonoids and polysaccharide phytochemicals in glioblastoma. *Molecules* 25(21): 4895.

Azizi, M., M. Tavana, M. Farsi, and F. Oroojalian. 2012. Yield performance of Lingzhi or Reishi medicinal mushroom, *Ganoderma lucidum* (W.Curt.:Fr.) P. Karst. (higher basidiomycetes), using different waste materials as substrates. *International Journal of Medicinal Mushrooms* 14(5): 521–527.

Baby, S., A.J. Johnson, and B. Govindan. 2015. Secondary metabolites from *Ganoderma*. *Phytochemistry* 114: 66–101.

Bambauer, R., C. Bambauer, B. Lehmann, R. Latza, and R. Schiel. 2012. LDL-apheresis: Technical and clinical aspects. *The Scientific World Journal* 2012: 1–19.

Berger, A., D. Rein, E. Kratky, I. Monnard, H. Hajjaj, I. Meirim, C. Piguet-Welsch, J. Hauser, K. Mace, and P. Niederberger. 2004. Cholesterol-lowering properties of *Ganoderma lucidum in vitro*, *ex vivo*, and in hamsters and minipigs. *Lipids in Health and Disease* 3(1): 2.

Betters, J.L., and L. Yu. 2010. NPC1L1 and cholesterol transport. *FEBS Letters* 584(13): 2740–2747.

Boarescu, P.-M., I. Boarescu, R.M. Pop, Ş.H. Roşian, I.C. Bocşan, V. Rus, R.O. Mada, et al. 2022. Evaluation of oxidative stress biomarkers, pro-inflammatory cytokines, and histological changes in experimental hypertension, dyslipidemia, and type 1 diabetes mellitus. *International Journal of Molecular Sciences* 23(3): 1438.

Bradberry, J.C., and D.E. Hilleman. 2013. Overview of omega-3 fatty acid therapies. *P & T: A Peer-Reviewed Journal for Formulary Management* 38(11): 681–691.

Caz, V., A. Gil-Ramírez, C. Largo, M. Tabernero, M. Santamaría, R. Martín-Hernández, F.R. Marín, G. Reglero, and C. Soler-Rivas. 2015. Modulation of cholesterol-related gene expression by dietary fiber fractions from edible mushrooms. *Journal of Agricultural and Food Chemistry* 63(33): 7371–7380.

Chan, S.W., B. Tomlinson, P. Chan, and C.W.K. Lam. 2021. The beneficial effects of *Ganoderma lucidum* on cardiovascular and metabolic disease risk. *Pharmaceutical Biology* 59(1): 1159–1169.

Chaudhary, R., J. Garg, N. Shah, and A. Sumner. 2017. PCSK9 inhibitors: A new era of lipid lowering therapy. *World Journal of Cardiology* 9(2): 76.

Chen, B., J. Tian, J. Zhang, K. Wang, L. Liu, B. Yang, L. Bao, and H. Liu. 2017. Triterpenes and meroterpenes from *Ganoderma lucidum* with inhibitory activity against HMGs reductase, aldose reductase and α-glucosidase. *Fitoterapia* 120: 6–16.

Chen, Y., M.-Y. Xie, S.-P. Nie, C. Li, and Y.-X. Wang. 2008. Purification, composition analysis and antioxidant activity of a polysaccharide from the fruiting bodies of *Ganoderma atrum*. *Food Chemistry* 107(1): 231–241.

Chugh, R.M., P. Mittal, N. Mp, T. Arora, T. Bhattacharya, H. Chopra, S. Cavalu, and R.K. Gautam. 2022. Fungal mushrooms: A natural compound with therapeutic applications. *Frontiers in Pharmacology* 13: 925387.

Cicero, A.F.G., F. Fogacci, and M. Banach. 2019. Red yeast rice for hypercholesterolemia. *Methodist DeBakey Cardiovascular Journal* 15(3): 192.

Daniels, T.F., K.M. Killinger, J.J. Michal, R.W. Wright Jr., and Z. Jiang. 2009. Lipoproteins, cholesterol homeostasis and cardiac health. *International Journal of Biological Sciences* 474–488.

De Carvalho, C., and M. Caramujo. 2018. The various roles of fatty acids. *Molecules* 23(10): 2583.

Di Stasi, S.L., T.D. MacLeod, J.D. Winters, and S.A. Binder-Macleod. 2010. Effects of statins on skeletal muscle: A perspective for physical therapists. *Physical Therapy* 90(10): 1530–1542.

Duan, Y., K. Gong, S. Xu, F. Zhang, X. Meng, and J. Han. 2022. Regulation of cholesterol homeostasis in health and diseases: From mechanisms to targeted therapeutics. *Signal Transduction and Targeted Therapy* 7(1): 265.

Eilam, Y., N. Pintel, H. Khattib, N. Shagug, R. Taha, and D. Avni. 2022. Regulation of cholesterol metabolism by phytochemicals derived from algae and edible mushrooms in non-alcoholic fatty liver disease. *International Journal of Molecular Sciences* 23(22): 13667.

Ekiz, E., E. Oz, A.M. Abd El-Aty, C. Proestos, C. Brennan, M. Zeng, I. Tomasevic, et al. 2023. Exploring the potential medicinal benefits of *Ganoderma lucidum*: From metabolic disorders to coronavirus infections. *Foods* 12(7): 1512.

Elisaf, M.S., M. Florentin, E.N. Liberopoulos, and D.P. Mikhailidis. 2023. Fibrate-associated adverse effects beyond muscle and liver toxicity. *Current Pharmaceutical Design* 14(6): 574–587.

Elkhateeb, W., G. Daba, D. Sheir, A. El-Dein, W. Fayad, E. Elmahdy, M. Shaheen, P. Thomas, and T. Wen. 2019. GC-MS analysis and *in-vitro* hypocholesterolemic, anti-rotavirus, anti-human colon carcinoma activities of the crude extract of a Japanese *Ganoderma* spp. *Egyptian Pharmaceutical Journal* 18(2): 102–110.

El Sheikha, A.F.E. 2022. Nutritional profile and health benefits of *Ganoderma lucidum* "Lingzhi, Reishi, or Mannentake" as functional foods: Current scenario and future perspectives. *Foods* 11(7): 1030.

Feldman, F., M. Koudoufio, Y. Desjardins, S. Spahis, E. Delvin, and E. Levy. 2021. Efficacy of polyphenols in the management of dyslipidemia: A focus on clinical studies. *Nutrients* 13(2): 672.

Feng, N., Y. Wei, J. Feng, Q. Tang, Z. Zhang, J. Zhang, and W. Han. 2018. Preparative isolation of Ganoderic Acid S, Ganoderic Acid T and Ganoderol B from *Ganoderma lucidum* mycelia by high-speed counter-current chromatography. *Biomedical Chromatography* 32(10): e4283.

Ferreira, I.C.F.R., S.A. Heleno, F.S. Reis, D. Stojkovic, M.J.R.P. Queiroz, M.H. Vasconcelos, and M. Sokovic. 2015. Chemical features of *Ganoderma* polysaccharides with antioxidant, antitumor and antimicrobial activities. *Phytochemistry* 114: 38–55.

Florentin, M., E.N. Liberopoulos, and M.S. Elisaf. 2007. Ezetimibe-associated adverse effects: What the clinician needs to know: Ezetimibe and side effects. *International Journal of Clinical Practice* 62(1): 88–96.

Friesen, J.A., and V.W. Rodwell. 2004. The 3-hydroxy-3-methylglutaryl coenzyme-A (HMG-CoA) reductases. *Genome Biology* 5(11): 248.

Frigerio, J., E. Tedesco, F. Benetti, V. Insolia, G. Nicotra, V. Mezzasalma, S. Pagliari, M. Labra, and L. Campone. 2021. Anticholesterolemic activity of three vegetal extracts (Artichoke, Caigua, and Fenugreek) and their unique blend. *Frontiers in Pharmacology* 12: 726199.

Galappaththi, M.C.A., N.M. Patabendige, B.M. Premarathne, K.K. Hapuarachchi, S. Tibpromma, D.-Q. Dai, N. Suwannarach, S. Rapior, and S.C. Karunarathna. 2022. A review of *Ganoderma* triterpenoids and their bioactivities. *Biomolecules* 13(1): 24.

Gao, J.L., K.S.Y. Leung, Y.T. Wang, C.M. Lai, S.P. Li, L.F. Hu, G.H. Lu, Z.H. Jiang, and Z.L. Yu. 2007. Qualitative and quantitative analyses of nucleosides and nucleobases in *Ganoderma* spp. by HPLC-DAD-MS. *Journal of Pharmaceutical and Biomedical Analysis* 44(3): 807–811.

Grienke, U., T. Kaserer, F. Pfluger, C.E. Mair, T. Langer, D. Schuster, and J.M. Rollinger. 2015. Accessing biological actions of *Ganoderma* secondary metabolites by in silico profiling. *Phytochemistry* 114: 114–124.

Grimm, D., and H.A.B. Wösten. 2018. Mushroom cultivation in the circular economy. *Applied Microbiology and Biotechnology* 102(18): 7795–7803.

Grundy, S.M., N.J. Stone, A.L. Bailey, C. Beam, K.K. Birtcher, R.S. Blumenthal, L.T. Braun, et al. 2019. 2018 AHA/ACC/AACVPR/AAPA/ABC/ACPM/ADA/AGS/APhA/ASPC/NLA/PCNA Guideline on the management of blood cholesterol: Executive summary: A report of the American college of cardiology/American heart association task force on clinical practice guidelines. *Journal of the American College of Cardiology* 73(24): 3168–3209.

Guo, W.-L., Y.-J. Cao, S.-Z. You, Q. Wu, F. Zhang, J.-Z. Han, X.-C. Lv, P.-F. Rao, L.-Z. Ai, and L. Ni. 2022. Ganoderic acids-rich ethanol extract from *Ganoderma lucidum* protects against alcoholic liver injury and modulates intestinal microbiota in mice with excessive alcohol intake. *Current Research in Food Science* 5: 515–530.

Guo, W.-L., J.-B. Guo, B.-Y. Liu, J.-Q. Lu, M. Chen, B. Liu, W.-D. Bai, P.-F. Rao, L. Ni, and X.-C. Lv. 2020. Ganoderic acid A from *Ganoderma lucidum* ameliorates lipid metabolism and alters gut microbiota composition in hyperlipidemic mice fed a high-fat diet. *Food & Function* 11(8): 6818–6833.

Guo, W.-L., Y.-Y. Pan, L. Li, T.-T. Li, B. Liu, and X.-C. Lv. 2018. Ethanol extract of *Ganoderma lucidum* ameliorates lipid metabolic disorders and modulates the gut microbiota composition in high-fat diet fed rats. *Food & Function* 9(6): 3419–3431.

Gupta, R., R.S. Rao, A. Misra, and S.K. Sharma. 2017. Recent trends in epidemiology of dyslipidemias in India. *Indian Heart Journal* 69(3): 382–392.

Gürgöze, M.T., A.H.G. Muller-Hansma, M.M. Schreuder, A.M.H. Galema-Boers, E. Boersma, and J.E. Roeters Van Lennep. 2019. Adverse events associated with PCSK 9 inhibitors: A real-world experience. *Clinical Pharmacology & Therapeutics* 105(2): 496–504.

Hajjaj, H., C. Macé, M. Roberts, P. Niederberger, and L.B. Fay. 2005. Effect of 26-oxygenosterols from *Ganoderma lucidum* and their activity as cholesterol synthesis inhibitors. *Applied and Environmental Microbiology* 71(7): 3653–3658.

Hong, T., J. Zou, X. Jiang, J. Yang, Z. Cao, Y. He, and D. Feng. 2022. Curcumin supplementation ameliorates bile cholesterol supersaturation in hamsters by modulating gut microbiota and cholesterol absorption. *Nutrients* 14(9): 1828.

Hsu, C.-L., and G.-C. Yen. 2014. Ganoderic acid and lucidenic acid (triterpenoid). In *The Enzymes*, 36, 33–56. Elsevier.

Hu, G., M. Zhai, R. Niu, X. Xu, Q. Liu, and J. Jia. 2018. Optimization of culture condition for ganoderic acid production in *Ganoderma lucidum* liquid static culture and design of a suitable bioreactor. *Molecules* 23(10): 2563.

Hu, J., Z. Zhang, W.-J. Shen, and S. Azhar. 2010. Cellular cholesterol delivery, intracellular processing and utilization for biosynthesis of steroid hormones. *Nutrition & Metabolism* 7(1): 47.

Huang, C.-H., W.-K. Lin, S. Chang, and G.-J. Tsai. 2021. *Ganoderma lucidum* culture supplement ameliorates dyslipidemia and reduces visceral fat accumulation in type 2 diabetic rats. *Mycology* 12(2): 94–104.

Huang, Z., F. Fang, and C.-W. Wong. 2011. *Ganoderma lucidum* spore lipid induces peroxisome proliferator-activated receptor alpha activity. *Journal of Food Biochemistry* 35(5): 1508–1513.

Huh, S., S. Lee, S.J. Choi, Z. Wu, J.-H. Cho, L. Kim, Y.S. Shin, et al. 2019. Quercetin synergistically inhibit EBV-associated gastric carcinoma with *Ganoderma lucidum* extracts. *Molecules* 24(21): 3834.

Hung, W.-T., S.-H. Wang, C.-H. Chen, and W.-B. Yang. 2008. Structure determination of β-glucans from *Ganoderma lucidum* with matrix-assisted laser desorption/ionization (MALDI) mass spectrometry. *Molecules* 13(8): 1538–1550.

Insull, W. 2006. Clinical utility of bile acid sequestrants in the treatment of dyslipidemia: A scientific review. *Southern Medical Journal* 99(3): 257–273.

Iqbal, J., and M.M. Hussain. 2009. Intestinal lipid absorption. *American Journal of Physiology-Endocrinology and Metabolism* 296(6): E1183–E1194.

Ison, H.E., S.L. Clarke, and J.W. Knowles. 1993. Familial Hypercholesterolemia. In *GeneReviews®*, ed. M.P. Adam, G.M. Mirzaa, R.A. Pagon, S.E. Wallace, L.J. Bean, K.W. Gripp, and A. Amemiya. University of Washington.

Jacobson, T.A., and F.H. Zimmerman. 2006. Fibrates in combination with statins in the management of dyslipidemia. *The Journal of Clinical Hypertension* 8(1): 35–41.

Jayachandran, M., J. Xiao, and B. Xu. 2017. A critical review on health promoting benefits of edible mushrooms through gut microbiota. *International Journal of Molecular Sciences* 18(9): 1934.

Jefferson, A. 2009. Role of oat beta-glucan in lowering cholesterol. *Practice Nursing* 20(4): 180–184.

Jeong, Y.-U., and Y.-J. Park. 2020. Ergosterol peroxide from the medicinal mushroom *Ganoderma lucidum* inhibits differentiation and lipid accumulation of 3T3-L1 adipocytes. *International Journal of Molecular Sciences* 21(2): 460.

Jesch, E.D., and T.P. Carr. 2017. Food ingredients that inhibit cholesterol absorption. *Preventive Nutrition and Food Science* 22(2): 67–80.

Jiang, Y., Y. Chang, Y. Liu, M. Zhang, H. Luo, C. Hao, P. Zeng, Y. Sun, H. Wang, and L. Zhang. 2017. Overview of *Ganoderma sinense* polysaccharide: An adjunctive drug used during concurrent chemo/radiation therapy for cancer treatment in China. *Biomedicine & Pharmacotherapy* 96: 865–870.

Jing, Y.-S., Y.-F. Ma, F.-B. Pan, M.-S. Li, Y.-G. Zheng, L.-F. Wu, and D.-S. Zhang. 2022. An insight into antihyperlipidemic effects of polysaccharides from natural resources. *Molecules* 27(6): 1903.

Johnson, B.M. 2010. Ganoderic acid DM: An alternative agent for the treatment of advanced prostate cancer. *The Open Prostate Cancer Journal* 3(1): 78–85.

Jose, J. 2016. Statins and its hepatic effects: Newer data, implications, and changing recommendations. *Journal of Pharmacy and Bioallied Sciences* 8(1): 23.

Karlic, H., and F. Varga. 2017. Mevalonate pathway. In *Reference Module in Biomedical Sciences*, B9780128012383650006. Elsevier.

Keong, C.Y. 2015. Medicinal values of selected mushrooms with special reference to anti-hypercholestero lemia. In *Hypercholesterolemia*, ed. S. Ashok Kumar. InTech.

Kim, M., and Y. Kim. 2010. Hypocholesterolemic effects of curcumin via up-regulation of cholesterol 7a-hydroxylase in rats fed a high fat diet. *Nutrition Research and Practice* 4(3): 191.

Kim, S.D. 2010. Isolation and structure determination of a cholesterol esterase inhibitor from *Ganoderma lucidum*. *Journal of Microbiology and Biotechnology* 20(11): 1521–1523.

Klimek, M., S. Wang, and A. Ogunkanmi. 2009. Safety and efficacy of red yeast rice (*Monascus purpureus*) as an alternative therapy for hyperlipidemia. *P & T: A Peer-Reviewed Journal for Formulary Management* 34(6): 313–327.

Klupp, N.L., H. Kiat, A. Bensoussan, G.Z. Steiner, and D.H. Chang. 2016. A double-blind, randomised, placebo-controlled trial of *Ganoderma lucidum* for the treatment of cardiovascular risk factors of metabolic syndrome. *Scientific Reports* 6(1): 29540.

Klupp, N.L., D. Chang, F. Hawke, H. Kiat, H. Cao, S.J. Grant, and A. Bensoussan. 2015. *Ganoderma Lucidum* for the Treatment of Cardiovascular Risk Factors. *Cochrane Database of Systematic Reviews*. 2015(2): CD007259.

Kolniak-Ostek, J., J. Oszmiański, A. Szyjka, H. Moreira, and E. Barg. 2022. Anticancer and antioxidant activities in *Ganoderma lucidum* wild mushrooms in poland, as well as their phenolic and triterpenoid compounds. *International Journal of Molecular Sciences* 23(16): 9359.

Komoda, Y., M. Shimizu, Y. Sonoda, and Y. Sato. 1989. Ganoderic acid and its derivatives as cholesterol synthesis inhibitors. *Chemical and Pharmaceutical Bulletin* 37(2): 531–533.

Lai, P., X. Cao, Q. Xu, Y. Liu, R. Li, J. Zhang, and M. Zhang. 2020. *Ganoderma lucidum* spore ethanol extract attenuates atherosclerosis by regulating lipid metabolism via upregulation of liver x receptor alpha. *Pharmaceutical Biology* 58(1): 760–770.

Lattimer, J.M., and M.D. Haub. 2010. Effects of dietary fiber and its components on metabolic health. *Nutrients* 2(12): 1266–1289.

Lee, H.A., J.-H. Cho, Q. Afinanisa, G.-H. An, J.-G. Han, H.J. Kang, S.H. Choi, and H.-A. Seong. 2020. *Ganoderma lucidum* extract reduces insulin resistance by enhancing ampk activation in high-fat diet-induced obese mice. *Nutrients* 12(11): 3338.

Lee, I., H. Kim, U. Youn, J. Kim, B. Min, H. Jung, M. Na, M. Hattori, and K. Bae. 2010. Effect of Lanostane triterpenes from the fruiting bodies of *Ganoderma lucidum* on adipocyte differentiation in 3T3-L1 Cells. *Planta Medica* 76(14): 1558–1563.

Lee, K.-H., S.L. Morris-Natschke, X. Yang, R. Huang, T. Zhou, S.-F. Wu, Q. Shi, and H. Itokawa. 2012. Recent progress of research on medicinal mushrooms, foods, and other herbal products used in traditional Chinese medicine. *Journal of Traditional and Complementary Medicine* 2(2): 1–12.

Lepor, N.E., and D.J. Kereiakes. 2015. The PCSK9 inhibitors: A novel therapeutic target enters clinical practice. *American Health & Drug Benefits* 8(9): 483–489.

Li, C., Y. Li, and H.H. Sun. 2006. New ganoderic acids, bioactive triterpenoid metabolites from the mushroom *Ganoderma lucidum*. *Natural Product Research* 20(11): 985–991.

Li, H., Y. Du, H. Ji, Y. Yang, C. Xu, Q. Li, L. Ran, C. Wu, Q. Zhou, and S. Wu. 2022. Adenosine-rich extract of *Ganoderma lucidum*: A safe and effective lipid-lowering substance. *iScience* 25(11): 105214.

Li, M., and J. Hu. 2014. Study on survival strategies of farmers engage in small-scale household cultivation of edible mushrooms: Take Shandong province as an example. *Modern Economy* 5(12): 1092–1100.

Li, T., M. Matozel, S. Boehme, B. Kong, L.-M. Nilsson, G. Guo, E. Ellis, and J.Y.L. Chiang. 2011. Overexpression of cholesterol 7α-hydroxylase promotes hepatic bile acid synthesis and secretion and maintains cholesterol homeostasis. *Hepatology* 53(3): 996–1006.

Liang, C., D. Tian, Y. Liu, H. Li, J. Zhu, M. Li, M. Xin, and J. Xia. 2019. Review of the molecular mechanisms of *Ganoderma lucidum* triterpenoids: Ganoderic acids A, C2, D, F, DM, X and Y. *European Journal of Medicinal Chemistry* 174: 130–141.

Liu, J.-Q., K. Shimizu, A. Tanaka, W. Shinobu, K. Ohnuki, T. Nakamura, and R. Kondo. 2012. Target proteins of ganoderic acid dm provides clues to various pharmacological mechanisms. *Scientific Reports* 2(1): 905.

Liu, J.-Q., C.-F. Wang, X.-R. Peng, and M.-H. Qiu. 2011. New alkaloids from the fruiting bodies of *Ganoderma* Sinense. *Natural Products and Bioprospecting* 1(2): 93–96.

Lopez-Jimenez, F., W. Almahmeed, H. Bays, A. Cuevas, E. Di Angelantonio, C.W. Le Roux, N. Sattar, et al. 2022. Obesity and cardiovascular disease: Mechanistic insights and management strategies. A joint position paper by the world heart federation and world obesity federation. *European Journal of Preventive Cardiology* 29(17): 2218–2237.

Lui, D.T.W., A.C.H. Lee, and K.C.B. Tan. 2021. Management of familial hypercholesterolemia: Current status and future perspectives. *Journal of the Endocrine Society* 5(1): bvaa122.

Mach, F., K.K. Ray, O. Wiklund, A. Corsini, A.L. Catapano, E. Bruckert, G. De Backer, et al. 2018. Adverse effects of statin therapy: Perception vs. the evidence—focus on glucose homeostasis, cognitive, renal and hepatic function, haemorrhagic stroke and cataract. *European Heart Journal* 39(27): 2526–2539.

Maíz, A., A. Arteaga, C.L. Villanueva, N. Velasco, and A.M. Acosta. 1990. Cholestyramine in the treatment of hypercholesterolemia. Our experience in 11 cases. *Revista Medica De Chile* 118(9): 1009–1013.

Mannu, G.S., M.J.S. Zaman, A. Gupta, H.U. Rehman, and P.K. Myint. 2013. Evidence of lifestyle modification in the management of hypercholesterolemia. *Current Cardiology Reviews* 9(1): 2–14.

Martínez-Montemayor, M.M., T. Ling, I.J. Suárez-Arroyo, G. Ortiz-Soto, C.L. Santiago-Negrón, M.Y. Lacourt-Ventura, A. Valentín-Acevedo, W.H. Lang, and F. Rivas. 2019. Identification of biologically active *Ganoderma lucidum* compounds and synthesis of improved derivatives that confer anti-cancer activities *in vitro*. *Frontiers in Pharmacology* 10: 115.

Meneses, M.E., D. Martínez-Carrera, L. González-Ibáñez, N. Torres, M. Sánchez-Tapia, C.C. Márquez-Mota, G. Rendón, et al. 2023. Effects of Mexican *Ganoderma lucidum* extracts on liver, kidney, and the gut microbiota of wistar rats: A repeated dose oral toxicity study. *PLoS One* 18(4): e0283605.

Meneses, M.E., D. Martínez-Carrera, N. Torres, M. Sánchez-Tapia, M. Aguilar-López, P. Morales, M. Sobal, et al. 2016. Hypocholesterolemic properties and prebiotic effects of Mexican *Ganoderma lucidum* in C57BL/6 mice. *PLoS One* 11(7): e0159631.

Meng, J., Sai-Zhen Wang, J. He, S. Zhu, B. Huang, S. Wang, M. Li, H. Zhou, S. Lin, and B. Yang. 2020. Ganoderic acid A is the effective ingredient of *Ganoderma* triterpenes in retarding renal cyst development in polycystic kidney disease. *Acta Pharmacologica Sinica* 41(6): 782–790.

Mikhailidis, D.P., R.W. Lawson, A.-L. McCormick, G.C. Sibbring, A.M. Tershakovec, G.M. Davies, and K. Tunceli. 2011. Comparative efficacy of the addition of ezetimibe to statin *vs* statin titration in patients with hypercholesterolaemia: Systematic review and meta-analysis. *Current Medical Research and Opinion* 27(6): 1191–1210.

Mirończuk-Chodakowska, I., K. Kujawowicz, and A.M. Witkowska. 2021. Beta-glucans from fungi: Biological and health-promoting potential in the COVID-19 pandemic era. *Nutrients* 13(11): 3960.

Morzycki, J.W. 2014. Recent advances in cholesterol chemistry. *Steroids* 83: 62–79.

Nelson, R.H. 2013. Hyperlipidemia as a risk factor for cardiovascular disease. *Primary Care: Clinics in Office Practice* 40(1): 195–211.

Nes, W.D. 2011. Biosynthesis of cholesterol and other sterols. *Chemical Reviews* 111(10): 6423–6451.

Oke, M.A., F.J. Afolabi, O.O. Oyeleke, T.A. Kilani, A.R. Adeosun, A.A. Olanbiwoninu, and E.A. Adebayo. 2022. *Ganoderma Lucidum*: Unutilized Natural Medicine and Promising Future Solution to Emerging Diseases in Africa. *Frontiers in Pharmacology* 13(22): 952027.

Olsen, L., E. Thum, and N. Rohner. 2021. Lipid metabolism in adaptation to extreme nutritional challenges. *Developmental Cell* 56(10): 1417–1429.

Pan, R., J. Lou, and L. Wei. 2021. Significant effects of *Ganoderma lucidum* polysaccharide on lipid metabolism in diabetes may be associated with the activation of the FAM3C-HSF1-CAM signaling pathway. *Experimental and Therapeutic Medicine* 22(2): 820.

Peng, H., L. Zhong, L. Cheng, L. Chen, R. Tong, J. Shi, and L. Bai. 2023. *Ganoderma lucidum*: Current advancements of characteristic components and experimental progress in anti-liver fibrosis. *Frontiers in Pharmacology* 13: 1094405.

Peters, S.A.E., Y. Singhateh, D. Mackay, R.R. Huxley, and M. Woodward. 2016. Total cholesterol as a risk factor for coronary heart disease and stroke in women compared with men: A systematic review and meta-analysis. *Atherosclerosis* 248: 123–131.

Plat, J., S. Baumgartner, T. Vanmierlo, D. Lütjohann, K.L. Calkins, D.G. Burrin, G. Guthrie, et al. 2019. Plant-based sterols and stanols in health & disease: "Consequences of human development in a plant-based environment?" *Progress in Lipid Research* 74: 87–102.

Radwan, F.F.Y. 2012. Apoptotic and immune restoration effects of ganoderic acids define a new prospective for complementary treatment of cancer. *Journal of Clinical & Cellular Immunology* 01(S3).

Rafieian-Kopaei, M., M. Setorki, M. Doudi, A. Baradaran, and H. Nasri. 2014. Atherosclerosis: Process, indicators, risk factors and new hopes. *International Journal of Preventive Medicine* 5(8): 927–946.

Ramkumar, S., A. Raghunath, and S. Raghunath. 2016. Statin therapy: Review of safety and potential side effects. *Acta Cardiologica Sinica* 32(6): 631–639.

Romero-Córdoba, S.L., I. Salido-Guadarrama, M.E. Meneses, G. Cosentino, M.V. Iorio, E. Tagliabue, N. Torres, et al. 2020. Mexican *Ganoderma lucidum* extracts decrease lipogenesis modulating transcriptional metabolic networks and gut microbiota in c57bl/6 mice fed with a high-cholesterol diet. *Nutrients* 13(1): 38.

Rossini, E., F. Biscetti, M.M. Rando, E. Nardella, A.L. Cecchini, M.A. Nicolazzi, M. Covino, A. Gasbarrini, M. Massetti, and A. Flex. 2022. Statins in high cardiovascular risk patients: Do comorbidities and characteristics matter? *International Journal of Molecular Sciences* 23(16): 9326.

Sang, T., C. Guo, D. Guo, J. Wu, Y. Wang, Y. Wang, J. Chen, et al. 2021. Suppression of obesity and inflammation by polysaccharide from sporoderm-broken spore of *Ganoderma lucidum* via gut microbiota regulation. *Carbohydrate Polymers* 256: 17594.

Santini, A., and E. Novellino. 2017. Nutraceuticals in hypercholesterolaemia: An overview: Nutraceuticals and hypercholesterolaemia. *British Journal of Pharmacology* 174(11): 1450–1463.

Sarat Chandra, K., M. Bansal, T. Nair, S.S. Iyengar, R. Gupta, S.C. Manchanda, P.P. Mohanan, et al. 2014. Consensus statement on management of dyslipidemia in Indian subjects. *Indian Heart Journal* 66: S1–S51.

Seethapathy, P., S. Sankaralingam, I.K. Muniraj, M. Perumal, and N. Pandurangan. 2023. Mass multiplication, economic analysis, and marketing of *Ganoderma* Sp. (Reishi Mushroom). In *Food Microbiology Based Entrepreneurship*, ed. N. Amaresan, D. Dharumadurai, and O.O. Babalola, 89–113. Springer Nature Singapore.

Seweryn, E., A. Ziała, and A. Gamian. 2021. Health-promoting of polysaccharides extracted from *Ganoderma lucidum*. *Nutrients* 13(8): 2725.

Sharifi-Rad, J., Y.E. Rayess, A.A. Rizk, C. Sadaka, R. Zgheib, W. Zam, S. Sestito, et al. 2020. Turmeric and its major compound Curcumin on health: Bioactive effects and safety profiles for food, pharmaceutical, biotechnological and medicinal applications. *Frontiers in Pharmacology* 11: 01021.

Sherratt, S.C.R., P. Libby, M.J. Budoff, D.L. Bhatt, and R.P. Mason. 2023. Role of omega-3 fatty acids in cardiovascular disease: The debate continues. *Current Atherosclerosis Reports* 25(1): 1–17.

Shiao, M.-S. 2003. Natural products of the medicinal fungus *Ganoderma lucidum*: Occurrence, biological activities, and pharmacological functions. *The Chemical Record* 3(3): 172–180.

Sizar, O., S. Khare, R.T. Jamil, and R. Talati. 2023. Statin medications. In *StatPearls*. StatPearls Publishing.

Skulas-Ray, A.C., P.W.F. Wilson, W.S. Harris, E.A. Brinton, P.M. Kris-Etherton, C.K. Richter, T.A. Jacobson, et al. 2019. Omega-3 fatty acids for the management of hypertriglyceridemia: A science advisory from the American heart association. *Circulation* 140(12): e673–e691.

Soliman, G.A. 2018. Dietary cholesterol and the lack of evidence in cardiovascular disease. *Nutrients* 10(6): 780.

Soliman, G.A. 2019. Dietary fiber, atherosclerosis, and cardiovascular disease. *Nutrients* 11(5): 1155.

Soran, H., S. Adam, J.B. Mohammad, J.H. Ho, J.D. Schofield, S. Kwok, T. Siahmansur, et al. 2018. Hypercholesterolaemia—practical information for non-specialists. *Archives of Medical Science* 1: 1–21.

Srivastava, R.A.K., S.L. Pinkosky, S. Filippov, J.C. Hanselman, C.T. Cramer, and R.S. Newton. 2012. AMP-activated protein kinase: An emerging drug target to regulate imbalances in lipid and carbohydrate metabolism to treat cardio-metabolic diseases. *Journal of Lipid Research* 53(12): 2490–2514.

Staels, B., J. Dallongeville, J. Auwerx, K. Schoonjans, E. Leitersdorf, and J.-C. Fruchart. 1998. Mechanism of action of fibrates on lipid and lipoprotein metabolism. *Circulation* 98(19): 2088–2093.

Staels, B., Y. Handelsman, and V. Fonseca. 2010. Bile acid sequestrants for lipid and glucose control. *Current Diabetes Reports* 10(1): 70–77.

Stoekenbroek, R.M., M.L. Hartgers, R. Rutte, D.D. De Wijer, E.S.G. Stroes, and G.K. Hovingh. 2018. PCSK9 inhibitors in clinical practice: Delivering on the promise? *Atherosclerosis* 270: 205–210.

Sun, P., L. Zhao, N. Zhang, J. Zhou, L. Zhang, W. Wu, B. Ji, and F. Zhou. 2021. Bioactivity of dietary polyphenols: The role in LDL-C lowering. *Foods* 10(11): 2666.

Toth, P.P., and T. Dayspring. 2012. Ezetimibe therapy: Mechanism of action and clinical update. *Vascular Health and Risk Management* 415: 415–427.

Trapani, L. 2012. Regulation and deregulation of cholesterol homeostasis: The liver as a metabolic "power station." *World Journal of Hepatology* 4(6): 184.

Trautwein, E.A., and S. McKay. 2020. The role of specific components of a plant-based diet in management of dyslipidemia and the impact on cardiovascular risk. *Nutrients* 12(9): 2671.

Trautwein, E.A., M. Vermeer, H. Hiemstra, and R. Ras. 2018. LDL-Cholesterol lowering of plant sterols and stanols—which factors influence their efficacy? *Nutrients* 10(9): 1262.

Troesch, B., M. Eggersdorfer, A. Laviano, Y. Rolland, A.D. Smith, I. Warnke, A. Weimann, and P.C. Calder. 2020. Expert opinion on benefits of long-chain omega-3 fatty acids (DHA and EPA) in aging and clinical nutrition. *Nutrients* 12(9): 2555.

Vourakis, M., G. Mayer, and G. Rousseau. 2021. The role of gut microbiota on cholesterol metabolism in atherosclerosis. *International Journal of Molecular Sciences* 22(15): 8074.

Wang, K., L. Bao, K. Ma, J. Zhang, B. Chen, J. Han, J. Ren, H. Luo, and H. Liu. 2017. A novel class of α-glucosidase and hmg-coa reductase inhibitors from *Ganoderma leucocontextum* and the anti-diabetic properties of Ganomycin I in KK-A y mice. *European Journal of Medicinal Chemistry* 127: 1035–1046.

Wang, K., L. Bao, W. Xiong, K. Ma, J. Han, W. Wang, W. Yin, and H. Liu. 2015. Lanostane triterpenes from the tibetan medicinal mushroom *Ganoderma leucocontextum* and their inhibitory effects on HMG-CoA reductase and α-glucosidase. *Journal of Natural Products* 78(8): 1977–1989.

Wang, L., J. Li, J. Zhang, Z. Li, H. Liu, and Y. Wang. 2020. Traditional uses, chemical components and pharmacological activities of the genus *Ganoderma* P. Karst.: A review. *RSC Advances* 10(69): 42084–42097.

Wang, W., Y. Zhang, Z. Wang, J. Zhang, and L. Jia. 2023. *Ganoderma lucidum* polysaccharides improve lipid metabolism against high-fat diet-induced dyslipidemia. *Journal of Ethnopharmacology* 309: 116321.

Ward, N.C., G.F. Watts, and R.H. Eckel. 2019. Statin toxicity: Mechanistic insights and clinical implications. *Circulation Research* 124(2): 328–350.

Wickman, B.E., B. Enkhmaa, R. Ridberg, E. Romero, M. Cadeiras, F. Meyers, and F. Steinberg. 2021. Dietary management of heart failure: DASH diet and precision nutrition perspectives. *Nutrients* 13(12): 4424.

Wider, B., M.H. Pittler, J. Thompson-Coon, and E. Ernst. 2013. Artichoke leaf extract for treating hypercholesterolaemia. In *Cochrane Database of Systematic Reviews*, ed. The Cochrane Collaboration, CD003335. pub3. John Wiley & Sons, Ltd.

Wong, T.Y., Y.Q. Tan, S. Lin, and L.K. Leung. 2017. Apigenin and Luteolin display differential hypocholesterolemic mechanisms in mice fed a high-fat diet. *Biomedicine & Pharmacotherapy* 96: 1000–1007.

World Heart Federation. 2023. Cholesterol. *World Heart Federation*. Accessed May 30.

Xia, Q., H. Zhang, X. Sun, H. Zhao, L. Wu, D. Zhu, G. Yang, et al. 2014. A comprehensive review of the structure elucidation and biological activity of triterpenoids from *Ganoderma* Spp. *Molecules* 19(11): 17478–17535.

Xiong, Z., X. Cao, Q. Wen, Z. Chen, Z. Cheng, X. Huang, Y. Zhang, C. Long, Y. Zhang, and Z. Huang. 2019. An overview of the bioactivity of Monacolin K/Lovastatin. *Food and Chemical Toxicology* 131: 110585.

Xu, J., C. Xiao, H. Xu, S. Yang, Z. Chen, H. Wang, B. Zheng, B. Mao, and X. Wu. 2021. Anti-inflammatory effects of *Ganoderma lucidum* sterols via attenuation of the P38 MAPK and NF-κB pathways in LPS-induced RAW 264.7 macrophages. *Food and Chemical Toxicology* 150: 112073.

Xu, Z., X. Chen, Z. Zhong, L. Chen, and Y. Wang. 2011. *Ganoderma lucidum* polysaccharides: Immunomodulation and potential anti-tumor activities. *The American Journal of Chinese Medicine* 39(01): 15–27.

Yang, S.-T., A.J.B. Kreutzberger, J. Lee, V. Kiessling, and L.K. Tamm. 2016. The role of cholesterol in membrane fusion. *Chemistry and Physics of Lipids* 199: 136–143.

Ye, L., S. Liu, F. Xie, L. Zhao, and X. Wu. 2018. Enhanced production of polysaccharides and triterpenoids in *Ganoderma lucidum* fruit bodies on induction with signal transduction during the fruiting stage. Ed. Olaf Kniemeyer. *PLoS One* 13(4): e0196287.

Yeh, Y.-Y., and L. Liu. 2001. Cholesterol-lowering effect of garlic extracts and organosulfur compounds: Human and animal studies. *The Journal of Nutrition* 131(3): 989S–993S.

Žák, A. 2015. Niacin in the treatment of hyperlipidemias in light of new clinical trials: Has niacin lost its place? *Medical Science Monitor* 21: 2156–2162.

Zhang, J., K. Ma, J. Han, K. Wang, H. Chen, L. Bao, L. Liu, et al. 2018. Eight new triterpenoids with inhibitory activity against HMG-CoA reductase from the medical mushroom *Ganoderma leucocontextum* collected in Tibetan plateau. *Fitoterapia* 130: 79–88.

Zhang, Z., X. Li, S. Sang, D.J. McClements, L. Chen, J. Long, A. Jiao, Z. Jin, and C. Qiu. 2022. Polyphenols as plant-based nutraceuticals: Health effects, encapsulation, nano-delivery, and application. *Foods* 11(15): 2189.

Zhao, P., M. Guan, W. Tang, N. Walayat, Y. Ding, and J. Liu. 2023. Structural diversity, fermentation production, bioactivities and applications of triterpenoids from several common medicinal fungi: Recent advances and future perspectives. *Fitoterapia* 166: 105470.

Zheng, C., P. Rangsinth, P.H.T. Shiu, W. Wang, R. Li, J. Li, Y.-W. Kwan, and G.P.H. Leung. 2023a. A review on the sources, structures, and pharmacological activities of Lucidenic acids. *Molecules* 28(4): 1756.

Zheng, G., Y. Zhao, Z. Li, Y. Hua, J. Zhang, Y. Miao, Y. Guo, et al. 2023b. GLSP and GLSP-derived triterpenes attenuate atherosclerosis and aortic calcification by stimulating ABCA1/G1-mediated macrophage cholesterol efflux and inactivating RUNX2-mediated VSMC osteogenesis. *Theranostics* 13(4): 1325–1341.

Zhou, Q., and J.K. Liao. 2023. Statins and cardiovascular diseases: From cholesterol lowering to pleiotropy. *Current Pharmaceutical Design* 15(5): 467–478.

Zhou, X.-W. 2017. Cultivation of *Ganoderma lucidum*. In *Edible and Medicinal Mushrooms*, ed. C.Z. Diego and A. Pardo-Giménez, 385–413. John Wiley & Sons, Ltd.

Zhou, X.-W., K.-Q. Su, and Y.-M. Zhang. 2012. Applied modern biotechnology for cultivation of *Ganoderma* and development of their products. *Applied Microbiology and Biotechnology* 93(3): 941–963.

Zhu, J., J. Jin, J. Ding, S. Li, P. Cen, K. Wang, H. Wang, and J. Xia. 2018. Ganoderic acid A improves high fat diet-induced obesity, lipid accumulation and insulin sensitivity through regulating SREBP pathway. *Chemico-Biological Interactions* 290: 77–87.

Index

G

H

I

L

T

V

W

X

Z